D0498054

François Cardarelli

Scientific Unit Conversion

A Practical Guide to Metrication

English translation by M.J. Shields, FIInfSc, MITI

Springer

François Cardarelli, PhD, MSc, BSc
Université Paul-Sabatier (UPS),
Laboratoire de Génie Chimique et Électrochimie,
118 route de Narbonne, F-31062 Toulouse Cedex, France

ISBN 3-540-76022-9 Springer-Verlag Berlin Heidelberg New York

British Library Cataloguing in Publication Data
Cardarelli, François
Scientific unit conversion: practical guide to metrication
1. Mensuration – Conversion tables 2. Metric system – Conversion tables
I. Title
502.1′2
ISBN 3540760229

Library of Congress Cataloging-in-Publication Data
Cardarelli, François, 1966–
Scientific unit conversion: practical guide to metrication/ François Cardarelli; English translation by M.J. Shields. – 1st ed.
p. cm.
Includes bibliographical references.
ISBN 3-540-76022-9 (alk. paper)
1. Metric system—Conversion tables—Handbooks, manuals, etc. 2. Units—Conversion tables—Handbooks, manuals, etc. I. Title.
QC94.C295 1996 96-33377
530.8′12—dc20 CIP

Typeset by T & A Typesetting Services, Rochdale, England
Printed by Interprint Limited, Malta
69/3830-543210 Printed on acid-free paper

Dedication

Scientific Unit Conversion is dedicated to my parents *Antonio* and *Claudine*, to my sister *Elsa*, and to *Véronique*, for their love and support.

François Cardarelli

Acknowledgments

Mr. Jean-Étienne Mittelmann, *Mr. Nicholas Pinfield*, *Mrs. Imke Mowbray*, and *Mr. Christopher Greenwell* are gratefully acknowledged for their valued assistance, patience, and advice.

Contents

List of Tables

1 Introduction

1.1 Why a Conversion Handbook?

Books on the conversion of scientific units into their SI equivalents are relatively rare in scientific literature. There are several specialized treatises (*see Bibliography*) on the subject as applied to certain areas of science and technology, which contain sections on the subject, supported by conversion tables. However, these tables are anything but exhaustive, and it is often necessary to consult sources in several very different areas in order to obtain the desired information.

This practical manual aims to be the most comprehensive work on the subject of unit conversion. It contains more than 10 000 precise conversion factors, and around 2000 definitions of the units themselves. The units included, and their conversions, are grouped into imperial and US units, conventional metric units, older or out-of-date units, ancient units, and SI units. The subject areas involved are: pure and applied science, technology, medicine, and economics. Some examples of individual sciences covered are mechanical, electrical, chemical, and nuclear engineering, civil and mining engineering, chemistry, physics, biology, medicine, economics, and computer sciences. In other words, this book places unit conversion at the disposal of everyone. It saves working time, and should be available in all research libraries and design offices. It has been kept as small as possible in order to facilitate consultation in all circumstances, whether in the office, on the production line, or on the move.

The aim of this book is to ensure rapid and accurate conversion of scientific units to their SI equivalents. However, the reader should be aware that it does not provide rules and advice for writing the names, nor the recommended symbols for physical quantities used in science and technology. Several specialized works already exist for this purpose.[1]

This book is suitable for researchers, scientists, engineers and technologists, economists, doctors, pharmacists, and patent lawyers, but is equally suitable for teachers and students.

Scientific Unit Conversion is the product of many years spent collecting information spread across scientific and technical literature. Each item of information has been carefully checked and verified. Additionally, certain pieces of information have been extracted from books or standards for the most part published by national and international bodies (e.g. ISO, AFNOR, BSI, DIN, IEC, ANSI, NACE, NIST). Every effort has been made to supply conversion factors as precisely as possible to an accuracy of nine decimal places.

---

[1] Mills, I., and Cvitas, T. (Eds.), *Quantities, Units and Symbols in Physical Chemistry, 2nd ed.*, IUPAC, Blackwell Scientific Publications, Oxford, 1993.

1.2 How to Use This Book

Chapter 2 contains a brief history of the metric system, including th organization and a complete description of SI Units (*Système Internation d'Unités*).

Chapter 3 gives a detailed description of a considerable number of oth systems of measurement. This includes several alternative modern systems o measurement, some of which are still in widespread use (e.g. imperial, US, cg MTS, FPS). Finally, there is a description of systems used in antiquity (e., ancient Chinese, Indian, Egyptian, Persian, Hebrew, Greek, Roman, Arabic), a well as older national or regional systems (e.g. French, Italian, Germa Japanese).

Chapter 4, which forms the most important part of the book, consists of a exhaustive set of conversion tables. This chapter covers the units in alphabe tical order. Each unit is fully described as follows: name, symbol(s), physica quantity, dimension, conversion factor, notes and definitions. The sectio covers some 2000 units, each with a precise conversion factor.

Chapter 5 enables a unit to be identified from its area of application. For th purpose, units are classed in groups. It contains thirty five conversion table ranging from mass to nuclear quantities.

In order to facilitate use of this manual, several supplementary sections hav been added to aid the researcher. These include tables of fundamental math ematical and physical constants to allow very precise calculation of conve sions. These form the sixth chapter of the book.

Appendices contain a list of many national and international organization in the area of standardization, rules of nomenclature for large number notation for times and dates, and a brief French-English glossary of name of units and associated physical quantities.

Finally, a detailed bibliography (e.g. national and international standard textbooks, specialized engineering handbooks) is presented at the end of th book in order to allow the reader to go further in their investigations.

This practical manual provides rapid answers to all questions concernin the conversion of scientific units. Some examples of the sort of questions tha can be answered more rapidly thanks to this manual, along with the chapter where their answers can be found, are given below:

- What is the history of SI units? (2
- What are the base units of the SI? (2
- What were the ancient systems of measurement? (3
- How is the imperial system organized? (3
- What is a kip? (4
- Which unit has the abbreviation pcu? (4
- What are the dimensions of the röntgen? (4
- What is the conversion factor from spats to steradians? (4
- What is the conversion factor from density in $lb.ft^{-3}$ to $kg.dm^{-3}$? (4
- What are the different kinds of units of pressure and stress? (4
- What is the exact value of the velocity of light in vacuum? (5
- What are the old symbols for imperial units? (6
- What are the addresses of standards organizations in the US? (6
- What are the ISO and AFNOR standards for quantities in nuclear physics? (7

2 The International System of Units

2.1 History

The origin of the metric system, and of its later version, the International System (*Système International*, SI) of units, goes back a long way into French history. Before that, the old French measures had presented two serious problems:

- units with the same name varied from one region to the next and had to be defined accordingly (pied de Paris, etc.)
- subdivisions were not decimal, which increased the complexity of commercial transactions

As early as 1670, the Abbé Gabriel Mouton proposed one minute of arc at the meridian as a rational standard of length. This represents a length of about 1.80 m. He gave this unit the name *milliare*, from the Latin for one thousand, and divided it decimally into three multiple units, named respectively *centuria*, *decuria*, and *virga*, and three submultiples *decima*, *centessima*, and *millesima*. Unfortunately, however, the Abbé died before seeing his ideas adopted.

Over the years, the English mathematician and architect Sir Christopher Wren (1667), the French astronomer Abbé Jean Picard (1671), the Dutch scientist Christiaan Huygens (1673), and the French geodesist Charles Marie de la Condamine (1746) proposed the length of the seconds pendulum as a unit of length. Finally, in 1789, there came a general call for the use of the same measures throughout France.

Soon afterwards, on March 9th, 1790, at the instigation of Charles Maurice de Talleyrand, Bishop of Autun and National Assembly Deputy, the *Constituante*[2] initiated a unification project for weights and measures. The project was adopted on May 8th, 1790, and the Academy of Sciences was given the task of studying the matter. A commission of French mathematicians was made responsible for establishing the base unit; its members were Count Louis de Lagrange, Gaspard Monge, Charles de Borda, and Marie Jean Antoine de Condorcet, plus the astronomer the Marquis Pierre Simon de Laplace. On March 19th, 1791, this commission decided on a unit of length equal to one ten millionth of the distance between the equator and the pole. This unit was called the *mètre*, from the Greek word, *metron*, for measure, and it is of course the origin of the name to the system itself.

---

2 Name of the French Revolutionary Assembly from July 9th, 1789 until September 30th, 1791.

In 1795, according to the text of the organic law of 18 Germinal year III (i terms of the revolutionary calendar then in force – equivalent to April 7t 1795), the Système Métrique Décimal was instituted in France. This decim metric system established a new set of units, the **are** (Latin *area*) for area, th **litre** (Greek *litra*, a 12-ounce weight) for volume, and for mass the **gra** (Greek *gramma*, the weight of a pea).

In 1799, Jean-Baptiste Delambre and Pierre Méchain, who had spent almo seven years measuring the geodetic distances between Rodez and Barcelon and Rodez and Dunkerque, respectively, determined the length of the quart meridian through Paris. The law established the length of the metre 0.513074 *toises de Paris*, and prototype standards of dense pure platinu for the metre and kilogram, made by Jeannetty from agglomerated platinu sponge, were adopted in June 1799. It is easy to imagine that the substitutic of these new metric measures for the old units in use until then was n achieved without a lot of problems and objections.

In 1812, the former units were re-established by the Emperor Napoleo However, metric units were reinstated by the Law of July 4th, 1837 whic declared the Decimal Metric System obligatory in France from January 184 and instituted penalties for the use of other weights and measures.

After that, the system slowly extended its application beyond the borders France, and even became legal, though not compulsory, in the United States i 1866. However, the main launching pad for its internationalization was th meeting of the *Commission Internationale du Mètre* (International Metr Commission) in Paris on August 8–13, 1872. The treaty known as the *Co vention du Mètre* (International Metric Convention) was signed on May 20t 1875 by an assembly of representatives of 17 countries including the USA. established the *Conférence Générale des Poids et Mesures* (General Conferen on Weights and Measures, CGPM), and the *Bureau International des Poids Mesures* (International Bureau of Weights and Measures, BIPM). The hea quarters of the International Bureau, which is maintained by all the nation members, was established at the Pavillon de Breteuil, at Sèvres, near Paris, i consideration of the role of France as the birthplace of the metric system. Th first General Conference on Weights and Measures (1st CGPM), held in 188 organized the distribution of copies of the international standard prototy metre to the 21 member states of the International Metric Convention. Th copies of the new standard prototype called *mètre international* were bu from platinum-iridium alloy (Pt90-Ir10) which is an outcome of works Sainte-Claire Deville et al[3,4]. The secondary standards were a typical bar with cross-section having a side of 2 cm.

The *Système International d'Unités* (SI) is the ultimate development of th metric system. Previous versions included the cgs (centimetre-gram-second the MTS (metre-tonne-second), the MKS (metre-kilogram-force-second), an the MKSA (metre-kilogram-second-ampere) or Giorgi systems.

In 1954, the 10th CGPM adopted a set of base units for the following physic quantities: length, mass, time, electric current, thermodynamic temperatur and luminous intensity. The 11th CGPM, in 1960, by its Resolution 2, adopte the name Système International d'Unités, to be known by its internation abbreviation SI. This system established rules for prefixes, for derived unit for supplementary units, as well as an overall control of units of measuremen Since then, the SI has evolved and developed via the various CGPMs.

---

[3] Sainte-Claire Deville, H., et Stas, *Ann. Chim. Phys.*, **22** (1881) 120.

[4] Sainte-Claire Deville, H., et Macart, *Ann. École Normale*, **8** (1879) 9.

In France, the SI became mandatory in 1961[5]. The 14th CGPM, in 1972, defined a new unit, the **mole**, for amount of substance, to be adopted as the seventh SI base unit.

The SI possesses several advantages. First, it is both metric and decimal. Second, fractions have been eliminated, multiples and submultiples being indicated by a system of standard prefixes, thus greatly simplifying calculations. Each physical quantity is expressed by one unique unit, and derived SI units are defined by simple equations relating two or more base units. Some derived units have been given individual names. In the interests of clarity, SI provides a direct relationship between mechanical, electrical, nuclear, chemical, thermodynamic, and optical units, thus forming a coherent system. There is no duplication of units for the same physical quantity, and all derived units are obtained by direct one-to-one relationships between base or other derived units. The same system of units can be used by researchers, engineers, or technologists.

2.2 The General Conference on Weights and Measures

The General Conference on Weights and Measures (CGPM) is an international organization made up of the delegates of all member states. In October 1985, the number of member states was 47. The remit of this organization is to take all necessary measures to ensure the propagation and the development of the SI, and to adopt various international scientific resolutions relative to new and fundamental developments in metrology.

Under the authority of the CGPM, the International Committee for Weights and Measures (CIPM) is responsible for the establishment and control of units of measurement. A permanent organization, the International Bureau of Weights and Measures (BIPM) created by the Metric Convention and signed by the 17 nations in Paris in 1875, operates under the supervision of the CIPM. The BIPM, which is located at the Pavillon of Breteuil[6], in the St-Cloud Park at Sèvres, has the remit of ensuring unification of measurements throughout the world, specifically:

- to establish fundamental standards and scales of the main physical quantities, and to preserve international prototypes
- to carry out comparisons of national and international standards
- to ensure co-ordination of appropriate measurement techniques
- to carry out and co-ordinate determination of physical constants involved in the above activities

2.3 Organization of the SI

The International System of Units (*Système International d'Unités*, SI) consists of three classes of units:

- seven base units
- two supplementary units
- a number of derived units

[5] Subsequent to the statutory order no. 61-501 of May 3rd, 1961, which appeared in the *Journal Officiel de la République Française* of May 20th, 1961 (Lois et Décrets, Ministère de l'Industrie, pp 4584–4593).

[6] Le Pavillon de Breteuil at Sèvres is an enclave which has international status.

In total, they form a coherent system of units officially known as **SI units**. Those units which do not form part of this system are known as **out-of-system units**.

It is recommended that only SI units should be used in scientific and technological applications, with SI prefixes where appropriate. The use of some out-of-system units (e.g. nautical mile, hectare, litre, hour, ampere-hour, bar) remains legal and they are temporarily retained because of their importance or their usefulness in certain specialized fields. Nevertheless, they should always be defined in terms of SI units, and SI units should be used wherever possible in order to maintain coherence in calculations.

2.3.1 SI Base Units

The seven SI base units are listed in ***Table 2-1*** below:

Table 2-1 The seven SI base units

Physical quantity	Dimension	Name	Symbol
Mass	M	kilogram	kg
Length	L	metre	m
Time	T	second	s
Temperature	Θ	kelvin	K
Amount of substance	N	mole	mol
Electric current intensity	I	ampere	A
Luminous intensity	J	candela	cd

These seven SI base units are officially and bilingually (French-English) defined as follows:

mètre Le mètre est la longueur du trajet parcouru dans le vide par la lumière pendant une durée de 1/299 792 458 s. [$17^{\text{ème}}$ CGPM (1983), Résolution 1].
The metre is the length of the path travelled by light in vacuum during a time interval of 1/299 792 458 s. [17th CGPM (1983), Resolution 1].

kilogramme Le kilogramme est l'unité de masse; il est égal à la masse du prototype international du kilogramme [$1^{\text{ère}}$ CGPM (1889), $3^{\text{ème}}$ CGPM (1901)].
The kilogram is the unit of mass; it is equal to the mass of the international prototype of the kilogram [1st CGPM (1889), 3rd CGPM (1901)].

seconde La seconde est la durée de 9 192 631 770 périodes de la radiation correspondant à la transition entre les deux niveaux hyperfins ($F = 4$, $m_F = 0$ à $F = 3$, $m_F = 0$) de l'état fondamental de l'atome de césium 133 [$13^{\text{ème}}$ CGPM (1967)].

The second is the duration of 9 192 631 770 periods of the radiation corresponding to the transition between the two hyperfine levels ($F=4$, $m_F=0$ to $F=3$, $m_F=0$) of the ground state of the cesium 133 atom [13th CGPM (1967)].

ampère

L'ampère est l'intensité d'un courant électrique constant qui maintenu dans deux conducteurs parallèles, rectilignes, de longueur infinie, de section circulaire négligeable et placés à une distance de 1 mètre l'un de l'autre dans le vide, produit entre ces conducteurs une force égale à 2×10^{-7} newton par mètre de longueur [9ème CGPM (1948), Résolution 2 et 7].
The ampere is that constant current which, if maintained in two straight parallel conductors of infinite length, of negligible circular cross-section, and placed 1 metre apart in vacuum, would produce between these conductors a force equal to 2×10^{-7} newton per metre of length [9th CGPM (1948), Resolution 2 and 7].

kelvin

Le kelvin, unité de température thermodynamique, est la fraction 1/273,16 de la température thermodynamique du point triple de l'eau [13ème CGPM (1967), Résolution 4].
The kelvin, unit of thermodynamic temperature, is the fraction 1/273.16 of the thermodynamic temperature of the triple point of water [13th CGPM (1967), Resolution 4].

mole

(i) La mole représente la quantité de matière totale d'un système qui contient autant d'entités élémentaires que 0,012 kg de carbone 12. (ii) Lorsque l'on emploie la mole, les entités élémentaires doivent être spécifiées et pêuvent être des atomes, des ions, des électrons, d'autres particules ou des groupements spécifiés de telles particules [14ème CGPM (1971), Résolution 3]. Dans cette définition, il est sous-entendu que les atomes de carbone 12 sont libres, au repos et dans leur état fondamental.
(i) The mole is the amount of substance of a system which contains as many elementary entities as there are atoms in 0.012 kg of carbon 12. (ii) When the mole is used, the elementary entities must be specified and may be atoms, molecules, ions, electrons, other particules, or specified groups of such particles [14th CGPM (1971), Resolution 3]. In this definition, it is understood that the carbon 12 atoms are unbound, at rest and in their ground state.

candela

La candela est l'intensité lumineuse, dans une direction donnée, d'une source qui émet un rayonnement monochromatique de fréquence 540×10^{12} Hz et dont l'intensité énergétique dans cette direction est de 1/683 watt par stéradian. [16ème CGPM (1979), Résolution 3].
The candela is the luminous intensity, in a given direction, of a source that emits monochromatic radiation of frequency 540×10^{12} Hz and that has a radiant intensity in that direction of 1/683 watt per steradian [16th CGPM (1979), Resolution 3].

2.3.2 SI Supplementary Units

In addition to the seven base units, the SI has two supplementary units, the **radian** for plane angle and the **steradian** for solid angle (see *Table 2-2*). These two units are non-dimensional (i.e. in a dimension equation they have the value unity). However, for clarity, they are sometimes included in dimensional equations using an arbitrary dimensional symbol, for example the Greek letter α or a Roman capital 'A' for plane angle, and the Greek capital Ω for solid angle. Equally, because of the non-official nature of this notation, it is possible to omit these symbols from a dimensional equation in cases where this does not cause ambiguity.

As an example, the expression for angular velocity could be equally well be written as either $[\alpha.T^{-1}]$ or $[T^{-1}]$.

However, for physical quantities in specialist areas such as particle transfer in stastistical physics or luminous transfer in photometry, it is usual to include solid angle in dimensional equations in order to avoid confusion.

Example: depending on whether the area of photometric work involves measurement of energy, visible light, or particle emissions, luminous intensity can be defined in three ways (see *Table 2-3*). It is clear that confusion can be avoided by introduction of a symbol for the steradian in the dimensional equation.

Table 2-2 The two supplementary SI units

Physical quantity	Dimension	Name	Symbol
Plane angle	α	radian	rad
Solid angle	Ω	steradian	sr

Table 2-3 Comparison of dimension equation for several quantities in photometry with and without symbol of solid angle

Photometry	Quantity	Dimensions		SI unit
		without symbol	with symbol	
Energy	Radiant intensity	$[E.T^{-1}]$	$[E.T^{-1}.\Omega^{-1}]$	$W.sr^{-1}$
Visible	Luminous intensity	$[J]$	$[J]$	cd
Particle	Photon intensity	$[T^{-1}]$	$[T^{-1}.\Omega^{-1}]$	$s^{-1}.sr^{-1}$

2.3.3 SI Derived Units

The SI derived units are defined by simple equations relating two or more base units. The names and symbols of some derived units may be substituted by special names and symbols. The nineteen derived units with special names and symbols are listed in *Table 2-4*. These derived units may themselves by used in combination to form further derived units.

Table 2-4 SI derived units with special names and symbols

Name	Symbol	Physical quantity	Dimension	Equivalent in SI base units
becquerel	Bq	radioactivity	T^{-1}	$1\ Bq = 1\ s^{-1}$
coulomb	C	quantity of electricity, electric charge	IT	$1\ C = 1\ A.s$
farad	F	electric capacitance	$M^{-1}L^{-2}T^4I^2$	$1\ F = 1\ kg^{-1}.m^{-2}.s^4.A^2$
gray	Gy	absorbed dose of radiation, kerma	L^2T^{-2}	$1\ Gy = 1\ m^2.s^{-2}$
henry	H	electric inductance	$ML^2T^{-2}I^{-2}$	$1\ H = 1\ kg.m^2.s^{-2}.A^{-2}$
hertz	Hz	frequency	T^{-1}	$1\ Hz = 1\ s^{-1}$
joule	J	energy, work, heat	ML^2T^{-2}	$1\ J = 1\ kg.m^2.s^{-2}$
lumen	lm	luminous flux	$J\Omega$	$1\ lm = 1\ cd.sr$
lux	lx	illuminance	$J\Omega L^{-2}$	$1\ lx = 1\ cd.sr.m^{-2}$
newton	N	force, weight	MLT^{-2}	$1\ N = 1\ kg.m.s^{-2}$
ohm	Ω	electric resistance	$ML^2T^{-3}I^{-2}$	$1\ \Omega = 1\ kg.m^2.s^{-3}.A^{-2}$
pascal	Pa	pressure, stress	$ML^{-1}T^{-2}$	$1\ Pa = 1\ kg.m^{-1}.s^{-2}$
poiseuille (pascal-second)	Po	absolute viscosity, dynamic viscosity	$ML^{-1}T^{-1}$	$1\ Po = 1\ kg.m^{-1}.s^{-1}$
siemens	S	electric conductance	$M^{-1}L^{-2}T^3I^2$	$1\ S = 1\ kg^{-1}.m^{-2}.s^3.A^2$
sievert	Sv	dose equivalent, dose equivalent index	L^2T^{-2}	$1\ Sv = 1\ m^2.s^{-2}$
tesla	T	induction field, magnetic flux density	$MT^{-2}I^{-1}$	$1\ T = 1\ kg.A^{-1}.s^{-2}$
volt	V	electric potential, electromotive force	$ML^2T^{-3}I^{-1}$	$1\ V = 1\ kg.m^2.s^{-3}.A^{-1}$
watt	W	power, radiant flux	ML^2T^{-3}	$1\ W = 1\ kg.m^2.s^{-3}$
weber	Wb	induction magnetic flux	$ML^2T^{-2}I^{-1}$	$1\ Wb = 1\ kg.m^2.s^{-2}.A^{-1}$

2.3.4 Non-SI and SI Units Used in Combination

For consistency and clarity, it is a general rule of SI that the use of non-SI units should be discontinued. However, there are some important instances where this is either impracticable or inadvisable. The SI therefore recognizes four categories of out-of-system units which may be used in combination with SI.

2.3.4.1 Commonly Used Legal Non-SI Units

The CIPM (1969) recognized that users of SI would wish to employ certain units that are important and widely used, but which do not properly fall within the SI. The special names and symbols for these units, and their definitions in terms of SI units, are listed in ***Table 2-5***.

Table 2-5 Commonly used non-SI units

Name	Symbol	Physical quantity	Dimension	Equivalent in SI base units
ampere-hour	Ah	electric charge	IT	1 Ah = 3600 C
day	d	time, duration, period	T	1 d = 86 400 s
degree	°	plane angle	α	$1^\circ = \pi/180$ rad
dioptre	d	refractive power	L^{-1}	$1\ d = 1\ m^{-1}$
hour	h	time, duration, period	T	1 h = 3600 s
kilowatt-hour	kWh	energy, work, heat	ML^2T^{-2}	$1\ kWh = 3.6 \times 10^6$ J
litre	l, L	capacity, volume	L^3	$1\ L = 10^{-3}\ m^3$
minute	min	time, duration, period	T	1 min = 60 s
minute of angle	′	plane angle	α	$1' = \pi/10\,800$ rad
second of angle	″	plane angle	α	$1'' = \pi/648\,000$ rad
tex	tex	linear mass density	ML^{-1}	$1\ tex = 10^{-6}\ kg.m^{-1}$
tonne (metric)	t	mass	M	$1\ t = 10^3$ kg

2.3.4.2 Non-SI Units Defined by Experiment

This class incorporates units accepted for use, the values of which are obtained by experiment; they are listed in ***Table 2-6*** below. These are important units widely used for special problems, and were accepted by the CIPM (1969) for continuing use in parallel with SI units.

Table 2-6 Commonly used legal non-SI units defined by experiment

Name	Symbol	Physical quantity	Dimension	Equivalent in SI base units
electronvolt	eV	electric potential	$ML^2T^{-3}I^{-1}$	$1\ eV = \frac{e}{C}$ J
faraday	F	molar electric charge	ITN^{-1}	$1\ F = eN_A\ C.mol^{-1}$
unified atomic mass unit	u, u.m.a.	mass	M	$1\ u = \frac{m_{^{12}C}}{12}$ kg

2.3.4.3 Non-SI Units Temporarily Maintained

In view of existing practice, the CIPM (1978) considered it acceptable to retain for the time being a third class of non-SI units for use with those of SI. These temporarily-maintained units are listed in ***Table 2-7***. The use of any and all of these units may be abandoned at some time in the future. They should not therefore be introduced where they are not already in current use.

2.3.4.4 Non-SI Units Which Must be Discontinued

These units, listed in ***Table 2-8***, are to be avoided in favour of an appropriate SI unit or decimal multiples using the common SI prefixes listed in ***Table 2-9***.

Table 2.7 Temporarily-maintained non-SI units

Name	Symbol	Physical quantity	Dimension	SI base units
ångström	Å	length	L	1 Å = 10^{-10} m
are	a	surface, area	L^2	1 a = 10^2 m^2
bar	bar	pressure	$ML^{-1}T^{-2}$	1 bar = 10^5 Pa
barn	b	surface, area	L^2	1 b = 10^{-28} m^2
curie	Ci	radioactivity	T^{-1}	1 Ci = 3.7 × 10^{10} Bq
gal	Gal	acceleration	LT^{-2}	1 Gal = 10^{-2} $m.s^{-2}$
hectare	ha	surface, area	L^2	1 ha = 10^4 m^2
hogshead (tonneau de mer)	–	capacity, volume	L^3	1 tonneau = 2.83 m^3
nautical mile	naut. mi	length	L	1 naut. mi = 1852 m
noeud, knot	kn	linear velocity	LT^{-1}	1 knot = 5.14444 × 10^{-1} $m.s^{-1}$
rad	rad	absorbed dose of radiation, kerma	L^2T^{-2}	1 rad = 10^{-2} Gy
rem	rem	dose equivalent, dose equivalent index	L^2T^{-2}	1 rem = 10^{-2} Sv
röntgen	R	exposure	$M^{-1}TI$	1 R = 2.58 × 10^{-4} $C.kg^{-1}$

Table 2-8 Non-SI units to be discontinued

Name	Symbol	Physical quantity	Dimension	SI base units
calorie (15°C)	$cal_{15°C}$	energy, heat	ML^2T^{-2}	1 $cal_{15°C}$ = 4.1855 J
calorie (IT)	cal_{IT}	energy, heat	ML^2T^{-2}	1 cal_{IT} = 4.1868 J
calorie (therm.)	cal_{th}	energy, heat	ML^2T^{-2}	1 cal_{th} = 4.1840 J
carat (metric)	ct	mass	M	1 ct = 2 × 10^{-4} kg
fermi	F	length	L	1 F = 10^{-15} m
gamma (induction)	γ	magnetic induction	$MT^{-2}I^{-1}$	1 γ = 10^{-9} T
gamma (mass)	γ	mass	M	1 γ = 10^{-9} kg
grade (gon)	gr, ^{g}	plane angle	α	1 gon = π/200 rad
kilogram-force	kgf	force	MLT^{-2}	1 kgf = 9.80665 N
lambda	λ	capacity, volume	L^3	1 l = 10^{-9} m^3
micron	μ	length	L	1 μ = 10^{-6} m
standard atmosphere	atm	pressure	$ML^{-1}T^{-2}$	1 atm = 101 325 Pa
stère	st	capacity, volume	L^3	1 st = 1 m^3
revolution	rev	plane angle	α	1 rev = 2π rad
revolutions per minute	rpm	angular velocity	$αT^{-1}$	1 rpm = π/30 $rad.s^{-1}$
torr	torr	pressure	$ML^{-1}T^{-2}$	1 torr = 101 325/760 Pa
X-unit	XU	length	L	1 XU = 1.0023 × 10^{-10}

2.4 SI Prefixes

The SI is a decimal system of units. Fractions have been eliminated, multiples and sub-multiples being formed using a series of prefixes ranging from yotta (10^{24}) to yocto (10^{-24}). The 20 SI prefixes that are to be used for multiples and sub-multiples of an SI unit are shown in ***Table 2-9***. Each prefix beyond $10^{\pm 3}$ represents a change in magnitude of 10^3 (power of 10 notation).

Table 2-9 The 20 SI prefixes

SI prefixes: multiples and submultiples					
Multiple			**Submultiple**		
Prefix (Etymology)	Symbol	Multiply by	Prefix (Etymology)	Symbol	Multiply by
yotta	Y	10^{24}	deci (Latin *decimus* tenth)	d	10^{-1}
zetta	Z	10^{21}	centi (Latin *centum* hundredth)	c	10^{-2}
exa	E	10^{18}	milli (Latin *milli* thousandth	m	10^{-3}
peta	P	10^{15}	micro (Greek *μικρος* small)	μ	10^{-6}
tera (Greek *τεραζ* monster)	T	10^{12}	nano (Greek *νανος* dwarf)	n	10^{-9}
giga (Greek *γιγαζ* gigant)	G	10^{9}	pico (Italian *piccolo* small)	p	10^{-12}
mega (Greek *μεγαζ* big)	M	10^{6}	femto (Danish *femten* fifteen)	f	10^{-15}
kilo (Greek *κιλιοι* thousand)	k	10^{3}	atto (Danish *atten* eighteen)	a	10^{-18}
hecto (Greek *εκατον* hundred)	h	10^{2}	zepto	z	10^{-21}
deca (Greek *δεκα* ten)	da	10	yocto	y	10^{-24}

Important Note: in Computer Science, the prefixes kilo, mega, and giga are commonly used, although in that context they are only approximations to powers of 10. These multiples of the byte are not equal to the SI prefixes because they are equal to the power of two according to the binary digit numeration. In order to avoid confusion, it is therefore recommended that in this context they are expressed with an initial capital as shown in ***Table 2-10***:

Table 2-10 Prefixes for Computer Science units

Name of unit (English, French)	Symbol	Conversion factor
Kilobyte, Kilooctet	KB, Ko	1 KB = 2^{10} = 1024 bytes
Megabyte, Mégaoctet	MB, Mo	1 MB = 2^{20} = 1 048 576 bytes
Gigabyte, Gigaoctet	GB, Go	1 GB = 2^{30} = 1 073 741 824 bytes

Prefixes and symbols should be printed in roman (upright) type with no space between the prefix and the unit symbol.

Example: 1 millimetre = 1 mm = 10^{-3} m (not m/m or m m)

When a prefix symbol is used with a unit symbol, the combination should be considered as a single new symbol that can be raised to a positive or negative power of 10 without using brackets.

Example: 1 $\mathrm{cm}^3 = (10^{-2}\ \mathrm{m})^3 = 10^{-6}\ \mathrm{m}^3$ (not $\mu(\mathrm{m}^3)$ or $10^{-2}\ \mathrm{m}^3$)

Prefixes are not to be combined into compound prefixes.

Example: 1 nm = 10^{-9} m (not mμm)
1 GW = 10^9 W (not MkW)
1 pF = 10^{-12} F (not μμF)

A prefix should never be used alone.

Example: 1 μm = 10^{-6} m (not μ)

Prefixes used with the kilogram (which already has the prefix kilo) are constructed by adding the appropriate prefix to the word gram and the symbol g.

Example: 1 Mg = 10^6 g (not kkg)

The prefixes apply to all standard and associated SI units with the exception of the following: h (hour), d (day), min (minute), rev (revolution), ° (degree), ′ (plane angle minute), and ″ (plane angle second).

If the name of the unit begins with a vowel, the prefix may be fused with it:

Example: 1 Ω = one meg**o**hm rather than megaohm

3 Other Systems of Units

Despite the internationalization of SI units, and the fact that other units are actually forbidden by law in France and other countries, there are still some older or parallel systems remaining in use in several areas of science and technology.

Before presenting conversion tables for them, it is important to put these systems into their initial context. A brief review of systems is given ranging from the ancient and obsolete (e.g. Egyptian, Greek, Roman, Old French) to the relatively modern and still in use (e.g. UK imperial, US, cgs, FPS), since a general knowledge of these systems can be useful in conversion calculations. Most of the ancient systems are now totally obsolete, and are included for general or historical interest.

3.1 MTS, MKpS, MKSA

3.1.1 The MKpS System

The former system of units referred to by the international abbreviations MKpS, MKfS, or MKS (derived from the French titles mètre-kilogramme-poids-seconde or mètre-kilogramme-force-seconde) was in fact entitled *Système des Mécaniciens* (Mechanical Engineers' System). It was based on three fundamental units, the **metre**, the **second**, and a weight unit, the **kilogram-force**. This had the basic fault of being dependent on the acceleration due to gravity *g*, which varies on different parts of the Earth, so that the unit could not be given a general definition. Furthermore, because of the lack of a unit of mass, it was difficult, if not impossible to draw a distinction between weight, or force, and mass (see also 3.4). In addition, the mechanical units were not self-consistent, as for example the unit of power, the horsepower, which is equal to 75 $kg.m.s^{-1}$. Finally, there were no links with magnetic, electrical, or thermodynamic units.

3.1.2 The MTS System

The French MTS system was based on the **metre**, the metric **tonne**, and the **second**, and was in fact the only legal system used in France between 1919 and 1961, when SI units were formally adopted. Several derived units with special names were based on these three fundamental units, for example the **sthene** (sn) for force or weight, and the **pieze** (pz) for pressure. Like the MKpS system, it had no links with electrical, magnetic, or thermodynamic units.

3.1.3 The MKSA (Giorgi) System

In 1904, the Italian physicist Giovanni L.T. Giorgi proposed a system based on five fundamental units. It was adopted by the IEC during the period 1935–1950. The units of length was to be the **standard metre** maintained at Sèvres, the unit of mass the **standard kilogram**, the unit of time the **second**, plus two new base units, the **ampere** for electric current intensity, and **vacuum magnetic permeability** which was defined as $\mu_0 = 4\pi \times 10^{-7}$ H.m^{-1}. This linkage meant that all units in the system could be used in electromagnetic or electrostatic contexts. The introduction of the factor 4π in the expression for vacuum magnetic permeability meant that all units could be rationalized, i.e. a factor of 2π applied when a system had cylindrical symmetry, and of 4π if it had spherical symmetry.

The advantage – or, for some physicists, the disadvantage – of this system lay in the fact that it made a clear distinction between magnetic field strength H and magnetic flux density B, and similarly between electric field strength E and electric flux density D. This distinction results from the expression for vacuum permittivity and magnetic permeability which is not equal to unity as in the cgs system. The vector equations relating these four quantities are therefore:

$$\vec{B} = \mu\vec{H} = \mu_0\mu_r\vec{H}$$
$$\vec{D} = \varepsilon\vec{E} = \varepsilon_0\varepsilon_r\vec{E}$$

The Giorgi system only became common in electrical engineering from 1948. At that time, the 9th CGPM adopted the modern definition of the ampere. The MKSA system is thus the precursor of the SI, and, perhaps for this reason, there remains some confusion between the two systems among some scientists and engineers.

3.2 Cgs, Gauss, IEUS, a.u.

3.2.1 The cgs System

The cgs (centimetre-gram-second) system has as its three base units the **centimetre**, the **gram**, and the **second**. It was proposed in 1873 by the distinguished British scientists Lord Kelvin and James Clerk Maxwell, and the famous German electrical engineer Ernst Werner von Siemens. As a system, it was outstanding for its consistency and for its clear distinction between force and mass. There are also advantages in the use of equations in four basic dimensions, one of which is electrical, and two fundamental subsystems came into existence. As a result, the General Assembly of the IUPAP in Copenhagen, 1951, approved via its Resolution 5 the introduction of the following generalized cgs subsystems:

- the *electrostatic cgs system* (centimetre, gram, second, and franklin)
- the *electromagnetic cgs system* (centimetre, gram second, and biot)

The system met with wide acceptance among scientists in many countries and was rapidly extended to every branch of physics. However, many of its units are too small for most scientific and engineering purposes.

Although the use of cgs units is officially discouraged since the introduction of the SI in 1960, practitioners in some fields of physics, such as electricity, magnetism, and optics, have continued to use unofficial derived units (e.g. dyne, erg, poise, stokes, gauss, oersted, maxwell, stilb, phot). The main reason for this is that these units are often of the same order of magnitude as the physical phenomena they define.

The major disadvantage of the cgs system is its inherent subdivision into three subsystems: electromagnetic units (emu or ab units), electrostatic units, (esu or stat units), and the system of practical units for common use. The complications introduced by inter-conversion of these sub-units were yet another reason for its eventual abandonment in favour of the MKSA system and ultimately the SI.

3.2.1.1 The esu Subsystem

In this cgs subsystem, the electrostatic force $\vec{F}$ between two point charges q_1 and q_2 separated by a distance r in a medium of permittivity ε is given by Coulomb's law, i.e. $\vec{F} = \frac{q_1 q_2}{\varepsilon r^2} \vec{e}_r$ if F, r, ε, are made equal to unity and $q_1 = q_2 = q$, q_1, and q_2 are unit electric charge. The cgs system of electrostatic units is based on this definition of electric charge. This is the **franklin** (Fr), the cgs unit of electric charge, which is formally defined as follows:

The **franklin** is that charge which exerts on an equal charge at a distance of one centimetre in vacuo a force of one dyne (1941).

All these units are prefixed with the separate acronym **esu** or an international or attached indicator **stat**.

Example: **stat**coulomb or **esu** coulomb (=1 Fr)

3.2.1.2 The emu Subsystem

As with the electrostatic units, the electromagnetic subsystem defines the electromagnetic force F between two hypothetical isolated point magnetic poles of strengths m_1 and m_2 separated by a distance r in a medium of magnetic permeability μ by Coulomb's Law for Magnetism, i.e. $\vec{F} = \frac{m_1 m_2}{\mu r^2} \vec{e}_r$, setting F, r, and m equal to unity, and $m_1 = m_2 = m$, m_1 and m_2 are unit pole strengths. The cgs system of electromagnetic units is based on this definition of pole strength, analogous with the electrostatic system, and has the **biot** (Bi) as the cgs unit of magnetic pole strength, defined as follows:

The **biot** is that constant current which, if maintained in two straight parallel conductors of infinite length, of negligible circular cross-section, and placed one centimetre apart in vacuo, would produce between these conductors a force equal to two dynes per centimetre of length (1961).

All these units are prefixed with the separate acronym **emu**, or attached indicator **ab**.

Example: **ab**coulomb or **emu** coulomb (=1 Bi)

Important Notes:

(i) The emu and esu are interconnected by the fundamental equation $\varepsilon\mu c^2 = 1$ where c is the velocity of light in vacuum. Thus the ratio of any pair of emu-esu primary units is equal to c or its reciprocal.

$$\textit{Example:} \quad \frac{\text{abampere}}{\text{statampere}} = \frac{\text{statvolt}}{\text{abvolt}} = c$$

For esu or emu derived units, the ab/stat ratio is obtained by considering each of the primary units involved, thus:

$$\textit{Example:} \quad \frac{\text{abfarad}}{\text{statfarad}} = \frac{\text{abcoulomb}}{\text{abvolt}} = \frac{\text{statvolt}}{\text{statcoulomb}} = c^2$$

(ii) Since electromagnetic and electrostatic units vary so enormously, a third cgs subsystem was used for most practical purposes in electrical engineering. Clearly, however, this added considerable complication to its general structure.

(iii) The use of the cgs system in fields other than mechanics involves exact definition of the subsystem concerned, which again adds to the confusion and is a great source of error in conversion computations.

3.2.2 The Gauss System

Gaussian units are a combination of the emu and esu subsystems. With three base units, it uses em units in magnetism and es units in electrostatics. This involves using the constant c (the velocity of light in vacuum) to interrelate these sets of units, resulting in complex and error-prone conversions.

Table 3-1 below gives the classes of units used for equivalent electromagnetic and electrostatic quantities.

Table 3-1 Organization of Gaussian units

Electrostatic units (esu)	Electromagnetic units (emu)
electric charge Q (Fr)	magnetic mass m
electric current intensity I (Fr.s^{-1})	magnetic flux Φ
electric field strength $\vec{E}$ (dyne.Fr^{-1})	magnetic field strength $\vec{H}$ (Bi.cm^{-1})
electric displacement $\vec{D}$ (Fr.cm^{-2})	induction field $\vec{B}$ (dyn.Bi^{-1}.cm^{-1})
electric potential $\vec{D}$ (erg.Fr^{-1})	magnetic potential $\vec{A}$
polarization $\vec{P}$	magnetization $\vec{M}$
electric dipole moment $\vec{p}$ (Fr.cm)	magnetic dipole moment $\vec{m}$ (erg.g^{-1})
electric susceptibility χ_e	magnetic susceptibility χ_m
polarizability α	magnetizability ξ

3.2.3 International Electrical Units

This separate system of electrical units was used by US electrical engineers, and was adopted internationally until 1947, when it was declared obsolete and replaced first by the MKSA and then by SI, with which it should not be confused. Its base units are defined in concrete terms, as shown by the examples in ***Table 3-2***.

Table 3-2 Basic IEUS units

ampere (int. mean)	One ampere (int. mean) is equal to the unvarying electric current intensity which deposits, in one second, by electrolysis from an aqueous silver nitrate solution, 0.00111800 g of silver metal at the cathode (IEC, 1881)
ohm (int. mean)	One ohm (int. mean) is equal to the electric resistance, measured at the temperature of melting ice (0°C), of a mercury column of 106.300 cm length which has a mass equal to 14.4521 g (IEC, 1908)
volt (int. mean)	One volt (int. mean) is equal to the electromotive force (e.m.f.), measured at 20°C, of a Weston electrochemical cell. It is equal exactly to 1.0183 int. volt (IEC, 1908)

3.2.4 Atomic Units (a.u.)

The system of atomic units with the international acronym **a.u.** was proposed by D.R. Hartree in 1927 with a view to simplifying calculations in problems involving the basic structures of the atom and molecule, as well as in computations in quantum mechanics. The system was based on units of four fundamental quantities, mass, length, time, and electric charge.

The fundamental units of this system were based on five universal constants: the **electron rest mass** (m_0) represented the unit of mass, the **elementary electrostatic charge** (e) was the unit of electric charge, the **first orbit Bohr radius** (a_0) was the unit of length, the first ionizing energy of the hydrogen atom in its ground state, or **rydberg** (Ry) was the unit of energy, and the **rationalized Planck constant** ($\hbar$) was the unit of angular momentum.

In 1959, Shull and Hall proposed a new unit of energy which was approximately equal to two rydbergs, to be called the **hartree** in honor of the inventor of the system.

The importance of this system of units lay in the fact that the numerical results of calculations were expressed as combinations of fundamental atomic constants. The a.u. system is therefore regarded as 'natural units' for calculations involving electronic structure in quantum chemistry. For clarity, they are usually set in *italic* type to distinguish them from other units which should be set in roman type.

It is also usual in specialized litreature (e.g. quantum chemistry, mathematical physics, nuclear and molecular physics) to find the acronym **a.u.** in place of the appropriate unit irrespective of the physical quantity involved. This of course leads to considerable confusion in identification and conversion.

Table 3-3 (overleaf) summarizes the main units employed in the system.

3.3 British and American Systems of Units

Despite the increasing importance of SI units, the systems of units developed in the UK and USA are still commonly used or referred to in the British Isles (England, Ireland, Scotland, and Wales), North America (USA and Canada), and in some Commonwealth countries such as Australia and New Zealand. These systems are not only non-metric, but also non-decimal, which increased the complexity of calculations and proved a powerful argument against their continued use, especially in science and technology. However, a counter-argument, that was perhaps more cogent before the age of the electronic calculator, is that their quantities were evenly divisible by a greater number of basic prime factors, and did not so often result in lengthy or recurring decimals.

Table 3-3 Base and derived units of the a.u. system

a.u. quantity	Dimension	SI conversion factor	Equation
angular momentum, action	ML^2T^{-1}	$= 1.05457266\,(63) \times 10^{-34}$ J.s	$\hbar = h/2\pi$
charge density	ITL^{-3}	$= 1.08120262 \times 10^{-12}$ C.m^{-3}	$e/a_0{}^3$
electric charge	IT	$= 1.60217733\,(49) \times 10^{-19}$ C	e
electric current intensity	I	$= 6.6236211\,(20) \times 10^{-3}$ A	$eE_h/\hbar$
electric dipole moment	LTI	$= 8.4783579\,(26) \times 10^{-30}$ C.m	ea_0
electric field gradient	$MT^{-3}I^{-1}$	$= 9.71736459 \times 10^{21}$ V.m^{-2}	$E_h/ea_0{}^2 = e/4\pi\varepsilon_0 a_0{}^3$
electric field strength	$MLT^{-3}I^{-1}$	$= 5.1422082\,(15) \times 10^{11}$ V.m^{-1}	$E_h/ea_0 = e/4\pi\varepsilon_0 a_0{}^2$
electric potential	$ML^2T^{-3}I^{-1}$	$= 27.21139622$ V	$E_b/e = e/4\pi\varepsilon_0 a_0$
electric quadripole moment	L^2TI	$= 4.48655413 \times 10^{-40}$ C.m^2	$ea_0{}^2$
energy (hartree)	ML^2T^{-2}	$= 4.3597482\,(26) \times 10^{-18}$ J	$E_h = \hbar^2/m_0 a_0{}^2$
force	MLT^{-2}	$= 8.2387295\,(25) \times 10^{-8}$ N	E_h/a_0
induction magnetic field, magnetic flux density	$MT^{-2}I^{-1}$	$= 2.35051808\,(71) \times 10^{5}$ T	$\hbar/ea_0{}^2$
length (1st Bohr radius)	L	$= 5.29177249(24) \times 10^{-11}$ m	$a_0 = 4\pi\varepsilon_0\hbar^2/m_0 e^2$
linear momentum	MLT^{-1}	$= 1.9928534\,(12) \times 10^{-24}$ N.s	$\hbar/a_0$
magnetic dipole moment	L^2I	$= 1.85480308\,(62) \times 10^{-23}$ J.T^{-1} (A.m^2)	$e\hbar/m_0 = 2\mu_B$
magnetizability	$L^{-1}I$	$= 7.89104 \times 10^{-29}$ A.m^{-1}	$e^2 a_0{}^2/m_0$
mass (electron rest mass)	M	$= 9.1093897\,(54) \times 10^{-31}$ kg	m_0
polarizability	$L^{-2}TI$	$= 1.64878 \times 10^{-41}$ J^{-1}.C^2.m^2	$4\pi\varepsilon_0 a_0{}^3$
time	T	$= 2.418884338 \times 10^{-17}$ s	$\hbar/E_b = m_e a_0{}^2/\hbar$

Preliminary notes for number writing:

- **Cardinal number writing:** in France, Italy and other European countries, the space ( ) or the point (.) is used to separate hundreds from thousands. In Britain and North America, either the comma is used (sometimes omitted in four-figure numbers) or the space (usually omitted in four-figure numbers). The latter is used in this book.

  Example: 1,657 or 1657 instead of 1 657.

- **Decimal number writing:** in France, Italy and other European countries, the comma (,) is used to separate integer from decimal. In Britain and North America the point is used.

  Example: 3.14159 instead of 3,14159.

3.3.1 Imperial Units

The British system of units, known as imperial units, was established by the *Weights and Measures Act* (WMA, 1824) of June 17th, 1824. Its three base units are the **pound avoirdupois**, the **yard**, and the **second.** The yard was defined in 1878 as the distance at 62°F between a pair of lines etched in gold plugs set in a bronze bar. Earlier in 1856, the pound avoirdupois was defined in terms of the mass of a platinum cylinder, known as the Imperial Standard Pound, and both were kept in the Standards Department of the Board of Trade in London. The imperial unit of capacity was the **gallon,** which was defined as the volume of 10 pounds avoirdupois of distilled water, weighed in air against brass[7] weights at a temperature of 62°F and atmospheric pressure of 30 inches of mercury. These legal measures were then used in all countries of the British Commonwealth. To a close approximation, the metric equivalents of these imperial pound and yard standards are 0.45359243 kg and 0.91443992 m respectively.

However, the WMA of 1963 modified the nature of these standards, and redefined the units in terms of the kilogram and metre standards maintained in Paris (at the Pavillon de Breteuil, Sèvres). These new precise definitions of the pound and yard are respectively 0.45359237 kg and 0.9144 m. The unit of time, the second, has always been the same in both systems and was redefined by the 13th CGPM in 1968.

In 1980, Parliament approved a Statutory Instrument (1980/1070) which began the progressive phasing out of the imperial system by withdrawing authorization of a substantial number of units, such as the **British thermal unit** (Btu), the **cran,** the **furlong,** the **horsepower** (HP), the **hundredweight** (cwt), the **ton** and the **Fahrenheit degree** (°F). In 1985, the **curie, rem,** and **rad** were discarded in favour of the **becquerel, sievert,** and **gray** respectively, although the legal status of the pound and yard was reaffirmed by the WMA of 1985.

However, from November 1995, the United Kingdom officially adopted metric units for general use and business transactions. As a result, the metre and kilogram are now compulsory in place of the yard and pound avoirdupois.

3.3.1.1 Imperial Units of Length

The many units of length used at various times in the UK fall into several categories, depending on area of application.

---

[7] For official comparison between standard organizations through the UK, the density of bras alloy was taken as being equal to 8413 kg m^3.

Table 3-4 UK linear measures

UK stat. league (st. lg)	UK stat. mile (st. mi)	Pole (rd)	Yard (yd)	Cubit (cu)	Foot (ft,′)	Span (sp)	Hand (hd)	Palm (plm)	Inch (in,″)	UK line (line)
1	= 3	= 960	= 5280	= 10 560	= 15 840	= 21 120	= 47 520	= 63 360	= 190 080	= 2 280 960
	1	= 320	= 1760	= 3520	= 5280	= 7040	= 15 840	= 21 120	= 63 360	= 76 0320
		1	= 11/2	= 11	= 16. 5	= 22	= 49.5	= 66	= 198	= 2376
			1	= 2	= 3	= 4	= 9	= 12	= 36	= 432
				1	= 3/2	= 2	= 9/2	= 6	= 18	= 216
					1	= 4/3	= 3	= 4	= 12	= 144
						1	= 9/4	= 3	= 9	= 108
							1	= 4/3	= 4	= 48
								1	= 3	= 36
									1	= 12

1 rod = 1 perch = 1 pole
1 land = 1 statute mile

3.3.1.1.1 UK Linear Measure

Current linear units are the **line**, the **inch**, the **foot**, the **yard**, the **statute mile** (or **land mile**) and the **statute league** (or **land league**). They are employed for measuring distance, length, width, depth, height, and thickness, and are listed fully in ***Table 3-4***.

Other units of length which have been used in the UK are:

1 skein (UK)	= 360 feet
1 bolt (UK)	= 120 feet
1 shackle (UK)	= 90 feet
1 rope (UK)	= 20 feet
1 ell (UK)	= 45 inches
1 nail (UK)	= 9 inches
1 span (UK)	= 9 inches
1 barleycorn (UK)	= 1/3 inch
1 caliber	= 0.01 inch
1 mil	= 0.001 inch
1 mil	= 1 thou

3.3.1.1.2 UK Nautical Measure

The units of length used in navigation are the **fathom**, and the **nautical mile**. The **cable length** and the **nautical league** are now obsolete.

Table 3-5 UK nautical measures of length
[1 mile (UK naut.)=1853.184 m (E)]

UK nautical league (UK, naut. lg)	**UK nautical mile** (UK, naut. mi)	**UK cable length** (UK, naut. cbl)	**UK nautical chain**	**Fathom** (fath)	**Yard** (yd)	**Foot** (ft)	**Inch** (in)
1	= 3	= 30	= 1216	= 3040	= 6080	= 18 240	= 218 880
	1	= 10	= 1216/3	= 3 040/3	= 6080/3	= 6080	= 72 960
		1	= 608/15	= 304/3	= 608/3	= 608	= 7296
			1	= 5/2	= 5	= 15	= 180
				1	= 2	= 6	= 72
					1	= 3	= 36
						1	= 12

3.3.1.1.3 UK Surveyors' Measure

Some other units of length were once used by UK land surveyors, the main ones being the **link**, the **chain** (**Gunter's chain**), the **rod**, and the **furlong**. All these units are now obsolete, except for the **furlong**, which is still commonly used to define distances in a UK horse-race.

Table 3-6 UK surveyors' measures of length

[1 chain (Gunter's) = 20.1168 m (E)]

Statute mile (st. mi)	Furlong (fur)	Gunter's chain (ch)	Rod (rd)	Yard (yd)	Foot (ft)	Gunter's link (lk)
1	= 8	= 80	= 320	= 1760	= 5280	= 8000
	1	= 10	= 40	= 220	= 660	= 1000
		1	= 4	= 22	= 66	= 100
			1	= 5.5	= 16.5	= 25
				1	= 3	= 50/11
					1	= 50/33

1 rod = 1 perch = 1 pole
1 land = 1 statute mile

3.3.1.2 Imperial Units of Area

As with linear measures, units of surface area fall into groups based on their application.

Notes

- It is important to note that in North America and Britain, the prefix *sq.* (an abbreviation of the word *square*) is sometimes used before an area unit instead of raising the unit to the power of two.

  Example: 1666.66 sq. ft = 1666.66 ft^2

- In some technical reports or scientific textbooks written before the 1960s, the term 'square' is represented by a square symbol drawn before the unit (thus □″ should be read as 'square inch').

  Example: 2.56 □′ = 2.56 ft^2 (obsolete)

3.3.1.2.1 UK Measures of Area

The conventional imperial units of surface area are the **square inch**, the **square foot**, the **square yard**, and the **square mile**. They are related as follows:

Table 3-7 UK measures of area

[1 $\text{ft}^2 = 9.290304 \times 10^{-2}\ \text{m}^2$ (E)]

Square mile (sq. mi)	Square rod (sq. rd)	Square yard (sq. yd)	Square foot (sq. ft)	Square inch (sq. in)
1	= 102 400	= 3 097 600	= 27 878 400	= 4 014 489 600
	1	= 121/4	= 1089/9	= 39 204
		1	= 9	= 1296
			1	= 144

1 square rod = 1 square perch = 1 square pole

3.3.1.2.2 *UK Surveyors' Measure*

The now obsolete units of area once used by surveyors are listed below. The **square foot** and the **acre** are the only measures to survive into recent times.

Table 3-8 UK surveyors' measure
[1 ft^2 = 9.290304 × 10^{-2} m^2 (E)]

Section of land (sq. mi)	Acre (ac)	Rood (ro)	Square chain (sq. yd)	Square rod (sq. rd)	Square foot (sq. ft)	Square link (sq. lk)
1	= 640	= 1600	= 6400	= 102 400	= 27 878 400	= 64 000 000
	1	= 4	= 10	= 160	= 43 560	= 100 000
		1	= 5/2	= 40	= 10 890	= 25 000
			1	= 16	= 4356	= 10 000
				1	= 1089/4	= 625
					1	= 2500/1089

1 rod = 1 perch = 1 pole
1 section of land = 1 square statute mile

3.3.1.2.3 *Circular Units*

Very rarely, so-called 'circular' units have been used, mainly for wire sizes and then only as the **circular mil** in the USA. In an analogy to square measure, circular units represent the area of a circle of a diameter equal to the equivalent linear unit. A circular mil is the area of a circle one mil (0.001 in) in diameter, and hence equal to 7.854 × 10^{-7} in^2 or 5.067 × 10^{-10} m^2. Circular units should not be confused with **circular measure**, which refers to the expression of angle in radians.

Table 3-9 Circular units
[1 cin = 5.067074791 × 10^{-4} m^2]

Circular inch (cin)	Circular millimetre (cmm)	Circular mil (cmil)
= 4.014 × 10^9	= 2.590 × 10^{12}	= 4.015 × 10^{15}
= 144	= 92 903.04	= 144 000 270
1	= 645.16	= 1 000 000
	1	= 1550

3.3.1.3 Imperial Units of Volume and Capacity

As with the UK units of length and area, units of volume and capacity are often associated with specific trades or fields of application. They do however, in common with many older systems, fall into three categories: geometric measure, expressed in cubic linear units (e.g. **cubic foot**), dry measure, and liquid measure. The latter two categories are known as units of capacity and consist of arbitrary volumes given specific names (e.g. **bushel, gallon**), with the same name sometimes being used for measuring solids or liquids and thus having different volumes (e.g. the UK dry pint = 0.5506 l, while a UK liquid pint = 0.473 l).

Note

- It is important to note that in North America and Britain, the prefix *cu.* (an abbreviation of the word *cubic*) is sometimes used before a volume unit instead of raising the unit to the power of three.

Example: 33.75 cu. ft = 33.75 ft^3

3.3.1.3.1 UK Measures of Volume

The volumetric units chiefly used in the UK are the **cubic inch, cubic foot, cubic yard**, and **cubic statute mile**. They are used for volumes of containers, tanks, boxes, etc. as well as for solids such as stone, concrete, woods, etc. and are summarized in the following table.

Table 3-10 UK geometric measures of volume
[1 ft^3 = 2.831684659 × 10^{-2} m^3 (E)]

Cubic mile (cu. st. mi)	Register ton	Rod	Cubic yard (cu. yd)	Cubic foot (cu. ft)	Cubic inch (cu. in)
1	= 5 451 770	= 54 517 760	= 5 451 776 000	= 1.472 × 10^{11}	= 2.543 × 10^{14}
	1	= 10	= 1000	= 27 000	= 46 602 000
		1	= 100	= 2700	= 4 660 200
			1	= 27	= 46 656
				1	= 1728

1 land = 1 statute mile

3.3.1.3.2 UK Liquid Measure

Practical commercial measurement of liquids meant that vessels of a common size had to be used, with the result that certain standard (if arbitrary) volumes came into use. Of those listed in ***Table 3-11***, the fluid ounce, pint, quart, and gallon are still in common use. Although the litre is now official in the UK, it is likely that the pint will continue in use for some time because of its popularity as a measure of alcoholic beverages, chiefly beer.

3.3.1.3.3 UK Dry Measure

Over the years, British trades and professions developed standard measures of dry volume for commercial purposes. They were based on liquid measures and are now identically equal to them. The measures in the ***Table 3-12*** are all either obsolete or becoming so, with the exception of the register ton used to specify the size (gross register tonnage) of cargo ships.

Table 3-11 UK liquid measures of capacity

[1 gallon (UK) = 4.546092 × 10^{-3} m^3 (E)]

Chaldron (UK chal)	Bucket (UK bk)	Gallon (UK gal)	Pottle (UK pot)	Quart (UK qt)	Pint (UK pt)	Gill (UK gi)	Fluid ounce (UK fl. oz)	Fluid dram (UK fl. dr)	Minim (UK min)
= 10/9	= 80	= 320	= 640	= 1280	= 2560	= 10 240	= 51 200	= 409 600	= 24 576 000
1	= 72	= 288	= 576	= 1152	= 2304	= 9216	= 46 080	= 368 640	= 22 118 400
	= 6	= 24	= 48	= 96	= 188	= 752	= 3760	= 30 080	= 1 804 800
	1	= 4	= 8	= 16	= 32	= 128	= 640	= 5120	= 307 200
		1	= 2	= 4	= 8	= 32	= 160	= 1 280	= 76 800
			1	= 2	= 4	= 16	= 80	= 640	= 38 400
				1	= 2	= 8	= 40	= 320	= 19 200
					1	= 4	= 20	= 160	= 9600
						1	**= 5**	= 40	= 2400
							1	**= 8**	= 480
								1	**= 60**

Table 3-12 UK dry measures

[1 gallon (UK) = 4.546092×10^{-3} m^3 (E)]

Wey (UK wy)	**Chaldron** (UK chal)	**Sack** (UK sk)	**Bushel** (UK bu)	**Bucket** (UK bk)	**Peck** (UK pk)	**Gallon** (UK gal)	**Quart** (UK qt)	**Pint** (UK pt)
1	= 10/9	= 40/3	= 40	= 80	= 160	= 320	= 1280	= 2560
	1	= 12	= 36	= 72	= 144	= 288	= 1152	= 2304
		1	= 3	= 6	= 12	= 24	= 96	= 188
			1	= 2	= 4	= 8	= 32	= 64
				1	= 2	= 4	= 16	= 32
					1	= 2	= 8	= 16
						1	= 4	= 8
							1	= 2
								1

Other UK capacity units are as follows:

1 last (UK)	= 640 gallons (UK)
1 butt (UK)	= 108 gallons (UK)
1 puncheon (UK)	= 70 gallons (UK)
1 seam (UK)	= 64 gallons (UK)
1 hogshead (UK)	= 63 gallons (UK)
1 quarter (UK)	= 1 seam (UK)
1 coomb (UK)	= 32 gallons (UK)
1 kilderkin (UK)	= 18 gallons (UK)
1 strike (UK)	= 16 gallons (UK)
1 firkin (UK)	= 9 gallons (UK)
1 roquille (UK)	= 1 gill (UK)
1 noggin (UK)	= 1 roquille (UK)
1 quartern (UK)	= 1 gill (UK)
1 drop (UK)	= 1 minim (UK)

3.3.1.4 Imperial Units of Weight

3.3.1.4.1 UK Avoirdupois Weight

About AD 1300, London merchants adopted a system of weights known as 'avoirdupois', from the Old French *aver de peis* (goods of weight). This system, used for wholesale weighing, was based on a pound of 7000 grains. The **pound avoirdupois** was established under the WMA of 1856 (see note under yard above) and until 1963 was defined as the mass of the Imperial Standard Pound, a platinum cylinder kept in the Standards Department of the Board of Trade in London. Over that period, it was imposed and used in all countries of the British Commonwealth.

Now, most countries are converting to metric units and the avoirdupois system is being phased out. It was however very widely used and there is still considerable resistance to its replacement. Its base unit is the pound of 16 oz, which under the WMA of 1963 was defined as 0.45359237 kg. A standard New Imperial Pound was created in the form of a platinum cylinder maintained at the Standards Office, Westminster, but, from 1995, Britain officially adopted metric units for all general and business purposes. The avoirdupois units of mass were indicated by the abbreviation avoir., avdp. or av.

Of the units listed in ***Table 3-13***, only the **ounce, pound, stone, hundredweight**, and **ton** have been in anything like regularly use in recent years.

3.3.1.4.2 UK Apothecaries' Weight

The units of weight of the apothecaries' system were formerly used by pharmacists for drugs and medicinal preparations, presumably because their small magnitudes better represented the quantities used in this profession. In this system, the base unit was the **apothecaries' ounce** and was equal to the Troy ounce (see below). The units were indicated by the abbreviation *ap.* or *apth.* These units were legalized by the *Medical Education Acts* of 1858 and 1862, but are now completely obsolete and their use is prohibited by the British and American Pharmacopoeia.

Table 3-13 UK avoirdupois weight

[1 lb av. (WMA, 1963) = 0.45359237 kg (E)]

Ton (UK ton)	Wey, Load (wy av.)	Hundredweight (cwt)	Quarter (qr av.)	Stone (st av.)	Clove (cv av.)	Pound (lb av.)	Ounce (oz av.)	Dram (dr av.)	Grain (gr av.)
1	= 80/9	= 20	= 80	= 160	= 280	= 2240	= 35 840	= 573 440	= 15 680 000
	1	= 9/4	= 9	= 18	= 63/2	= 252	= 4032	= 64 512	= 1 764 000
		1	= 4	= 8	= 28	= 112	= 1792	= 28 672	= 784 000
			1	= 2	= 7/2	= 28	= 448	= 7168	= 196 000
				1	= 7/4	= 14	= 224	= 3584	= 98 000
					1	= 8	= 128	= 2048	= 56 000
						1	= 16	= 256	= 7000
							1	= 16	= 437.5
								1	= 27.34375

Table 3-14 UK apothecaries' weight [1 lb apoth. = 0.3732417216 kg]

Pound (lb apoth.)	Ounce (oz apoth.)	Drachm (dr apoth.)	Scruple (scr apoth.)	Grain (gr apoth.)
1	= 12	= 96	= 288	= 5760
	1	= 8	= 24	= 480
		1	= 3	= 60
			1	= 20

Note: There was also apothecaries' measure, for liquid quantities, which can be adequately covered by the following brief statement regarding its units. The parallel system of apothecaries' measure was used for liquid quantities. In this system, analogous to imperial capacity measure, 60 minims was equal to one fluid drachm, which was one eighth of a fluid ounce.

3.3.1.4.3 UK Troy Weight

The troy system was formerly the UK legal system for weighing precious metals and gems. The name is derived from the town of Troyes in Northern France, famous in medieval times for its commercial fairs. It is now obsolete except in the USA. Its base unit is the **troy pound** of 5760 grains defined as 0.3732417216 kg, with other units as in the table below. The troy grain is identical to the grain avoirdupois.

Table 3-15 UK troy weight [1 lb troy = 0.3732417216 kg (E)]

Pound (lb troy)	Ounce (oz troy)	Pennyweight (dwt)	Grain (gr troy)
1	= 12	= 240	= 5760
	1	= 20	= 480
		1	= 24

Note: Another unit still in common use in jewellery for weighing gems is the **carat** the old version of this unit of weight was equal to about 205 mg (3.164 grains). The **metric carat** was standardized at 200 mg (= 3.086 grains) in 1932. The **carat** (occasionally **karat**) is also a unit of purity of gold and other precious metal, with 24 carats as 100% wt pure, and other measures in proportion (e.g. 18 carat gold is 75% wt pure).

3.3.2 The American System of Measures (US Customary Units)

This system is substantially the same as the imperial system, the main differences being mentioned below, and the same criticisms of inconsistency and complexity of calculations apply. These units are however still fully legal in the USA and Canada. Where there are differences between them and imperial units, the designation (US) is normally applied.

American weights and measures are based on units used in Britain prior to 1824, when the imperial system was officially established. The US law of 1866 established a relationship with the metric system by defining the metre as equal to 39.37 in, and, in 1883, the yard was also defined in terms of the metre (=3600/3937 m). Under the *Menenhall Order* of 1893, the **US yard, pound** and all other derived units were redefined in terms of metric units of length and mass, so that from then on there was no longer a direct relationship with UK units, though the differences were often minute. In 1959, an agreement between English-speaking countries unified metric definitions of units for scientific and technical uses, with the yard defined as 0.9144 m and the pound as 0.45359237 kg. In order to accommodate data from the US geodetic surveys, however, the old standard of 1 ft = 1200/3937 m was retained with the name **US survey foot**. It has the following relationships with other units:

1 rod (perch)	= 16.5 ft
1 chain	= 66 ft
1 US mile	= 5280 ft

3.3.2.1 US Customary Units of Length

As with their British equivalents, the many units of length used at various times in the USA fall into several categories depending on area of application

3.3.2.1.1 US Linear Measure

Current linear units are the **inch**, the **foot**, the **yard**, and the **statute mile** (or **land mile**). They are employed for measuring distance, length, width, depth, height, and thickness, and are listed fully in the table below.

Table 3-16 US linear measure

[1 ft = 0.3048 m (E)]

Statute league (US st. leag)	Statute mile (US st. mi)	Pole (rd)	Yard (yd)	Foot (ft)	Inch (in)	Line (US line)
1	= 3	= 960	= 5280	= 15 840	= 190 080	= 7 603 200
	1	= 320	= 1760	= 5280	= 63 360	= 2 534 400
		1	= 5.5	= 16.5	= 198	= 7920
			1	= 3	= 36	= 1440
				1	= 12	= 480
					1	= 40

1 rod = 1 perch = 1 pole
1 land = 1 statute mile

3.3.2.1.2 US Nautical Measure

US nautical measures are substantially the same as those of the UK, and are summarized in ***Table 3-17***.

Table 3-17 US nautical measure [1 mile (US naut.) = 1853.184 m (E)]						
US nautical league (US naut. leag)	**US nautical mile** (US naut. mi)	**US cable length** (US cbl)	**Fathom** (fath)	**Yard** (yd)	**Foot** (ft)	**Inch** (in)
1	= 3	= 30	= 3040	= 6080	= 18 240	= 218 880
	1	= 76/9	= 9120/9	= 18240/9	= 6080	= 72 960
		1	= 120	= 240	= 720	= 8640
			1	= 2	= 6	= 72
				1	= 3	= 36
					1	= 12

.3.2.1.3 *US Surveyors' Measure*

nits of length in the US Customary system are the **link**, the **chain** (**Ramsden's hain**), the **rod**, and the **furlong**.

Table 3-18 US surveyors' measure [1 chain (Ramsden's or Engineer's) = 30.48 m (E)]					
Statute mile (mi)	**Furlong** (fur)	**Ramsden's chain** (ch)	**Rod** (rod)	**Foot** (ft)	**Ramsden's link** (lk)
1	= 8	= 52.8	= 320	= 8000	= 8000
	1	= 6.6	= 40	= 1000	= 1000
		1	= 6	= 100	= 100
			1	= 16.5	= 16.5
				1	= 1

rod = 1 perch = 1 pole
land = 1 statute mile

.3.2.2 US Customary Units of Area

s with linear units, US square measure is virtually identical to the British ystem.

.3.2.2.1 *US Measures of Area*

he conventional US units of area are directly related to the linear measures n which they are based, thus the **square inch, square foot, square yard,** and **quare mile**. They are related as follows:

Table 3-19 US measures of area [1 ft^2 = 9.290304 × 10^{-2} m^2 (E)]				
Square mile (sq. st. mi)	**Square rod** (sq. rd)	**Square yard** (sq. yd)	**Square foot** (sq. ft)	**Square inch** (sq. in)
1	= 102 400	= 3 097 600	= 27 878 400	= 4 014 489 600
	1	= 121/4	= 1089/4	= 39 204
		1	= 9	= 1296
			1	= 144

rod = 1 perch = 1 pole
section of land = 1 square statute mile

3.3.2.2.2 US Surveyors' Measure

The now obsolete units of area once used by surveyors are listed below i relation to the **square foot** and the **acre**, the only two such measures to surviv into relatively recent times.

Table 3-20 US surveyors' measure
[1 $ft^2 = 9.290304 \times 10^{-2}$ m^2 (E)]

Square mile (sq. mi)	Square chain (sq. yd)	Square rod, perch, pole (sq. rd)	Square foot (sq. ft)	Square link (sq. lk)
1	= 6400	= 230 400	= 62 726 400	= 143 999 742
	1	= 36	= 9801	= 22 500
		1	= 1089/3	= 625
			1	= 22500/9801

1 rod = 1 perch = 1 pole
1 section of land = 1 square statute mile

3.3.2.2.3 Circular Units

Very rarely, so-called 'circular' units have been used, mainly for wire sizes an then only as the **circular mil** in the USA (see also ***Table 3-9***). In an analogy t square measure, circular units represent the area of a circle of a diamete equal to the equivalent linear unit. A circular mil is the area of a circle one m (0.001 in) in diameter, and hence equal to $7.853981634 \times 10^{-7}$ in^2 c $5.067074791 \times 10^{-10}$ m^2. Circular units should not be confused with *circul measure*, which refers to the expression of angle in radians.

3.3.2.3 US Units of Volume and Capacity

As with the UK units of length and area, units of volume and capacity are ofte associated with specific trades or fields of application. They do however, i common with many older systems, fall into three categories: geometr measure, expressed in cubic linear units (e.g. **cubic foot**), dry measure, an liquid measure. The latter two categories are known as units of capacity an consist of arbitrary volumes given specific names (e.g. **bushel, gallon**), wit the same name sometimes being used for measuring solids or liquids and th having different volumes (e.g. the US dry pint ≈ 0.551 l, while a US liquid pi ≈ 0.473 l).

3.3.2.3.1 US Measures of Volume

The volumetric units chiefly used in the US are the **cubic inch, cubic foo cubic yard**, and **cubic statute mile** identical to imperial units.

Table 3-21 US measures of volume
[1 $ft^3 = 2.831684659 \times 10^{-2}$ m^3 (E)]

Cubic mile (cu. mi)	Cubic yard (cu. yd)	Cubic foot (cu. ft)	Cubic inch (cu. in)
1	= 5451776000	$= 1.472 \times 10^{11}$	$= 2.543 \times 10^{14}$
	1	= 27	= 46 656
		1	= 1728

3.3.2.3.2 *US Liquid Measure*

Liquid measures in the USA are based on the old wine measures in use in Britain before 1824 and differ considerably from their imperial equivalents. The fluid ounce is identical, but there are only 16 to the pint as compared with 20 in the UK system. Thus the US quart, and gallon, both of which are still in common use, are 20% smaller than their British equivalents. The most important multiple is the **barrel** (=42 US gallons, 35 imperial gallons, 0.158987295 m^3) which is extensively used in the international oil industry (*Table 3.22* overleaf).

3.3.2.3.3 *US Dry Measure*

US measures of dry volume are related to liquid measures but differ from them in some respects. They are used for measuring the volume of powdered or granular materials (e.g. flour, sand, grain, powdered ore, etc.).

Table 3-23 US dry measure
[1 gallon (US, dry) = 4.404884 × 10^{-3} m^3 (E)]

Bushel (US bu)	**Peck** (US pk)	**Gallon** (US gal)	**Dry quart** (US dr. qt)	**Dry pint** (US dr. pt)
1	=4	=8	=32	=64
	1	=2	=8	=16
		1	=4	=8
			1	=2

3.3.2.3.4 *US Apothecaries' Measures of Capacity*

The US Customary Units commonly used to measure volumes of medicinal liquids are the **tablespoon** and the **teaspoon.** They are defined as in the table below:

Table 3-24 US Apothecaries' measures of capacity
[1 US fl. oz = 2.957350 × 10^{-5} m^3]

Fluid ounce (US fl. oz)	**Tablespoon**	**Teaspoon**
1	=2	=6
	1	=3

Tablespoon (metric)	**Teaspoon** (metric)	**Cubic centimetre** (cm^3)
1	=3	=15
	1	=5

3.3.2.4 US Customary Units of Weight

3.3.2.4.1 *US Avoirdupois Weight*

The US pound avoirdupois was defined by Act of Congress in 1866 as 1/2.2046 kg, but was more accurately related to its metric equivalent in 1895 as 0.45359224277 kg. Today, for engineering purposes, the US and British pounds can be regarded as identical, defined as equal to 0.45359237 kg by the WMA, 1963 in Britain, and in North America by the *US Metric Board* in 1959 (USMB, 1959).

Table 3-22 US fluid measure

[1 gallon (US, liquid) = $3.785411784 \times 10^{-3}$ m^3 (E)]

Tun (US tu)	**Hogshead** (US hhd)	**Barrel** (US bbl)	**Gallon** (US gal)	**Quart** (US qt)	**Pint** (US pt)	**Gill** (US gi)	**Fluid ounce** (US fl oz)	**Fluid dram** (US fl dr)	**US minim** (US min)
1	**= 4**	= 8	= 252	= 1008	= 2016	= 8064	= 32 256	= 258 048	= 15 482 880
	1	**= 2**	= 63	= 252	= 504	= 2016	= 8064	= 64 512	= 3 870 720
		1	**= 31.5**	= 126	= 252	= 1008	= 4032	= 32 256	= 1 935 360
			1	= 4	= 8	= 32	= 128	= 1024	= 61 440
				1	= 2	= 8	= 32	= 256	= 15 360
					1	= 4	= 16	= 128	= 7680
						1	= 4	= 32	= 1920
							1	= 8	= 480
								1	**= 60**

Now, when most countries are converting to metric units and the avoirdupois system is being phased out, there is pressure in the US for alignment with the rest of the world. So far, however, the pressure has been resisted on the grounds of economic and social upheaval.

Table 3-25 US avoirdupois weight
[1 lb av. (USMB, 1959) = 0.45359237 kg (E)]

Short ton (US ton)	Long hundredweight (lg cwt)	Short hundredweight (sh cwt)	Pound (lb av)	Ounce (oz av.)	Dram (dram av.)	Grain (gr av.)
1	= 250/14	= 20	= 2000	= 32 000	= 512 000	= 14 000 000
	1	= 28/25	= 112	= 1792	= 28 672	= 784 000
		1	= 100	= 1600	= 25 600	= 700 000
			1	= 16	= 256	= 7000
				1	= 16	= 875/2
					1	= 875/32

3.3.2.4.2 *US Apothecaries' Weight*

As with the UK units, these are now completely obsolete and their use is prohibited by the American Pharmacopoeia.

Table 3-26 US apothecaries' weight
[1 lb apoth. = 0.3732417216 kg]

Pound (lb apoth.)	Ounce (oz apoth.)	Drachm (dr apoth.)	Scruple (scr apoth.)	Grain (gr apoth.)
1	= 12	= 96	= 288	= 5760
	1	= 8	= 24	= 480
		1	= 3	= 60
			1	= 20

3.3.2.4.3 *US Troy Weight*

The troy system was formerly the UK legal system for weighing precious metals and gems. The name is derived from the town of Troyes in Northern France, famous in medieval times for its commercial fairs. It is now obsolete except in the USA. Its base unit is the **troy pound** of 5760 grains defined as 0.3732417216 kg, with other units as in the table below. The troy grain is identical to the grain avoirdupois.

Table 3-27 US troy weight
[1 lb troy = 0.3732417216 kg (E)]

Pound (lb troy)	Ounce (oz troy)	Pennyweight (dwt)	Grain (gr troy)
1	= 12	= 240	= 5760
	1	= 20	= 480
		1	= 24

Note: Another unit still in common use in jewellery for weighing gems is the **carat**; the old version of this unit of weight was equal to about 205 mg (3.164 grains). The **metric carat** was standardized at 200 mg (=3.086 grains) in 1932. The **carat** (occasionally **karat**) is also a unit of purity of gold and other precious metal, with 24 carats as 100% wt pure, and other measures in proportion (e.g. 18 carat gold is 75% wt pure).

3.3.3 Obsolete Weight and Capacity Measures

In both the UK and the USA, some old measures of weight and capacity survive in a few specialized areas, chiefly with regard to beer, wines, and spirits, and agricultural products, though most are replaced by metric measures. The summary below is given mainly for historical interest.

3.3.3.1 UK Units for Beer, Wines, and Spirits

Table 3-28 UK units for alcohols and spirits

Unit (alcohol)	Approx. volume (UK gal)	Approx. volume (dm^3)
octave (whisky)	16	73
quarter	17–30	77–136
aum (hock)	30–32	136–145
hogshead	44–60	200–273
hogshead (madeira)	45–48	205–218
hogshead (brandy)	56–61	255–277
puncheon	70–120	318–546
pipe	90–120	409–546
butt	108–117	491–532
tonneau, tun	190–200	864–909
stuck	260–265	1182–1205

Units especially applied to measures of beer are given in the table below:

Table 3-29 UK units of capacity for beer [1 gallon (UK) = 4.546092 dm^3 (E)]

Tun	Butt	Puncheon	Hogshead	Barrel	Firkin	Pin	Gallon
1	=2	=3	=4	=6	=24	=48	=216
	1	=3/2	=2	=3	=12	=24	=108
		1	=4/3	=2	=8	=16	=72
			1	=3/2	=6	=12	=54
				1	=4	=8	=36
					1	=2	=9
						1	=9/2

Some bottle sizes, once used for wines in general, are now exclusively reserved for champagne, and then only in general parlance.

Table 3-30 UK units of capacity for wine

French name (English word)	Reputed quarts (No. of bottles)	Volume (UK gal)	Volume (dm^3)
1 nabuchodonosor (nabuchadnezzar)	20	10/3	15.153640
1 balthazar or balthasar (belshazzar)	16	8/3	12.122912
1 salmanazar (salmarazd)	12	2	9.092184
1 mathusalem (methuselah)	9	3/2	6.819138
1 rèhoboam (rehoboam)	6	1	4.546092
1 jèroboam (jeroboam)	4	2/3	3.030728
1 magnum (magnum)	2	1/3	1.515364
1 bouteille champenoise (reputed quart)	1	1/6	0.757682

3.3.2 British and American Hay and Straw Weights

Measures of agricultural fodder formerly used are:

Table 3-31 UK and US hay and straw weights [1 lb (UK, straw) = 0.453592338 kg]

Load	Truss	Pound (straw)	Pound (old hay)	Pound (new hay)
1	= 36	= 1296	= 2016	= 2160
	1	= 36	= 56	= 60
		1	= 14/9	= 5/3
			1	= 15/14

3.3.3 British Weights for Wool

Table 3-32 UK weight for wool [1 wey (UK, wool) = 114.30527724 kg (E)]

Load	Sack	Wey	Stone	Pound
1	= 6	= 12	= 168	= 2100
	1	= 2	= 28	= 350
		1	= 14	= 175
			1	= 12.5

3.4 The Foot-Pound-Second (FPS) System

In parallel with the development of the cgs system (see 3.2.1 above) came what was seen as its imperial equivalent, the foot-pound-second system, proposed by W. Stroud in 1880 and sometimes called the Stroud system in commemoration. It became very widely employed in all branches of engineering, and most technical papers written in Britain, the USA, and other parts of the English-speaking world before about 1960 would have used these units, although scientific papers tended to use cgs units.

Its popularity in engineering was due not only to the use of imperial units as its base, but also because the pound and the foot were felt to be more convenient for engineers than the too-small centimetre and gram, and the too-large metre and kilogram. Although it was, strictly speaking, a non-decimal system, this was technically irrelevant since quantities could be expressed in decimals of feet, pounds, etc., so that the criticism of complexity in calculations usually aimed at the imperial system did not necessarily apply. It must be admitted, however, that, engineers being what they are, measures such as tons-feet and inches per second were not uncommon.

Its main problem was that the *pound* had long been in common use as a unit of both *weight* and *mass*. This makes no difference in general and commercial usage, since, because of Earth's gravity, a mass of one pound weighs exactly one pound. For the non-technical reader, the difference may be illustrated by considering the same mass taken to another planet such as the Moon. Because the Moon's gravitational force is about one-sixth that of the Earth, the one-pound mass would weigh only one-sixth of a pound, although the mass itself would not have changed. Weight, therefore, is the *force* with which a mass is attracted by gravity, and, since it is an entirely different quantity, it requires a different unit.

Force was defined by Newton as *mass* × *acceleration*. In a coherent system of units, any derived unit must interrelate one-to-one with the system's base units, so that one force unit equals one mass unit times one acceleration unit. In the FPS system, with the pound as the unit of mass, one force unit is required to impart one acceleration unit (1 ft.s^{-2}) to a mass of one pound. The acceleration due to gravity is approximately 32 ft.s^{-2}, so that *weight* of one pound *mass* is in fact equal to 32 force units, and the force unit must therefore be 1/32 pounds (to be accurate, since g is 32.1740486 ft.s^2, it is $1/32.17$ or 0.031081 lbf, = 0.138255 N). This is termed the **poundal**.

However, because in general use the pound had always been appreciated as a unit of weight, there was a tendency among engineers to continue to use it in this way. In a variant of the FPS system, usually termed *technical*, *gravitational* or *engineers' units*, the **pound-force** (lbf) was taken as a base unit, and a unit of mass was derived from it by a reversal of the above considerations. This unit was named the *slug*, and was the mass which when acted upon by one pound-force experienced an acceleration of 1ft.s^{-2}, so was equal to 32.17 lb. This version of the FPS system was more commonly used in the United States than anywhere else.

The FPS system was never made fully coherent by the incorporation of electrical or molar units. It did however have derived units which were for the most part expressed clearly in terms of their base units and not given separate names as in the SI. It is true that in practice they were often abbreviated (e.g. *psi* for lbf.in^{-2}), and that they were often used in a non-standard way, or in a way that confused the two subsystems. The example just quoted shows this quite clearly: the 'p' is intended to mean 'pounds-force', the abbreviation should be written lbf.in^{-2}, and the correct FPS pressure unit should have been poundals per square foot (pdl.ft^{-2}). In fact, the pound-force and pound-weight were often used quite indiscriminately, with the acceleration due to gravity, '*g*', being used so commonly as a correction factor that it was humorously referred to as the 'engineers' constant'.

Abbreviations of derived units often became acronymic in that engineers spoke of *pee-ess-eye* or *ar-pee-em* rather than pounds force per square inch or revolutions per minute. Among the non-standard abbreviations were sq. ft and cu. ft for square and cubic feet respectively, with sometimes a small square being used with an abbreviation (e.g. □ft) in drawings and calculations. A selection of FPS derived units is given in the following table.

Table 3-33 FPS derived units in common use

Quantity	FPS unit	Abbreviation (other units)	Conversion factor in SI unit
acceleration	foot per square second	ft.s^{-2}	1 ft.s^{-2} = 0.3048 m.s^{-2} (E)
angular velocity	revolutions per second	rev.s^{-1} (also rps, rev/min, rpm)	1 rps = 2π rad.s^{-1} (E)
area	square foot	ft^2, sq. ft	1 ft^2 = 9.290304 × 10^{-2} m^2
energy, work	foot-poundal	ft.pdl	1ft.pdl = 4.21401101 × 10^{-2} J
force	poundal	pdl	1 pdl = 0.1382549544 N
frequency	cycles per second	cycle.s^{-1} (cps)	1 cps = 1 Hz
heat	foot-poundal	ft.pdl (also British thermal unit, Btu)	1 ft.pdl = 4.21401101 × 10^{-2} J 1 Btu = 1055.06 J
power	foot-poundal per second	ft.pdl.s^{-1} (also horsepower, hp)	1 ft.pdl.s^{-1} = 4.21401101 × 10^{-2} W 1 hp = 745.7 W
pressure, stress	poundal per square foot	pdl.ft^{-2} (also pound-force/ sq. in, psi)	1 pdl.ft^{-2} = 1.488163944 Pa 1 psi = 6894.75729 Pa
velocity	foot per second	ft.s^{-1} (also miles/hr, mph)	1ft.s^{-1} = 0.3048 m.s^{-1} (E)
volume	cubic foot	ft^3, cu. ft	1ft^3 = 2.83168466 × 10^{-2} m^{-3}

Note that psi, rpm, mph, Btu, and hp, although in common use in engineering calculations, were not derived FPS units.

With its various inconsistencies, inherent and imposed, and with the increasing internationalization of the metric system culminating in the creation of the SI, it was inevitable that the FPS system would become obsolete. Yet, in concluding this brief survey of the FPS system, it is worth noting that there was nothing inherently inconsistent in a system based on the foot and the pound in themselves. Decimalized, with a single set of force and mass units, and integrated with electrical and molar quantities, it could have been just as consistent and international as the metric-based SI. And, as the considerations of the following section on ancient units will show, there is a feeling among human beings that units based on the human body are somehow more comprehensible than those derived from the circumference of the Earth, or referred to the energy level of an atom.

3.5 Ancient and Obsolete Systems of Weights and Measures

From the very earliest times, human beings have found it necessary to weigh and measure the world around them, and the most ancient records include references to units of measurement. Most of these ancient units are now entirely obsolete. Our knowledge of them comes from texts and inscriptions which have survived, but the values of many have been quite reliably determined. This presentation of ancient systems of measurement is mainly for historical and general interest. Nevertheless, they illustrate the development of our modern systems, and the associated conversion tables could be of use to engineers and others reconstructing or evaluating ancient machinery, ships, or buildings.

Many early units, with an anthropocentrism which has persisted up to the present day, were based on the human body or its attributes. Units such as the **finger**, **hand**, **palm**, and **foot** are self-explanatory, while the **span** was the maximum width, thumb-tip to small fingertip, of the spread hand. The **inch** was the distance from the tip to the first joint of the thumb, and the surviving French word *pouce* in fact means 'thumb'. As a matter of incidental interest, the English word 'inch' comes from the Old German for 'one-twelfth', and has the same derivation as the word 'ounce'. The **cubit** was the distance from elbow to fingertips (from the Latin *cubitum*, elbow), and is also sometimes known as the *ell* from the Germanic word for 'forearm' ('elbow' being derived from Old German *elnboga*, arm-bend). Together with the **fathom**, derived from the span of the arms or the height of a man, these ancient units can still be seen in use today when builders, woodworkers, or other tradespeople make rough estimates of quantities.

Units of length greater than the human body itself were usually expressed in terms of walking distances. The yard was about the length of a pace, and one thousand paces was a **mile** (Latin *milia passuum*). Other units which expressed distance in terms of human activity were rough periods of time such as an **hour's walking**, or a **day's sailing**.

Measures of area and volume using square or cubic units are relatively recent inventions. Areas were first thought of in terms of reasonable-sized fields, that could be ploughed in a given time, and indeed the word acre is ultimately derived from the Sanskrit for 'field' (*ajra*). Likewise, volumes were thought of in terms of containers such as churns or barrels, or of human-sized portable units such as bundles of wood, bales of hay, sacks of grain, or pails of milk. The parallel between liquids and granular or powered solids was also noted, with the same measures being used for both. Sometimes, however, to equalize the 'feel' of measured quantities, a smaller measure was used for heavier materials, resulting in different sized 'pints', for example, in dry and liquid measures.

Weight units arose from these capacity measures, and there is a unit more or less equal to, and often called, a **pound** (Latin *pondus*, weight), in many ancient and obsolete systems. The relationship remains clear even in modern times. A pint of wine weighs about a pound, and a gallon of water weighs ten pounds. The word **ton** is from the same root as the name of the older unit, tun, a large wine cask of some 250 gallons, which would therefore weigh about 2000 pounds when full.

One of the oldest units of measurement is the **degree of arc**, which is usually supposed to have been invented by the Babylonians over 4000 years ago. It is curious that they should have chosen to divide the circle into 360 degrees: the ıplest way would have been to divide it into halves, quarters, and so on, ing 256, or another multiple of 2, degrees in the circle. It has been suggested t the number 360 arose in an early attempt to guess the number of days in year, but this is unlikely, since accurate astronomical data were known ore recorded history began. However, the lunar month of approximately 30 's, the division of the solar year into 12 months, and the solar day into 12 ırs, cannot fail to be related to the Babylonian number base of 60 and the ision of the circle into 360 degrees.

'inally, of course, it should be remembered that these early units were ınected with measuring time and constructing a calendar. In this respect it nteresting to consider that the second, determined by the Babylonians, is the fundamental unit of time in all current systems of measurement, even SI itself. It is therefore the unit in longest continuous official use, and, with degree of arc, one of only two to have been in use throughout recorded tory.

.1 Systems from Antiquity

this section, only the more important and the best known of these ancient tems have been included. It is also important to bear in mind that these tems were not consecutive, but were in a constant state of evolution and rlapped with one another to a large extent. It is therefore impossible to ıblish a time-scale over which any one system was used.

.1.1 The Chinese System

.1.1.1 Old Chinese Units of Length

ese are shown in ***Table 3.34*** (overleaf).

.1.1.2 Old Chinese Units of Area

ıble 3-35 Old Chinese units of area
[1 meou = 614.4 m^2]

hing	King	Meou	Kish	Fen	Lyi	Kung (sq. pou)	Hao
1	= 10	= 100	= 400	= 1000	= 10 000	= 24 000	= 100 000
	1	= 10	= 40	= 100	= 1000	= 2400	= 10 000
		1	= 4	= 10	= 100	= 240	= 1000
			1	= 5/2	= 25	= 60	= 250
				1	= 10	= 24	= 100
					1	= 12/5	= 10
						1	= 25/6

Table 3-34 Old Chinese units of length

[1 tchi = 0.23 m]

Tou	Thsan	Poû	Li	Kyo	fen	Yin (Yan)	Zhang	Pou	Tchi	Cun (tsouen)	Fen	Lô	Hao	Su	Hoé
	= 25/8	= 25	= 250	= 1 500	= 3750	= 4500	= 45 000	= 90 000	= 450 000	$= 4.5 \times 10^6$	$= 4.5 \times 10^7$	$= 4.5 \times 10^8$	$= 4.5 \times 10^9$	$= 4.5 \times 10^{10}$	$= 4.5 \times 10^{11}$
	1	= 8	= 80	= 480	= 1200	= 1440	= 14 400	= 28 800	= 144 000	$= 1.44 \times 10^6$	$= 1.44 \times 10^7$	$= 1.44 \times 10^8$	$= 1.44 \times 10^9$	$= 1.44 \times 10^{10}$	$= 1.44 \times 10^{11}$
		1	= 10	= 60	= 150	= 180	= 1800	= 3600	= 18 000	$= 1.8 \times 10^5$	$= 1.8 \times 10^6$	$= 1.8 \times 10^7$	$= 1.8 \times 10^8$	$= 1.8 \times 10^9$	$= 1.8 \times 10^{10}$
			1	= 6	= 15	= 18	= 180	= 360	= 1800	$= 1.8 \times 10^4$	$= 1.8 \times 10^5$	$= 1.8 \times 10^6$	$= 1.8 \times 10^7$	$= 1.8 \times 10^8$	$= 1.8 \times 10^9$
				1	= 5/2	= 3	= 30	= 60	= 300	$= 3 \times 10^3$	$= 3 \times 10^4$	$= 3 \times 10^5$	$= 3 \times 10^6$	$= 3 \times 10^7$	$= 3 \times 10^8$
					1	= 6/5	= 12	= 24	= 120	$= 1.2 \times 10^3$	$= 1.2 \times 10^4$	$= 1.2 \times 10^5$	$= 1.2 \times 10^6$	$= 1.2 \times 10^7$	$= 1.2 \times 10^8$
						1	= 10	= 20	= 100	$= 10^3$	$= 10^4$	$= 10^5$	$= 10^6$	$= 10^7$	$= 10^8$
							1	= 2	= 10	= 100	$= 10^3$	$= 10^4$	$= 10^5$	$= 10^6$	$= 10^7$
								1	= 5	= 50	= 500	= 5000	= 50 000	= 500 000	= 5 000 000
									1	= 10	= 100	$= 10^3$	$= 10^4$	$= 10^5$	$= 10^6$
										1	= 10	= 100	$= 10^3$	$= 10^4$	$= 10^5$
											1	= 10	= 100	$= 10^3$	$= 10^4$
												1	= 10	= 100	$= 10^3$
													1	= 10	= 100
														1	= 10

.5.1.1.3 *Old Chinese Units of Weight*

Table 3-36 Old Chinese units of weight
[1 jin = 0.250 kg]

)an shih)	Tan	Jun (kwan)	Jin (tchin)	Liang	Zhu	Shu
1	= 6/5	= 4	= 120	= 1920	= 46 080	= 4 608 000
	1	= 10/3	= 100	= 1600	= 38 400	= 3 840 000
		1	= 30	= 480	= 11 520	= 1 152 000
			1	= 16	= 384	= 38 400
				1	= 24	= 2400
					1	= 100

.5.1.1.4 *Old Chinese Units of Capacity*

hese are shown in ***Table 3-37***.

.5.1.2 The Indian System

.5.1.2.1 *Old Indian Units of Length*

Table 3-38 Indian units of length
[1 hasta = 0.457 m]

Yodjana	Gavyuti	Crosa	Dhanush (orgyla)	Hasta (cubit)	Vistati (span)	Angula (finger)
1	= 2	= 4	= 16 000	= 32 000	= 64 000	= 768 000
	1	= 2	= 4000	= 16 000	= 32 000	= 384 000
		1	= 2000	= 8000	= 16 000	= 192 000
			1	= 4	= 8	= 96
				1	= 2	= 24
					1	= 12

.5.1.2.2 *Indian Units of Weight*

ee ***Table 3-39*** (p. 47).

.5.1.2.3 *Indian Units of Capacity*

ee ***Table 3-40*** (p. 48).

.5.1.3 The Egyptian System

.5.1.3.1 *Egyptian Units of Length*

he earliest known unit of length was the **cubit**, which is the distance between ne elbow and the tip of the middle finger. It was used by the Sumerians, abylonians, Israelites, and Egyptians as a base unit. The Egyptian system of near measure, sometimes called *Pharaonic measurements*, included two inds of cubit, as shown in ***Table 3-41*** (p. 49).

Table 3-37 Old Chinese units of capacity

[1 cheng = 1.03544 dm^3]

Ping	Chei	Hou	To	Cheng	Yo	Khô	Chao	Ço	Quei
1	= 5	= 10	= 50	= 500	= 2500	= 5000	= 50 000	= 500 000	= 5 000 000
	1	= 2	= 10	= 100	= 500	= 1000	= 10 000	= 100 000	= 1 000 000
		1	= 5	= 50	= 250	= 500	= 5000	= 50 000	= 500 000
			1	= 10	= 50	= 100	= 1000	= 10 000	= 100 000
				1	= 5	= 10	= 100	= 1000	= 10 000
					1	= 2	= 20	= 200	= 2000
						1	= 10	= 100	= 1000
							1	= 10	= 100
								1	= 10

Table 3-39 Indian units of weight

[1 pala = 47 × 10^{-3} kg]

Achita	Bara	Hara	Tuba	Pala	Kharsha	Tola	Kona	Dharana	Tank-sala	Masha	Retti (ratica)	Yava
1	= 10	= 100	= 200	= 20 000	= 200 000/3	= 80 000	= 400 000/3	= 200 000	= 3 200 000/9	= 3 200 000/3	= 6 400 000	= 64 000 000
	1	= 10	= 20	= 2000	= 20 000/3	= 8000	= 40 000/3	= 20 000	= 320 000/9	= 320 000/3	= 640 000	= 6 400 000
		1	= 2	= 200	= 2000/3	= 800	= 4000/3	= 2000	= 32 000/9	= 32 000/3	= 64 000	= 640 000
			1	= 100	= 1000/3	= 400	= 2000/3	= 1000	= 16 000/9	= 16 000/3	= 32 000	= 320 000
				1	= 10/3	= 4	= 20/3	= 10	= 160/9	= 160/3	= 320	= 3200
					1	= 6/5	= 2	= 3	= 16/3	= 16	= 96	= 960
						1	= 5/3	= 5/2	= 40/9	= 40/3	= 80	= 800
							1	= 3/2	= 8/3	= 8	= 48	= 480
								1	= 16/9	= 16/3	= 32	= 320
									1	= 3	= 24	= 240
										1	= 6	= 60
											1	= 10

Table 3-40 Indian units of capacity (measured by weight)
[1 drona = 13.2 kg]

Baha	**Cumbha**	**Shari**	**Cumbha short**	**Drona**	**Adhaka**	**Prastha**	**Cudava**	**Musti** (pala)
1	**= 10**	= 50/4	= 100	= 200	= 800	= 3200	= 6400	= 51 200
	1	**= 5/4**	= 10	= 20	= 80	= 320	= 640	= 5120
		1	**= 8**	= 16	= 64	= 256	= 512	= 4096
			1	**= 2**	= 8	= 32	= 64	= 512
				1	**= 4**	= 16	= 32	= 256
					1	**= 4**	= 8	= 64
						1	**= 2**	= 16
							1	**= 8**

Table 3-41 Egyptian units of length

[1 derah (Royal cubit) = 0.5235 m]

Royal atour	Parasange	Shoëme	Atour	Mille	Stade	Senus	Canne	Orgye (pace)	Xilon	Long cubit	Derah (Royal cubit)	Pigon	Zereth (Royal foot)
1	= 3/2	= 5/3	= 2	= 6	= 50	= 200	= 18 000/7	= 5 000	= 20 000/3	= 15 000	= 20 000	= 24 000	= 30 000
	1	= 10/9	= 4/3	= 4	= 100/3	= 400/3	= 12 000/7	= 10 000/3	= 40 000/9	= 10 000	= 40 000/3	= 16 000	= 20 000
		1	= 6/5	= 18/5	= 30	= 120	= 10 800/7	= 3000	= 4000	= 9000	= 12 000	= 14 400	= 18 000
			1	= 3	= 25	= 100	= 9000/7	= 2500	= 10 000/3	= 7500	= 10 000	= 12 000	= 15 000
				1	= 25/3	= 100/3	= 3000/7	= 2500/3	= 10 000/9	= 2500	= 10 000/3	= 4000	= 5000
					1	= 4	= 360/7	= 100	= 400/3	= 300	= 400	= 480	= 600
						1	= 90/7	= 25	= 100/3	= 75	= 100	= 120	= 150
							1	= 35/18	= 70/27	= 35/6	= 70/9	= 28/3	= 35/3
								1	= 4/3	= 3	= 4	= 24/5	= 6
									1	= 9/4	= 3	= 18/5	= 9/2
										1	= 4/3	= 8/5	= 2
											1	= 6/5	= 3/2
												1	= 5/4

Table 3-41 Egyptian units of length (continued)

[1 derah (Royal cubit) = 0.5235 m]

Orgye (pace)	**Xilon**	**Long cubit**	**Derah** (Royal cubit)	**Pigon**	**Zereth** (Royal foot)	**Spithame**	**Dichas**	**Choryos** (palm)	**Thebs** (finger)	**Digits**
1	**= 4/3**	= 3	= 4	= 24/5	= 6	= 9/2	= 12	= 24	= 96	= 192
	1	**= 9/4**	= 3	= 18/5	= 9/2	= 27/8	= 9	= 18	= 72	= 144
		1	**= 4/3**	= 8/5	= 2	= 3/2	= 4	= 8	= 32	= 64
			1	**= 6/5**	= 3/2	= 9/8	= 3	= 6	= 24	= 48
				1	**= 5/4**	= 15/16	= 5/2	= 5	= 20	= 40
					1	**= 3/4**	= 2	= 4	= 16	= 32
						1	**= 8/3**	= 16/3	= 64/3	= 128/3
							1	**= 2**	= 8	= 16
								1	**= 4**	= 8
									1	= 2

3.5.1.3.2 *Egyptian Units of Area*

Table 3-42 Egyptian units of area
[1 pekeis = 27.405 m^2]

Setta	Aurure	Rema	Ten (dizaine)	Sû	Pekeis	Square cubit
1	= 10	= 20	= 100	= 160	= 1000	= 100 000
	1	= 2	= 10	= 16	= 100	= 10 000
		1	= 5	= 8	= 50	= 5000
			1	= 8/5	= 10	= 1000
				1	= 25/4	= 625
					1	= 100

3.5.1.3.3 *Egyptian Units of Weight*

Table 3-43 Egyptian units of weight
[1 deben = 13.65 × 10^{-3} kg]

Kedet	Deben	Sep	Grains
1	= 10	= 100	= 150
	1	= 10	= 15
		1	= 3/2

3.5.1.3.4 *Egyptian Units of Capacity*

Table 3-44 Egyptian units of capacity (measured by weight)
[1 khar = 34 kg]

Letech	Artabe	Metretes of Heron	Khar (keramion)	Apt	Hecte	Maân (mine)	Outen
1	= 27/16	= 27/8	= 135/32	= 135/8	= 135/2	= 675/4	= 675
	1	= 2	= 3/2	= 6	= 24	= 60	= 240
		1	= 5/4	= 5	= 20	= 50	= 200
			1	= 4	= 16	= 40	= 160
				1	= 4	= 10	= 40
					1	= 5/2	= 10
						1	= 4

3.5.1.4 The Assyrio-Chaldean-Persian System

3.5.1.4.1 *Persian Units of Length*

These are shown in ***Table 3-45***.

Table 3-45 Persian units of length

[1 foot (Persian) = 0.320 m]

Mansion (stathmos)	Schoëme	Parasang	Mille	Ghalva (stadion)	Chebel	Qasab (cane)	Pace	Cubit	Zereth (foot)	Palm	Finger
1	**= 100/27**	= 4	= 400/27	= 1000/9	= 1000	= 20 000/3	= 40 000/3	= 40 000	= 80 000	= 320 000	= 1 280 000
	1	**= 27/25**	= 4	= 30	= 270	= 1800	= 3600	= 10 800	= 21 600	= 86 400	= 345 600
		1	**= 100/27**	= 250/9	= 250	= 5000/3	= 10 000/3	= 10 000	= 20 000	= 80 000	= 320 000
			1	**= 15/2**	= 135/2	= 450	= 900	= 2700	= 5400	= 21 600	= 86 400
				1	**= 9**	= 60	= 120	= 360	= 720	= 2880	= 11 520
					1	**= 20/3**	= 40/3	= 40	= 80	= 320	= 1 280
						1	**= 2**	= 6	= 12	= 48	= 192
							1	**= 3**	= 6	= 24	= 96
								1	**= 2**	= 8	= 32
									1	**= 4**	= 16
										1	= 4

3.5.1.4.2 *Persian Units of Area*

Table 3-46 Persian units of area
[1 gar = 14.7 m^2]

Gur	Gan	Ten (dizaine)	Gar	Square foot
1	= 10	= 100	= 1000	= 144 000
	1	= 10	= 100	= 14 400
		1	= 10	= 1440
			1	= 144

3.5.1.4.3 *Persian Units of Capacity*

Table 3-47 Persian units of capacity (measured by weight)
[1 amphora = 32.60 kg]

Gariba	Long amphora	Long artaba	Short artaba	Amphora	Woëbe (modius)	Makuk	Cados
1	= 8/3	= 4	= 16/3	= 8	= 16	= 64	= 128
	1	= 3/2	= 2	= 3	= 6	= 24	= 96
		1	= 4/3	= 2	= 4	= 16	= 64
			1	= 3/2	= 3	= 12	= 48
				1	= 2	= 8	= 32
					1	= 4	= 16
						1	= 4

3.5.1.5 The Hebrew System

3.5.1.5.1 *Hebrew Units of Length*

Table 3-48 Hebrew units of length
[1 sacred cubit = 0.640 m]
[1 cubit = 0.555 m]

Cubit	Zereth	Palm	Finger
1	= 2	= 6	= 24
	1	= 3	= 12
		1	= 4

3.5.1.5.2 *Hebrew Units of Weight (Sacred System)*

Table 3-49 Hebrew units of weight (sacred system)
[1 mina = 0.850 kg]

Talent of Moses	Mina	Shekel	Bekah	Rabah	Gerah (Obol)
1	= 50	= 3000	= 6000	= 12 000	= 60 000
	1	= 60	= 120	= 240	= 1200
		1	= 2	= 4	= 20
			1	= 2	= 10
				1	= 5

3.5.1.5.3 *Hebrew Units of Weight (Talmudic System)*

Table 3-50 Hebrew units of mass (Talmudic or Rabbinical system)
[1 mina = 0.3542 kg]

Talent	Mina	Shekel	Zuzah (Drachm)	Mehah (Obol)	Pondiuscule
1	= 60	= 1500	= 6000	= 36 000	= 72 000
	1	= 25	= 100	= 600	= 1200
		1	= 4	= 24	= 48
			1	= 6	= 12
				1	= 2

3.5.1.5.4 *Hebrew Units of Capacity (Dry)*

Table 3-51 Hebrew units of capacity (measured by weight)
(Dry products)
[1 elphah (Old) = 29.376 kg]
[1 elphah (New) = 21.420 kg]

Cor	Elphah	Sath (modius)	Gomor	Cab	Log
1	= 10	= 100/3	= 100	= 180	= 720
	1	= 10/3	= 10	= 18	= 72
		1	= 3	= 27/5	= 108/5
			1	= 9/5	= 36/5
				1	= 4

3.5.1.5.5 *Hebrew Units of Capacity (Liquids)*

Table 3-52 Hebrew units of capacity (measured by weight) (Liquids)
[1 bath (Old) = 29.376 kg]
[1 bath (New) = 21.420 kg]

Cor	Bath	Hin	Log
1	= 10	= 60	= 720
	1	= 6	= 72
		1	= 12
			1

3.5.1.6 The Greek System (Attic)

3.5.1.6.1 *Greek Units of Length*

These are shown in ***Table*** **3-53**.

Table 3-53 Greek (Attic) units of length

[1 pous = 0.30856 m]

Mille (mile)	Stadion	Plethron	Amma (cord)	Akaina	Orguia (fathom)	Bema (pace)	Cubit	Pous (foot)	Spithane (span)	Dichas	Palestra (palm)	Condylos	Daktylos (finger)
1	= 15/2	= 45	= 75	= 500	= 750	= 1800	= 2250	= 4500	= 6000	= 9000	= 18 000	= 36 000	= 72 000
	1	= 6	= 10	= 200/3	= 100	= 240	= 300	= 600	= 800	= 1200	= 2400	= 4800	= 9600
		1	= 5/3	= 100/9	= 2000/27	= 16 000/9	= 20 000/9	= 100	= 400/3	= 200	= 400	= 800	= 1600
			1	= 20/3	= 10	= 24	= 30	= 60	= 80	= 120	= 240	= 480	= 960
				1	= 3/2	= 18/5	= 9/2	= 9	= 12	= 18	= 36	= 72	= 144
					1	= 12/5	= 3	= 6	= 8	= 12	= 24	= 48	= 96
						1	= 5/4	= 5/2	= 10/3	= 5	= 10	= 20	= 40
							1	= 2	= 8/3	= 4	= 8	= 16	= 32
								1	= 4/3	= 2	= 4	= 8	= 16
									1	= 3/2	= 3	= 6	= 12
										1	= 2	= 4	= 8
											1	= 2	= 4
												1	= 2

3.5.1.6.2 *Greek Units of Weight*

Table 3-54 Greek (Attic) units of weight
[1 talent = 25.920 kg]

Talent	Mine	Drachma	Diobol	Obol	Chalque
1	= 60	= 6000	= 18 000	= 36 000	= 432 000
	1	= 100	= 300	= 600	= 4800
		1	= 3	= 6	= 48
			1	= 2	= 16
				1	= 8

3.5.1.6.3 *Greek Units of Capacity (Dry)*

The Greek system of capacity measures, which was divided into two sub-systems, one for dry substances and one other for liquids, was the following:

Table 3-55 Greek (Attic) units of capacity (Dry products)
[1 cotyle = 0.27 dm^3]

Medimnos	Hektos (modius)	Chenica	Sexte	Cotyle	Oxybaphon	Cyanthos
1	= 6	= 48	= 96	= 192	= 768	= 1152
	1	= 8	= 16	= 32	= 128	= 192
		1	= 2	= 4	= 16	= 24
			1	= 2	= 8	= 12
				1	= 4	= 6
					1	= 3/2

3.5.1.6.4 *Greek Units of Capacity (Liquids)*

Table 3-56 Greek (Attic) units of capacity (Liquids)
[1 cotyle = 0.27 dm^3]

Metretes	Amphora	Maris	Khous (conge)	Cotyle
1	= 2	= 6	= 12	= 144
	1	= 3	= 6	= 64
		1	= 2	= 24
			1	= 12

3.5.1.7 The Roman System

The Roman system of weights and measures was among the many customs adopted by the peoples conquered by the Romans throughtout Europe and western Asia.

3.5.1.7.1 *Roman Units of Length*

These are shown in ***Table 3-57***.

Table 3-57 Roman units of length

[1 Roman mile = 1472 m]

Milliarum (mile)	**Actus** (chain)	**Decempeda** (perch)	**Passus** (double pace)	**Gradus** (simple pace)	**Cubitus** (cubit)	**Palmipes**	**Pes** (foot)	**Palmus** (span)	**Uncia** (inch)	**Digitus** (finger)
1	= 125/3	= 500	= 1000	= 2000	= 10 000/3	= 4000	= 5000	= 20 000	= 60 000	= 80 000
	1	= 12	= 24	= 48	= 80	= 96	= 120	= 480	= 1440	= 1920
		1	= 2	= 4	= 20/3	= 8	= 10	= 40	= 120	= 160
			1	= 2	= 10/3	= 4	= 5	= 20	= 60	= 80
				1	= 5/3	= 2	= 5/2	= 10	= 30	= 40
					1	= 6/5	= 3/2	= 6	= 18	= 24
						1	= 5/4	= 5	= 15	= 20
							1	= 4	= 12	= 16
								1	= 3	= 4
									1	= 4/3

3.5.1.7.2 Roman Units of Area

See ***Table 3-58***.

3.5.1.7.3 Roman Units of Weight

See ***Table 3-59*** (p. 60).

3.5.1.7.4 Roman Units of Capacity (Dry)

Table 3-60 Roman units of capacity (Dry materials) [1 modius = 8.788 dm^3]

Quadrantal	Modius (Muid)	Semodius	Sextarius (Setier)	Hemina
1	= 3	= 6	= 48	= 96
	1	= 2	= 16	= 32
		1	= 8	= 16
			1	= 2

3.5.1.7.5 Roman Units of Capacity (Liquids)

See ***Table 3-61*** (p. 61).

3.5.1.8 The Arabic System

3.5.1.8.1 Arabic Units of Length

See ***Table 3-62*** (p. 62).

3.5.1.8.2 Arabic Units of Area

See ***Table 3-63*** (p. 63).

3.5.1.8.3 Arabic Units of Weight (System of the Prophet)

See ***Table 3-64*** (p. 63).

3.5.1.8.4 Arabic Units of Capacity

See ***Table 3-65*** (p. 64).

Table 3-58 Roman units of area

[1 quadratus pes = 0.086 m^2]

Saltus	Centuria	Heredium	Jugerum	Actus	Versum	Clima	Short actus	Decempeda quadrata	Quadratus pes
1	= 4	= 400	= 800	= 1600	= 2304	= 6400	= 57 600	= 230 400	= 23 040 000
	1	= 100	= 200	= 400	= 576	= 1600	= 14 400	= 57 600	= 5 760 000
		1	= 2	= 4	= 144/25	= 16	= 144	= 576	= 57 600
			1	= 2	= 72/25	= 8	= 72	= 288	= 28 800
				1	= 36/25	= 4	= 36	= 144	= 14 400
					1	= 25/9	= 25	= 100	= 10 000
						1	= 9	= 36	= 3 600
							1	= 4	= 400
								1	= 100

Table 3-59 Roman units of weight

[1 uncia = 0. 02725 kg]

Libra (podium)	Deunx	Dextans	Dodrans	Bes	Septunx	Semis	Quicunx	Triens	Quadrans	Sextans	Uncia (ounce)	Semuncia	Scripulum
1	= 12/11	= 6/5	= 4/3	= 3/2	= 12/7	= 2	= 12/5	= 3	= 4	= 6	= 12	= 24	= 288
	1	**= 11/10**	= 11/9	= 11/8	= 11/7	= 11/6	= 11/5	= 11/4	= 11/3	= 11/2	= 11	= 22	= 264
		1	**= 10/9**	= 5/4	= 10/7	= 5/3	= 2	= 5/2	= 10/3	= 5	= 10	= 20	= 240
			1	**= 9/8**	= 9/7	= 3/2	= 9/5	= 9/4	= 3	= 9/2	= 9	= 18	= 216
				1	**= 8/7**	= 4/3	= 8/5	= 2	= 8/3	= 4	= 8	= 16	= 192
					1	**= 7/6**	= 7/5	= 7/4	= 7/3	= 7/2	= 7	= 14	= 168
						1	**= 6/5**	= 3/2	= 2	= 3	= 6	= 12	= 144
							1	**= 5/4**	= 5/3	= 5/2	= 5	= 10	= 120
								1	**= 4/3**	= 2	= 4	= 8	= 96
									1	**= 3/2**	= 3	= 6	= 72
										1	**= 2**	= 4	= 48
											1	**= 2**	= 24
												1	**= 12**

Table 3-59 Roman units of weight (continued) [1 uncia = 0.02725 kg]

Libra (podium)	Uncia (ounce)	Semuncia	Duella	Sicilium	Milliaresum	Solidus (sextula)	Denarius	Denier	Scripulum
1	= 12	= 24	= 36	= 48	= 60	= 72	= 84	= 96	= 288
	1	= 2	= 3	= 4	= 5	= 6	= 7	= 8	= 24
		1	= 3/2	= 2	= 5/2	= 3	= 7/2	= 4	= 12
			1	= 4/3	= 5/3	= 2	= 7/3	= 8/3	= 8
				1	= 5/4	= 3/2	= 7/4	= 2	= 6
					1	= 6/5	= 7/5	= 8/5	= 24/5
						1	= 7/6	= 4/3	= 4
							1	= 8/7	= 24/7
								1	= 3

Table 3-61 Roman units of capacity (liquids) [1 sextarius = 0.547 dm^3]

Culleus (dolium) (hogshead)	Amphora (metrete)	Urna (urn)	Congius (gallon)	Sextarius (setier)	Hemina	Quartus	Acetabulum	Cyathus
1	= 20	= 40	= 160	= 960	= 1920	= 3840	= 7680	= 11 520
	1	= 2	= 8	= 48	= 96	= 192	= 384	= 576
		1	= 4	= 24	= 48	= 96	= 192	= 288
			1	= 6	= 12	= 24	= 48	= 72
				1	= 2	= 4	= 8	= 12
					1	= 2	= 4	= 6
						1	= 2	= 3
							1	= 3/2

Table 3-62 Arabic units of length

[1 foot = 0.320 m]

Marhala	Barid (veredus)	Parasang	Mille	Ghalva	Seir (stadion)	Qasab	Orgye (pace)	Cubit (hachemic)	Cubit (new)	Foot	Cabda (palm)	Assbaa (finger)
1	= 2	= 8	= 24	= 200	= 240	= 12 000	= 24 000	= 72 000	= 96 000	= 144 000	= 576 000	= 2 304 000
	1	= 4	= 12	= 100	= 120	= 6000	= 12 000	= 36 000	= 48 000	= 72 000	= 2 88 000	= 1 152 000
		1	= 3	= 25	= 30	= 1500	= 3000	= 9000	= 12 000	= 18 000	= 72 000	= 288 000
			1	= 225/27	= 10	= 500	= 1000	= 3000	= 4000	= 6000	= 24 000	= 96 000
				1	= 1.2	= 60	= 120	= 360	= 480	= 720	= 2880	= 11 520
					1	= 50	= 100	= 300	= 400	= 600	= 2400	= 9600
						1	= 2	= 6	= 8	= 12	= 48	= 192
							1	= 3	= 4	= 6	= 24	= 96
								1	= 4/3	= 2	= 8	= 32
									1	= 1.5	= 6	= 24
										1	= 4	= 16
											1	= 4

Table 3-63 Arabic units of area [1 feddan = 5898.24 m^2]

Feddan	Djarib	Daneq	Qirat	Cafiz	Habbah	Qamha	Qasaba (Achir)	Square cubit
1	= 4	= 6	= 24	= 40	= 72	= 96	= 400	= 14 400
	1	= 3/2	= 6	= 10	= 18	= 24	= 100	= 3600
		1	= 4	= 20/3	= 12	= 16	= 200/3	= 2400
			1	= 5/3	= 3	= 4	= 50/3	= 600
				1	= 9/5	= 12/5	= 10	= 360
					1	= 4/3	= 50/9	= 300
						1	= 25/6	= 225
							1	= 36

Table 3-64 Arabic units of weight (so-called *System of the Prophet*) [1 rotl = 0.340 kg]

Kikkar	Quantha	Ocque	Man (mine)	Rotl (rotolo)	Oukia	Nasch	Nevat	Dihrem
1	= 5/4	= 125/4	= 125/2	= 125	= 375	= 750	= 2250	= 11 250
	1	= 25	= 50	= 100	= 300	= 600	= 1800	= 9000
		1	= 2	= 4	= 12	= 24	= 72	= 360
			1	= 2	= 6	= 12	= 36	= 180
				1	= 3	= 6	= 24	= 120
					1	= 2	= 6	= 30
						1	= 3	= 15
							1	= 5

Table 3-65 Arabic units of capacity (measured by weight)

[1 cafiz = 32.640 kg]

Gariba (den)	Artabe (amphora)	Modius	Cafiz	Khoull (woëbe)	Ferk	Makuk	Sâa	Caphite (kist, kiladja)	Mudd
1	= 4	= 32/5	= 8	= 16	= 32	= 64	= 96	= 192	= 384
	1	= 8/5	= 2	= 4	= 8	= 16	= 24	= 48	= 96
		1	= 5/4	= 5/2	= 5	= 10	= 15	= 30	= 60
			1	= 2	= 4	= 8	= 12	= 24	= 48
				1	= 2	= 4	= 6	= 12	= 24
					1	= 2	= 3	= 6	= 12
						1	= 3/2	= 3	= 6
							1	= 2	= 4
								1	= 2

3.5.2 Obsolete National and Regional Systems

3.5.2.1 Old French System (Ancien Régime)

In France, under the *Ancien Régime* (i.e., before the French Revolution of 1789), the old measures derived from the system of Charlemagne. However, units varied from one region to another; subdivisions were irregular and also suffered regional variations, which tended to complicate business transactions.

3.5.2.1.1 Old French Units of Length

Table 3-66 Old French units of length
[1 pied (de Paris) = 0.3248394167 m]
[1 toise (de Perou) = 1.9490365 m (E)]

Lieue (league)	Perche (perch)	Toise	Aune	Pied (foot)	Pouce (inch)	Ligne (line)	Point (point)
1	= 380/3	= 2 280	= 760	= 13 680	= 161 160	= 1 969 920	= 23 639 040
	1	= 18	= 54	= 108	= 1296	= 15 552	= 186 624
		1	= 3	= 6	= 72	= 864	= 10 368
			1	= 2	= 24	= 288	= 3456
				1	= 12	= 144	= 1728
					1	= 12	= 144
						1	= 12

3.5.2.1.2 Old French Units of Area

Table 3-67 Old French units of area
[1 pied carré (de Paris) = 0.105520646 m^2]

Arpent (Eaux et Forêts)	Arpent (de Paris)	Perche (Eaux et Forêts)	Perche (de Paris)	Toise carrée	Pied carré
1	= 121/81	= 100	= 12 100/81	= 12 100/9	= 48 400
	1	= 8100/121	= 100	= 900	= 32 400
		1	= 121/81	= 121/9	= 484
			1	= 9	= 324
				1	= 36

3.5.2.1.3 Old French Units of Capacity

Units of capacity for liquids are shown in ***Table 3-68*** (overleaf).

Table 3-69 Old French units of capacity (Dry materials)
[1 sétier = 152 dm^3]

Muid	Setier	Mine	Minot	Boisseau	Quart	Litron
1	= 12	= 24	= 48	= 144	= 576	= 2304
	1	= 2	= 4	= 12	= 48	= 192
		1	= 2	= 6	= 24	= 96
			1	= 3	= 12	= 48
				1	= 4	= 16
					1	= 4

Table 3-68 Old French units of capacity (Liquids)

[1 pinte (de Paris) = 0.931389 dm^3]

Muid	Feuillette	Quartaut	Velte	Pot (quade, cade)	Pinte	Chopine (sétier)	Demi-setier	Posson	Demi-posson	Roquille
1	= 2	= 4	= 36	= 144	= 288	= 1152	= 2304	= 4608	= 9216	= 18 432
	1	= 2	= 18	= 72	= 144	= 576	= 1152	= 2304	= 4608	= 9216
		1	= 9	= 36	= 72	= 288	= 576	= 1152	= 2304	= 4608
			1	= 4	= 8	= 16	= 32	= 64	= 128	= 256
				1	= 2	= 4	= 8	= 16	= 32	= 64
					1	= 2	= 4	= 8	= 16	= 32
						1	= 2	= 4	= 8	= 16
							1	= 2	= 4	= 8
								1	= 2	= 4
									1	= 2

Table 3-70 Old French units of weight

[1 livre (de Paris) = 0.48951 kg]

Quintal (quintal)	Livre (pound)	Marc (mark)	Quarteron	Once (ounce)	Lot	Gros (drachm)	Denier (scruple)	Grain (grain)
1	= 1000	= 2000	= 4000	= 16 000	= 32 000	= 128 000	= 384 000	= 9 216 000
	1	= 2	= 4	= 16	= 32	= 128	= 384	= 9216
		1	= 2	= 8	= 16	= 64	= 192	= 4608
			1	= 4	= 8	= 32	= 96	= 2304
				1	= 2	= 8	= 24	= 576
					1	= 4	= 12	= 288
						1	= 3	= 72
							1	= 24

3.5.2.1.4 Old French Units of Weight

These are shown in ***Table 3-70*** (p. 67).

3.5.2.2 Old French System (1812–1840)

In 1812, the old weights and measures used before 1789 were restored by th French Emperor Napoleon Bonaparte. These pseudo-metric units were a follows:

Table 3-71 Old French units of length (metric) (period 1812–1840) [1 pied (metric) = 0.33 m (E)]

Lieue (metric)	Toise (metric)	Pied (metric)	Pouce (metric)	Ligne (metric)
1	= 2000	= 12 120	= 145 440	= 1 745 280
	1	= 6060	= 72 727	= 872 727
		1	= 12	= 144
			1	= 12

However, the Law of July 4th, 1837 reinstated the metric system, making obligatory in France from January 1st, 1840, and banning the use of othe weights and measures from that date.

3.5.2.3 Old Italian System

These units were in use before the adoption of the metric system. The definition varies geographically, and some units have changed over the year The metric system became compulsory in Italy in 1861, but it was adopted i Milan as early as 1803.

3.5.2.3.1 Italian Units of Length

Table 3-72 Old Italian units of length [1 piede liprando = 0.51377 mm]

Miglio	Trabucco	Canna	Piede	Oncia	Punto
1	= 6500/9	= 3250/3	= 13 000/3	= 52 000	= 624 000
	1	= 3/2	= 6	= 72	= 864
		1	= 4	= 48	= 576
			1	= 12	= 144
				1	= 12

5.2.3.2 Old Italian Units of Weight

able 3-73 Old Italian units of weight
[1 libbra = 0.307 kg]

antaro	Rubbo	Libbra	Oncia	Ottavo	Denaro	Grano
1	= 6	= 150	= 4800	= 38 400	= 115 200	= 2 764 800
	1	= 25	= 300	= 2400	= 7200	= 172 800
		1	= 12	= 96	= 288	= 6912
			1	= 8	= 24	= 576
				1	= 3	= 72
					1	= 24

5.2.3.3 Old Italian measures (Regional Variations)

able 3-74 Old Italian measures (regional variations)

City	Measures of weight	Measures of length	Measures of capacity
enezia (enice)	1 libbra grossa = 12 once = 0.477 kg 1 libbra sottile = 0.301 kg	1 braccio = 0.683 m 1 piede = 0.348 m	1 moggio = 8 mezzeni = 333.3 l
ilano (ilan)	1 libbra grossa = 28 once = 0.763 kg	1 braccio = 12 once = 0.595 m 1 trabucco = 6 piedi	1 moggio = 8 staia = 146.2 l 1 brenta = 96 boccali = 75.6 l
orino (urin)	1 libbra = 12 once = 0.369 kg	1 trabucco = 6 piedi liprandi = 3.096 m 1 raso = 0.6 m 1 piede = 0.293 m	1 sacco = 5 mine = 115.3 l 1 carro = 10 brente = 493.11 l
ologna	1 libbra mercantile = 12 once = 0.362 kg	1 braccio = 0.64 m 1 piede = 0.38 m	1 corba = 2 staia = 60 boccali = 78.6 l
irenze (lorence)	1 libbra = 12 once = 0.3395 kg	1 braccio = 2 palmi = 0.583 m	1 moggio = 8 sacca = 584.7 l 1 barile (vino) = 20 fiaschi = 45.6 l 1 barile (olio) = 16 fiaschi = 33.43 l
enova (enoa)	1 libbra = 12 once = 0.317 kg	1 palmo = 0.248 m	1 mina = 116.5 l 1 barile = 70 l
oma (ome)	1 libbra = 12 once = 0.339 kg	1 canna = 10 palmi = 2.234 m	1 rubblo = 22 scorzi = 294.5 l 1 barile = 32 boccali = 75.5 l
apoli (aples)	1 rotolo = 0.861 kg 1 libbra = 12 once = 0.321 kg	1 canna = 10 palmi = 2.646 m	1 botte (vino) = 12 barili = 523.5 l 1 tomolo = 55.54 l
alermo	1 cantaro = 100 rotoli = 79.34 kg 1 libbra = 12 once = 0.317 kg	1 canna = 10 palmi = 2.065 m	1 salma = 4 bisace = 16 tomoli = 2.75 l

5.2.4 Old Spanish System (Castillian)

5.2.4.1 Old Spanish Units of Length

ese are shown in *Table 3-75*.

Table 3-75 Old Spanish units of length

[1 vara = 0.835905 m]

Legua	Milla	Estadal	Estado	Passo (pace)	Vara (yard)	Codoc	Pie (foot)	Palma (palm)	Sesma	Pulgada	Diedo (finger)	Linea (line)	Punto (point)
1	= 3	= 1250	= 2500	= 3000	= 5000	= 10 000	= 15 000	= 20 000	= 30 000	= 180 000	= 240 000	= 2 880 000	= 34 560 000
	1	= 1250/3	= 2500/3	= 1000	= 5000/3	= 10 000/3	= 5000	= 20 000/3	= 10 000	= 60 000	= 80 000	= 960 000	= 11 520 000
		1	= 2	= 12/5	= 4	= 8	= 12	= 16	= 24	= 144	= 192	= 2 304	= 27 648
			1	= 6/5	= 2	= 4	= 6	= 8	= 12	= 72	= 96	= 1 152	= 13 824
				1	= 5/3	= 10/3	= 5	= 20/3	= 10	= 60	= 80	= 960	= 11 520
					1	= 2	= 3	= 4	= 6	= 36	= 48	= 576	= 6912
						1	= 3/2	= 2	= 3	= 18	= 24	= 288	= 3546
							1	= 4/3	= 2	= 12	= 16	= 192	= 2304
								1	= 3/2	= 9	= 12	= 144	= 1728
									1	= 6	= 8	= 96	= 1152
										1	= 4/3	= 16	= 192
											1	= 12	= 144
												1	= 12

3.5.2.4.2 Old Spanish Units of Area

Table 3-76 Old Spanish units of area
[1 square vara = 0.6987372 m^2]

Yugada	Fanegada	Aranzada	Calemin	Cuatilla	Square vara
1	= 50	= 72	= 600	= 18 432	= 460 800
	1	= 36/25	= 12	= 9216/25	= 9216
		1	= 25/3	= 256	= 6400
			1	= 768/25	= 768
				1	= 25
					1

3.5.2.4.3 Old Spanish Units of Weight

These are shown in ***Table 3-77*** (overleaf).

3.5.2.4.4 Old Spanish Units of Capacity (Liquids)

See ***Table 3-78*** (p. 73).

3.5.2.4.5 Old Spanish Units of Capacity (Dry)

Table 3-79 Old Spanish units of capacity (Dry materials)
[1 fanega = 55.501 dm^3]

Cahiz	Fanega	Cuartilla	Almude (calemin)	Medio	Cuartillo	Racion	Ochavillo
1	= 12	= 48	= 144	= 288	= 576	= 2304	= 9216
	1	= 4	= 12	= 24	= 48	= 192	= 768
		1	= 3	= 6	= 12	= 48	= 192
			1	= 2	= 4	= 16	= 64
				1	= 2	= 8	= 32
					1	= 4	= 16
						1	= 4

3.5.2.5 Old Portuguese System

3.5.2.5.1 Old Portuguese Units of Length

These are shown in ***Table 3-80*** (p. 73).

Table 3-77 Old Spanish units of weight

[1 libra = 0.460093 kg]

Tonnelada	Quintalmacho	Quintal	Barril	Arroba	Libra (pound)	Marco	Onza (ounce)	Escrupolo (scruple)	Ochava	Adarme (drachm)	Dinero	Tomin	Arienzo	Grano (grain)
1	= 40/3	= 20	= 40	= 80	= 2000	= 4000	= 32 000	= 128 000	= 256 000	= 512 000	= 768 000	= 1 536 000	= 4 608 000	= 18 432 000
	1	= 3/2	= 3	= 6	= 150	= 300	= 2400	= 9600	= 19 200	= 38 400	= 57 600	= 115 200	= 345 600	= 1 382 400
		1	= 2	= 4	= 100	= 200	= 1600	= 6400	= 12 800	= 25 600	= 38 400	= 76 800	= 230 400	= 921 600
			1	= 2	= 50	= 100	= 800	= 3200	= 6400	= 12 800	= 19 200	= 38 400	= 115 200	= 460 800
				1	= 25	= 50	= 400	= 1600	= 3200	= 6400	= 9600	= 19 200	= 57 600	= 230 400
					1	= 2	= 16	= 64	= 128	= 256	= 384	= 768	= 2304	= 9216
						1	= 8	= 32	= 64	= 128	= 192	= 384	= 1152	= 4608
							1	= 4	= 8	= 16	= 24	= 48	= 144	= 576
								1	= 2	= 4	= 6	= 12	= 36	= 144
									1	= 2	= 3	= 6	= 18	= 72
										1	= 3/2	= 3	= 9	= 36
											1	= 2	= 6	= 24
												1	= 3	= 12
													1	= 4

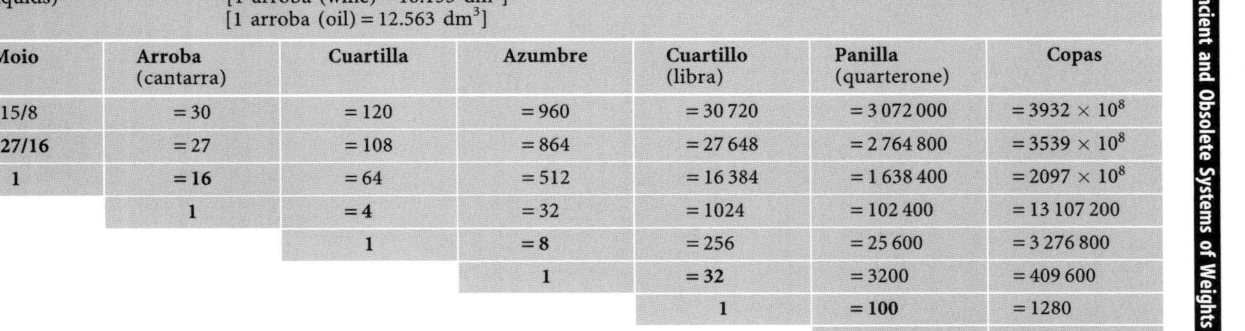

Table 3-78 Old Spanish units of capacity (Liquids) [1 arroba (wine) = 16.133 dm^3] [1 arroba (oil) = 12.563 dm^3]

Bota	Pipa	Moio	Arroba (cantarra)	Cuartilla	Azumbre	Cuartillo (libra)	Panilla (quarterone)	Copas
1	**= 10/9**	= 15/8	= 30	= 120	= 960	= 30 720	= 3 072 000	= 3932×10^8
	1	**= 27/16**	= 27	= 108	= 864	= 27 648	= 2 764 800	= 3539×10^8
		1	**= 16**	= 64	= 512	= 16 384	= 1 638 400	= 2097×10^8
			1	**= 4**	= 32	= 1024	= 102 400	= 13 107 200
				1	**= 8**	= 256	= 25 600	= 3 276 800
					1	**= 32**	= 3200	= 409 600
						1	**= 100**	= 1280
							1	**= 128**

Table 3-80 Old Portuguese units of length [1 pe = 0.3285 m]

Legoa	Milha	Estadio	Vara	Covada	Pe	Palmo	Pollegada	Linha
1	**= 3**	= 24	= 5664	= 9440	= 14 160	= 21 240	= 169 920	= 2 039 040
	1	**= 8**	= 1416	= 2360	= 4720	= 7080	= 56 640	= 679 680
		1	**= 236**	= 393	= 785	= 1177.5	= 9420	= 113 040
			1	**= 5/3**	= 10/3	= 10	= 80	= 960
				1	**= 2**	= 3	= 24	= 288
					1	**= 3/2**	= 12	= 144
						1	**= 8**	= 96
							1	**= 12**

3.5.2.5.2 *Old Portuguese Units of Weight*

Table 3-81 Old Portuguese units of weight [1 libra = 0.459 kg]

Quital	Arroba	Libra (arratel)	Meio (marco)	Onca	Outava	Escrupolo	Grao
1	=4	=128	=8192	=65 536	=524 288	=1 570 864	=37 748 736
	1	=32	=64	=512	=4096	=12 288	=249 912
		1	=2	=16	=128	=384	=9216
			1	=8	=64	=192	=4608
				1	=8	=24	=576
					1	=3	=72
						1	=24

3.5.2.5.3 *Old Portuguese Units of Capacity (Dry)*

Table 3-82 Old Portuguese units of capacity (Dry) [1 fanga = 54 dm^3]

Moio	Fanga	Alqueira	Meio	Quarto	Outava
1	=15	=60	=120	=240	=480
	1	=4	=8	=16	=32
		1	=2	=4	=8
			1	=2	=4
				1	=2

3.5.2.5.4 *Old Portuguese Units of Capacity (Liquids)*

Table 3-83 Old Portuguese units of capacity (Liquids) [1 almude = 16.5 dm^3]

Tonnelada	Bota (pipa)	Almude	Alqueira	Canada	Meio	Quartillo
1	=16/13	=32	=192	=384	=768	=1536
	1	=26	=156	=312	=624	=1248
		1	=6	=12	=24	=48
			1	=2	=4	=8
				1	=2	=4
					1	=2

3.5.2.6 Old Dutch System

3.5.2.6.1 *Old Dutch Units of Length*

Table 3-84 Old Dutch units of length [1 voeten (Amsterdam) = 0.2830594 m]

Roeden	Elle	Voeten	Duime	Lyne
1	=26/5	=13	=156	=1872
	1	=5/2	=30	=360
		1	=12	=144
			1	=12

.5.2.6.2 Old Dutch Units of Weight

Table 3-85 Old Dutch units of weight
[1 pond (Amsterdam) = 0.49409032 kg]
[1 pond (ordinary) = 0.49216772 kg]
[1 pond (apothecary) = 3/4 pond (ordinary) = 0.369126 kg]

Pond	Mark	Unze	Drachme	Engel	Vierling	Grein
1	= 2	= 16	= 128	= 320	= 1280	= 7680
	1	= 8	= 64	= 160	= 640	= 3840
		1	= 8	= 20	= 80	= 480
			1	= 5/2	= 10	= 60
				1	= 4	= 24
					1	= 6

.5.2.6.3 Old Dutch Units of Capacity (Dry)

Table 3-86 Old Dutch units of capacity (Dry)
[1 schepel = 27.26 dm^3]

Last	Mud	Zak	Schepel	Vierd	Kop
1	= 27	= 36	= 108	= 432	= 3456
	1	= 4/3	= 4	= 16	= 128
		1	= 3	= 12	= 96
			1	= 4	= 32
				1	= 8

.5.2.6.4 Old Dutch Units of Capacity (Liquids)

'hese are shown in ***Table 3-87*** (overleaf).

.5.2.7 Old German System (Prussian)

.5.2.7.1 Old German Units of Length

'hese Old German units were employed under the Prussian system.

Table 3-88 Old German units of length
[1 fuss = 0.313857 m]

Meile (mile)	Ruthe (yard)	Elle	Fuss (foot)	Zoll (inch)	Linie (line)
1	= 2000	= 192 000/17	= 24 000	= 288 000	= 3 456 000
	1	= 96/17	= 12	= 144	= 1728
		1	= 17/8	= 25.5	= 306
			1	= 12	= 144
				1	= 12

Table 3-87 Old Dutch units of capacity (Liquids)

[1 mingelen = 1.200 dm^3]

Vat	Oxhooft	Aam	Anker	Steekan	Stoop	Mingelen	Pint	Mutsje
1	= 4	= 6	= 24	= 48	= 384	= 768	= 1536	= 6144
	1	= 3/2	= 6	= 12	= 96	= 192	= 384	= 1536
		1	= 4	= 8	= 64	= 128	= 256	= 512
			1	= 2	= 16	= 32	= 64	= 256
				1	= 8	= 16	= 32	= 128
					1	= 2	= 4	= 16
						1	= 2	= 8
							1	= 4

3.5.2.7.2 *Old German Units of Weight*

Table 3-89 Old German units of weight
[1 pfund = 0.467711 kg (E)]

Schiffspfund	Doppelzentner	Zentner	Stein	Pfund (pound)	Loth	Quentchen
1	= 3/2	= 3	= 15	= 330	= 10 560	= 31 680
	1	= 2	= 10	= 220	= 7040	= 21 120
		1	= 5	= 110	= 3520	= 10 560
			1	= 22	= 704	= 2112
				1	= 32	= 96
					1	= 3

3.5.2.7.3 *Old German Units of Capacity (Dry)*

Table 3-90 Old German units of capacity (Dry)
[1 metzen = 37.0596 dm^3]

Scheffel	Metzen	Mässel	Dreissiger
1	= 6	= 48	= 192
	1	= 8	= 32
		1	= 4

3.5.2.7.4 *Old German Units of Capacity (Liquids)*

Table 3-91 Old German units of capacity (Liquids)
[1 quart = 1.145030 dm^3]

Fuder	Oxhoft	Ohm	Eimer	Anker	Quart
1	= 4	= 6	= 12	= 24	= 720
	1	= 3/2	= 3	= 6	= 180
		1	= 2	= 4	= 120
			1	= 2	= 60
				1	= 30

3.5.2.8 Old Russian System

All the following units were used before the 1917 Revolution and are now obsolete, although they will be encountered in 19th century Russian literature. The metric system was formally adopted as the national standard of measurement by the Soviet government in 1927.

3.5.2.8.1 *Old Russian Units of Length*

These are shown in ***Table 3-92***.

Table 3-92 Old Russian units of length

[1 foute = 0.3048 m (E)]

Vyorst	Saaschen	Arshin	Foute (foot)	Chevert	Vershok	Duîme	Sotka	Pal'ets	Line	Totchka
1	= 500	= 1500	= 3500	= 6000	= 24 000	= 42 000	= 50 000	= 84 000	= 420 000	= 4 200 000
	1	= 3	= 7	= 12	= 48	= 84	= 100	= 168	= 840	= 8400
		1	= 7/3	= 4	= 16	= 28	= 100/3	= 56	= 280	= 2800
			1	= 12/7	= 48/7	= 12	= 100/7	= 24	= 120	= 1200
				1	= 4	= 7	= 25/3	= 14	= 70	= 700
					1	= 7/4	= 25/12	= 7/2	= 35/2	= 175
						1	= 25/21	= 2	= 10	= 100
							1	= 42/25	= 42/5	= 84
								1	= 5	= 50
									1	= 10

5.2.8.2 *Old Russian Units of Weight (Ordinary)*

Table 3-93 Old Russian units of weight (Ordinary) [1 funt = 0.4095171792 kg]

Berkovets	Pood	Funt (pound)	Lana	Once	Loth (lot)	Solotnik (denier)	Doli (grain)
1	= 10	= 400	= 4 800	= 6400	= 12 800	= 38 400	= 3 686 400
	1	= 40	= 480	= 640	= 1280	= 3840	= 368 640
		1	= 12	= 16	= 32	= 96	= 9216
			1	= 4/3	= 8/3	= 8	= 768
				1	= 2	= 6	= 576
					1	= 3	= 288
						1	= 96

5.2.8.3 *Old Russian Units of Weight (Apothecary)*

Table 3-94 Old Russian units of weight (Apothecary) [1 doli = 44.4349403 mg]

Pound	Once	Drachme	Scrupule	Grain	Doli (grain)
1	= 120	= 960	= 2880	= 57 600	= 80 064
	1	= 8	= 24	= 480	= 672
		1	= 3	= 60	= 84
			1	= 20	= 28
				1	= 1.4

5.2.8.4 *Old Russian Units of Capacity (Dry)*

These are shown in ***Table 3-95*** (overleaf).

5.2.8.5 *Old Russian Units of Capacity (Liquids)*

See ***Table 3-96*** (p. 81).

5.2.9 Old Swedish System

5.2.9.1 *Old Swedish Units of Length*

See ***Table 3-97*** (p. 82).

5.2.9.2 *Old Swedish Units of Weight*

See ***Table 3-98*** (p. 83).

5.2.9.3 *Old Swedish Units of Capacity (Dry)*

See ***Table 3-99*** (p. 84).

Table 3-95 Old Russian units of capacity (Dry)

[1 garnetz = 3.279842 dm^3]

Tchevert	Osmini	Lof	Payok	Tcheverik	Vedro	Garnetz	Polou-garnetz	Krushky	Tchast
1	= 2	= 120/37	= 4	= 8	= 16	= 64	= 128	= 160	= 1920
	1	= 60/37	= 2	= 4	= 8	= 32	= 64	= 80	= 960
		1	= 37/30	= 74/30	= 148/30	= 592/30	= 1 184/30	= 148/3	= 592
			1	= 2	= 4	= 16	= 32	= 40	= 480
				1	= 2	= 8	= 16	= 20	= 240
					1	= 4	= 8	= 10	= 120
						1	= 2	= 5/2	= 30
							1	= 5/4	= 15
								1	= 12
									1

Table 3-96 Old Russian units of capacity (Liquids)

[1 vedro = 12.29941 dm^3]

Botchka	Pipe	Anker	Stekar	Vedro	Shtoff	Krouchka	Bottle (wine)	Bottle (vodka)	Tcharka	Chkalik
1	= 10/9	= 40/3	= 80/3	= 40	= 160	= 200	= 320	= 400	= 2000	= 4000
	1	= 12	= 24	= 36	= 144	= 180	= 288	= 360	= 1800	= 3600
		1	= 2	= 3	= 12	= 15	= 24	= 30	= 150	= 300
			1	= 3/2	= 6	= 15/2	= 12	= 15	= 75	= 150
				1	= 4	= 5	= 8	= 10	= 50	= 100
					1	= 5/4	= 2	= 5/2	= 25/2	= 25
						1	= 8/5	= 2	= 10	= 20
							1	= 5/4	= 25/4	= 25/2
								1	= 5	= 10
									1	= 2

Table 3-97 Old Swedish units of length

[1 fot = 0.29690 m]

Mil (mile)	Ref	Alnar (ell)	Stang	Famm (fathom)	Alm	Fot (foot)	Turn (inch)	Linie
1	= 225/2	= 1000	= 1125	= 3000	= 9000	= 18 000	= 216 000	= 2 592 000
	1	= 80/9	= 10	= 80/3	= 80	= 160	= 1920	= 23 040
		1	= 9/8	= 3	= 9	= 18	= 216	= 2592
			1	= 8/3	= 8	= 16	= 192	= 2304
				1	= 3	= 6	= 72	= 864
					1	= 2	= 24	= 288
						1	= 12	= 144
							1	= 12

Table 3-98 Old Swedish units of weight

[1 skålpund = 0.4250797024 kg]

Nyläst	Skippund	Waag	Centner	Sten	Liespund	Skålpund (pound)	Untz	Lod	Ort	Quintin	Korn	As
1	= 30	= 800/11	= 120	= 375	= 600	= 12 000	= 192 000	= 384 000	= 1 200 000	= 1 536 000	= 12 000 000	= 106 176 000
	1	= 80/33	= 4	= 25/2	= 20	= 400	= 6400	= 12 800	= 40 000	= 51 200	= 400 000	= 3 539 200
		1	= 33/20	= 165/32	= 33/4	= 165	= 2640	= 5280	= 16 500	= 21 120	= 165 000	= 1 459 920
			1	= 25/8	= 5	= 100	= 1600	= 3200	= 10 000	= 12 800	= 100 000	= 884 800
				1	= 8/5	= 32	= 512	= 1024	= 3200	= 4096	= 32 000	= 283 136
					1	= 20	= 320	= 640	= 2000	= 2560	= 20 000	= 176 960
						1	= 16	= 32	= 100	= 128	= 1000	= 8848
							1	= 2	= 25/4	= 8	= 125/2	= 553
								1	= 25/8	= 4	= 125/4	= 553/2
									1	= 32/25	= 10	= 2212/25
										1	= 125/16	= 553/8
											1	= 1106/125

Table 3-99 Old Swedish units of capacity (Dry)

[1 kanna = 2.617162 dm^3]

Kolläst	Koltunna	Tunna	Spanna	Fjerdingar	Kappar	Kanna	Stop	Quarter	Junkfra (ort)
1	= 12	= 27/2	= 27	= 108	= 432	= 756	= 1512	= 4536	= 24 192
	1	= 9/8	= 9/4	= 9	= 36	= 63	= 126	= 378	= 2016
		1	= 2	= 8	= 32	= 56	= 112	= 336	= 1792
			1	= 4	= 16	= 28	= 52	= 156	= 832
				1	= 4	= 7	= 14	= 42	= 224
					1	= 7/4	= 7/2	= 21/2	= 56
						1	= 2	= 8	= 32
							1	= 3	= 16
								1	= 16/3

3.5.2.9.4 Old Swedish Units of Capacity (Liquids)

Table 3-100 Old Swedish units of capacity (Liquids)
[1 kanna = 2.617162 dm^3]

Fuder	Oxhoft	Ohm	Eimer	Ankar	Kanna
1	= 4	= 6	= 12	= 24	= 360
	1	= 3/2	= 3	= 6	= 90
		1	= 2	= 4	= 60
			1	= 2	= 30
				1	= 15

3.5.2.10 Old Japanese System

3.5.2.10.1 Old Japanese Units of Length

These are shown in ***Table 3-101*** (overleaf).

3.5.2.10.2 Old Japanese Units of Area

Table 3-102 Old Japanese units of area
[1 tsubo = 100/30.25 = 3.30578512 m^2]

Square ri	Chô	Tan	Se	Tsubo	Gô
1	= 46 656	= 466 560	= 4 665 600	= 46 656 000	= 466 560 000
	1	= 10	= 100	= 3000	= 30 000
		1	= 10	= 300	= 3000
			1	= 30	= 300
				1	= 10

3.5.2.10.3 Old Japanese Units of Capacity

Table 3-103 Old Japanese units of capacity
[1 sho = (2401/1331) = 1.8039068 dm^3]

Koku	To	Sho	Gô	Shaku
1	= 10	= 100	= 1000	= 10 000
	1	= 10	= 100	= 1000
		1	= 10	= 100
			1	= 10

3.5.2.10.4 Old Japanese Units of Weight

These are shown in ***Table 3-104*** (p. 87).

Table 3-101 Old Japanese units of length

[1 shaku = 10/33 m]

Ri	Chô	Jô	Ken	Hiro	Yabiki	Shaku	Sun	Bu	Rin	Mô	Shi
1	= 36	= 1296	= 2160	= 2592	= 5184	= 12 960	= 129 600	= 1 296 000	= 12 960 000	= 120 960 000	= 1 209 600 000
	1	= 36	= 60	= 72	= 144	= 360	= 3600	= 36 000	= 360 000	= 3 600 000	= 36 000 000
		1	= 5/3	= 2	= 4	= 10	= 100	= 1000	= 10 000	= 100 000	= 1 000 000
			1	= 6/5	= 12/5	= 6	= 60	= 600	= 6000	= 60 000	= 600 000
				1	= 2	= 5	= 50	= 500	= 5000	= 50 000	= 500 000
					1	= 5/2	= 25	= 250	= 2500	= 25 000	= 250 000
						1	= 10	= 100	= 1000	= 10 000	= 100 000
							1	= 10	= 100	= 1000	= 10 000
								1	= 10	= 100	= 1000
									1	= 10	= 100
										1	= 10

Table 3-104 Old Japanese units of weight

[1 kwan = (15/4) = 3.75 kg]

Komma-ichi-da	Karus hiri-ichi-da	Kiyak-kin	Ninsoku-ichi-nin	Kwan	Kin	Hyaku-mé	Niyo	Mommé	Candareen (fun)	Rin	Mô
1	= 20/9	= 5/2	= 40/7	= 40	= 250	= 400	= 10 000	= 40 000	= 400 000	= 4 000 000	= 40 000 000
	1	= 9/8	= 18/7	= 18	= 225/2	= 180	= 4500	= 18 000	= 180 000	= 1 800 000	= 18 000 000
		1	= 16/7	= 16	= 100	= 160	= 4000	= 16 000	= 160 000	= 1 600 000	= 16 000 000
			1	= 7	= 175/4	= 70	= 1750	= 7000	= 70 000	= 700 000	= 7 000 000
				1	= 25/4	= 10	= 250	= 1000	= 10 000	= 100 000	= 1 000 000
					1	= 8/5	= 40	= 160	= 1600	= 16 000	= 160 000
						1	= 25	= 100	= 1000	= 10 000	= 100 000
							1	= 4	= 40	= 400	= 4000
								1	= 10	= 100	= 1000
									1	= 10	= 100
										1	= 10

4 Conversion Tables

4.1 Units in Alphabetical Order

In this conversion table, the units are listed in alphabetical order. Each unit is completely described as follows: name, symbol, physical quantity, dimension, conversion factor, and notes and definitions. The table includes about 2000 units with accurate conversion factors. The main abbreviations used in the table are listed below:

(E)	Exact numerical value
@	Astronomical units
a.u.	Atomic system of units
apoth., ap.	UK and US apothecaries' weights
avdp., av.	UK and US avoirdupois weights
BIH	Bureau International de l'Heure
BIPM	Bureau International des Poids et Mesures
CAN	Canadian units
CGPM	Conférence Générale des Poids et Mesures
cgs	Centimetre-gram-second units
FPS	Foot-pound-second units
INT	International use
MKpS	MKpS or MKfS units
MTS	MTS units
SI	SI units
IEUS	IEUS units
troy	UK and US Troy weights
UK	UK imperial units
US	US customary units
USMB	US Metric Board
WMA	Weights and Measures Act
Arabic	Old Arabian units (used in ancient times)
Attic	Old Greek units (used in ancient times)
Austrian	Old Austrian units

Chinese	Old Chinese units (used in ancient times)
Dutch	Old Dutch units
Egyptian	Old Egyptian units (used in ancient times)
French	Old French units
German	Old German (Prussian) units
Hebrew	Old Hebrew units (used in ancient times)
Hungarian	Old Hungarian units
Italian	Old Italian units
Japanese	Old Japanese
Portuguese	Old Portuguese units
Roman	Old Roman units (used in ancient times)
Russian	Old Russian (pre-Revolutionary) units
Spanish	Old Spanish units
Swedish	Old Swedish units
Swiss	Old Swiss units

Important Note

A precise conversion between two units needs first of all an exact knowledge of the unit's origin (see Chapters 2 and 3) and secondly numerical data from laboratory experiments. These basic values are given below, but the reader is also referred to Chapter 5 of this book where more information on fundamental constants may be obtained.

- the *density of pure water at 4°C (39.2°F)*[8] $\rho^{w}_{4°C} = 999.972\ \text{kg.m}^{-3}$
- the *density of pure water at 15.56°C (60°F)*[9] $\rho^{w}_{60°F} = 999.022\ \text{kg.m}^{-3}$
- the *density of mercury at 0°C (32°F)*[10] $\rho^{Hg}_{0°C} = 13\,595.08\ \text{kg.m}^{-3}$
- the *density of mercury at 15.56°C (60°F)* $\rho^{Hg}_{60°F} = 13\,558.14\ \text{kg.m}^{-3}$
- the *standard acceleration of gravity* $g_n = 9.80665\ \text{m.s}^{-1}$
- the *velocity of light in vacuum* $c_0 = 2.99792458 \times 10^8\ \text{m.s}^{-1}$
- the *permeability of vacuum* $\mu_0 = 4\pi \times 10^{-7}\ \text{H.m}^{-1}$

[8] This temperature has been accepted because the maximum density of water is measured at 3.98°C with $\rho^{w}_{3.98°C} = 999.973\ \text{kg.m}^{-3}$.

[9] From 'Density of water from 0 to 100°C' in Perry, R.H. and Green, D.W. (Eds.), *Perry's Chemical Engineer's Handbook*, 6th ed., McGraw-Hill Book Company, New York, 1984, pp. 3–75 to 3-76.

[10] From 'Density of mercury from 0 to 350°C' in Perry, R.H. and Green, D.W. (Eds.), *Perry's Chemical Engineer's Handbook*, 6th ed., McGraw-Hill Book Company, New York, 1984, pp. 3–77.

Unit (synonym, acronym)	Symbol	Physical quantity	Dimension	Conversion factor (SI equivalent unit)	Notes, definitions, other conversion factors	System
a.u. of action	a.u.	action, angular momentum	ML^2T^{-1}	1 a.u. of action $= 1.05457266 \times 10^{-34}$ J.s	1 a.u. of action $= \hbar = h/2\pi$	a.u.
a.u. of angular momentum	a.u.	action, angular momentum	ML^2T^{-1}	1 a.u. of angular momentum $= 1.05457266 \times 10^{-34}$ J.s	1 a.u. of angular momentum $= \hbar = h/2\pi$	a.u.
a.u. of charge	a.u., e	quantity of electricity, electric charge	IT	1 a.u. of charge $= 1.6021773349 \times 10^{-19}$ C	1 a.u. of charge $= e$	a.u.
a.u. of charge density	a.u.	electric charge density	ITL^{-3}	1 a.u. of charge density $= 1.081202621 \times 10^{12}$ C	1 a.u. of charge density $= e/a_0{}^3$	a.u.
a.u. of electric current	a.u.	electric current intensity	I	1 a.u. of current $= 6.62362108 \times 10^{-3}$ A	1 a.u. of current $= eE_h/\hbar$	a.u.
a.u. of electric dipole moment	a.u.	electric dipole moment	LTI	1 a.u. of electric dipole moment $= 8.478357919 \times 10^{-30}$ C.m	1 a.u. of electric dipole moment $= ea_0$	a.u.
a.u. of electric field strength	a.u.	electric field strength	$MLT^{-3}I^{-1}$	1 a.u. of electric field strength $= 5.142208245 \times 10^{11}$ V.m^{-1}	1 a.u. of electric field strength $= E_h/a_0e = e/4\pi\varepsilon_0 a_0{}^2$	a.u.
a.u. of electric potential	a.u.	electric potential, electric potential difference, electromotive force	$ML^2T^{-3}I^{-1}$	1 a.u. of electric potential $= 27.21139613$ V	1 a.u. of electrical potential $= E_h/e = e/4\pi\varepsilon_0 a_0$	a.u.
a.u. of electric quadrupole moment	a.u.	electric quadrupole moment	L^2TI	1 a.u. of electric quadrupole moment $= 4.486554120 \times 10^{-40}$ C.m^2	1 a.u. of electric quadrupole moment $= ea_0{}^2$	a.u.
a.u. of energy (hartree)	E_h	energy, work, heat	ML^2T^{-2}	1 a.u. of energy $= 4.3597482 \times 10^{-18}$ J	The unit is named after D.R. Hartree (1897–1958). 1 a.u. of energy = 1 hartree (E) 1 a.u. of energy = 2 rydbergs (E) 1 a.u. of energy $= \hbar^2/m_0 a_0{}^2$	a.u.

Unit (synonym, acronym)	Symbol	Physical quantity	Dimension	Conversion factor (SI equivalent unit)	Notes, definitions, other conversion factors	System
a.u. of force	a.u.	force, weight	MLT^{-2}	1 a.u. of force $= 8.238729477 \times 10^{-8}$ N	1 a.u. of force $= E_h/a_0$	a.u.
a.u. of gradient of electric field strength	a.u.	gradient of electric field strength	$MT^{-3}I^{-1}$	1 a.u. of gradient of electric field strength $= 9.717364560 \times 10^{21}$ $V.m^{-2}$	1 a.u. of gradient of electric field strength $= E_h/ea_0{}^2 = e/4\pi\varepsilon_0 a_0{}^3$	a.u.
a.u. of length (1st Bohr radius, Bohr)	a_0, b, a.u.	length, distance	L	1 a.u. of length $= 5.29177249 \times 10^{-11}$ m	Fundamental physical constant. Base unit of a.u. system. 1 a.u. of length $= 4\pi\varepsilon_0\hbar^2/m_0e^2$	a.u.
a.u. of linear momentum	a.u.	linear momentum, momentum	MLT^{-1}	1 a.u. of linear momentum $= 1.992853362 \times 10^{-24}$ N.s	1 a.u. of linear momentum $= \hbar/a_0$	a.u.
a.u. of magnetic dipole moment	a.u.	magnetic dipole moment	L^2I	1 a.u. of magnetic dipole moment $= 1.854803086 \times 10^{-23}$ $J.T^{-1}$	1 a.u. of magnetic dipole moment $= e\hbar/m_0 = 2\mu_B$	a.u.
a.u. of magnetic flux density	a.u.	magnetic flux density	$MT^{-2}I^{-1}$	1 a.u. of magnetic flux density $= 2.350518086 \times 10^5$ T	1 a.u. of magnetic flux density $= \hbar/ea_0{}^2$	a.u.
a.u. of magnetizability	a.u.	magnetizability	$M^{-1}L^2I^2T^2$	1 a.u. of magnetizability $= 7.891039396 \times 10^{-29}$ $J.T^{-2}$	1 a.u. of magnetizability $= e^2a_0{}^2/m_0$	a.u.
a.u. of mass (electron rest mass)	m_0, m_e	mass	M	1 a.u. of mass $= 9.1093897 \times 10^{-31}$ kg	Mass base unit in the a.u. system. 1 a.u. of mass $= m_0$	a.u.
a.u. of polarizability	a.u.	polarizability	$M^{-1}T^4I^2$	1 a.u. of polarizability $= 1.648777621 \times 10^{-41}$ $J^{-1}.C^2.m^2$	1 a.u. of polarizability $= 4\pi\varepsilon_0 a_0{}^3$	a.u.
a.u. of time	a.u.	time, period, duration	T	1 a.u. of time $= 2.418884338 \times 10^{-17}$ s	1 a.u. of time $= \hbar/E_h = m_e a_0{}^2/\hbar$	a.u.

aam (Dutch)	–	capacity, volume	L^3	1 aam (Dutch) = 153.600×10^{-3} m^3	Obsolete Dutch unit of capacity used for liquid substances. 1 aam (Dutch) = 128 Mingelen (E)	Dutch
abampere (emu of electric current intensity)	aA, Bi	electric current intensity	I	1 abampere = 10 A (E)	Obsolete cgs unit of electric current in the emu subsystem. 1 abampere = 1 Bi (E)	cgs
abbe	–	linear frequency, spatial frequency	$L^{-1}T^{-1}$	1 abbe = 10^3 $Hz.m^{-1}$ (E)	The unit is named after Ernst Abbe (1840-1901). It was suggested in 1973 as the unit of linear spatial frequency. 1 abbe = 1 $Hz.mm^{-1}$ (E)	
abcoulomb (emu of charge)	aC	quantity of electricity, electric charge	IT	1 abcoulomb = 10 C (E)	Obsolete cgs unit of electric charge in the emu subsystem.	cgs
abfarad (emu of electric capacitance)	aF	electric capacitance	$M^{-1}L^{-2}T^4I^2$	1 abfarad = 10^9 F (E)	Obsolete cgs unit of electric capacitance in the emu subsystem.	cgs
abhenry (emu of electric inductance)	aH	electric inductance	$ML^2T^{-2}I^{-2}$	1 abhenry = 10^{-9} H (E)	Obsolete cgs unit of electric inductance in the emu subsystem.	cgs
abmho (emu of conductance)	aS, $(a\Omega)^{-1}$	electric conductance	$M^{-1}L^{-2}T^3I^2$	1 abmho = 10^9 S (E)	Obsolete cgs unit of electric conductance in the emu subsystem.	cgs
abohm (emu of resistance)	aΩ	electric resistance	$ML^2T^{-3}I^{-2}$	1 abohm = 10^{-9} Ω (E)	Obsolete cgs unit of electric resistance in the emu subsystem.	cgs
abvolt (emu of electric potential)	aV	electric potential, electric potential difference, electromotive force	$ML^2T^{-3}I^{-1}$	1 abvolt = 10^{-8} V (E)	Obsolete cgs unit of electric potential in the emu subsystem.	cgs
abwatt (emu of power)	aW	power	ML^2T^{-3}	1 abwatt = 10^{-8} W (E)	Obsolete cgs unit of power in the emu subsystem.	cgs

Unit (synonym, acronym)	Symbol	Physical quantity	Dimension	Conversion factor (SI equivalent unit)	Notes, definitions, other conversion factors	System
acetabulum (Roman)	–	capacity, volume	L^3	1 acetabulum (Roman) $= 6.837500 \times 10^{-3}$ m^3	Obsolete Roman unit of volume employed in ancient times. It was used for capacity measurements of liquids. 1 acetabulum (Roman gallon) = 1/8 sextarius (E)	Roman
achir (qasaba)	–	surface, area	L^2	1 achir (qasaba) = 14.750 m^2	Obsolete Arabic unit of area used in ancient times. 1 achir (qasaba) = 1/400 feddan (E)	Arabic
achita (Indian)	–	mass	M	1 achita (Indian) = 940.800 kg	Obsolete Indian unit of mass used in ancient times. 1 achita (Indian) = 20 000 pala (E)	Indian
achtel (Austrian, dry)	–	capacity, volume	L^3	1 achtel (Austrian, dry) $= 7.686125 \times 10^{-3}$m^3	Obsolete Austrian unit of capacity used for dry substances. 1 achtel (Austrian, dry) = 1/8 metzel (E)	Austrian
acoustic ohm (cgs)	–	acoustic impedance	$ML^{-4}T^{-1}$	1 cgs acoustic ohm $= 10^{-5}$ Pa.m^{-3}.s (E)	Obsolete cgs unit of acoustic impedance. An acoustic impedance (including acoustic resistance and reactance) has a magnitude of 1 cgs acoustic ohm when a sound pressure of 1 barye produces a volume velocity of 1 cm^3/s. 1 cgs acoustic ohm = 1 dyne.s.cm^{-5} (E) 1 cgs acoustic ohm = 1 barye.s.cm^{-3} (E)	cgs
acoustic ohm (SI)	–	acoustic impedance	$ML^{-4}T^{-1}$	1 SI acoustic ohm = 1 Pa.s.m^{-3} (E)	SI unit of acoustic impedance. An acoustic impedance (including acoustic resistance and reactance) has a magnitude of 1 SI acoustic ohm when a sound pressure	SI

					of 1 Pa produces a volume velocity of 1 $m^3.s^{-1}$. 1 SI acoustic ohm = 1 N.s. m^{-5} (E)	
acre	ac, acre	surface, area	L^2	1 acre = 4.0468564224 × 10^3 m^2	Obsolete British unit of area employed in surveyor's measurements. It was first defined in England in the reign of Edward I (1272–1307) and is reputed to be the area which a yoke of oxen could plough in a day. 1 acre = 1/640 square mile (E) 1 acre = 4 roods (E) 1 acre = 10 square chains (Gunter's) (E) 1 acre = 160 square rods (E) 1 acre = 4840 square yards (E)	UK, US, CAN
acre (Cunningham)	ac, acre	surface, area	L^2	1 acre (Cunningham) = 5.188277465 × 10^3 m^2	Obsolete British unit of area employed in surveyor's measurements. 1 acre (Cunningham) = (1/0.78) acres (E)	UK
acre (Ireland)	ac, acre	surface, area	L^2	1 acre (Ireland) = 6.52787778 × 10^3 m^2	Obsolete Irish unit of area employed in surveyor's measurements. 1 acre (Ireland) = (1/0.62) acres (E)	UK, Ireland
acre (Plantation)	ac, acre	surface, area	L^2	1 acre (Plantation) = 5.188277465 × 10^3 m^2	Obsolete British unit of area employed in surveyor's measurements. 1 acre (Plantation) = (1/0.78) acres (E)	UK
acre (Scotland)	ac, acre	surface, area	L^2	1 acre (Scotland) = 4.935190759 × 10^3 m^2	Obsolete Scottish unit of area employed in surveyor's measurements. 1 acre (Scotland) = (1/0.82) acre (E)	UK, Scotland

Unit (synonym, acronym)	Symbol	Physical quantity	Dimension	Conversion factor (SI equivalent unit)	Notes, definitions, other conversion factors	System
acre (US, Survey)	ac (US, Surv.)	surface, area	L^2	1 acre (US, Survey) $= 4.046872610 \times 10^3$ m^2	Obsolete American unit of area used in geodetic measurements. 1 acre (US, Survey) = 4840 square yards (US, Survey) (E)	US
acre-foot	ac-ft, acre-ft	capacity, volume	L^3	1 acre-foot $= 1.233481837 \times 10^3$ m^3	Obsolete British unit of capacity employed to express volume of water in surveyor's measurements. It is equal to the product of covered area in acres multiply by depth in feet. 1 ac-ft = 43 560 ft^3 (E)	UK
acre-foot (US, Survey)	ac.-ft (US, Surv.)	capacity, volume	L^3	1 acre-foot (US, Survey) $= 1.23348677 \times 10^3$ m^3	Obsolete American unit of capacity employed to express volume of water in surveyor's measurements. It is equal to the product of covered area in acres multiplied by depth in feet. 1 ac-ft (US, Survey) = 43 560 ft^3 (E) 1 ac-ft (US, Survey) $= 3.259 \times 10^5$ gal (US)	US
acre-inch	ac-in	capacity, volume	L^3	1 acre-inch = 102.7901531 m^3	Obsolete British unit of capacity employed to express volume of water in surveyor's measurements. It is equal to the product of covered area in acres multiplied by depth in inches. 1 acre-inch = 3630 cu. ft (E)	UK

actus (Roman) [Roman chain]	–	length, distance	L	1 actus (Roman) = 35.328 m	Obsolete Roman unit of length employed in ancient times. 1 actus (Roman) = 120 pes (E)	Roman
actus (Roman)	–	surface, area	L^2	1 actus (Roman) = 1263.375936 m^2	Obsolete Roman unit of area employed in ancient times. 1 actus (Roman) = 14 400 quadratus pes (E)	Roman
adarme (Spanish)	–	mass	M	1 adarme (Spanish) $= 1.797238281 \times 10^{-3}$ kg	Obsolete Spanish unit of mass. 1 adarme (Spanish) = 1/256 libra (E)	Spanish
adhaka (Indian)	–	capacity, volume	L^3	1 adhaka (Indian) $= 3.300 \times 10^{-3}$ m^3 (of water)	Obsolete Indian unit of capacity used in ancient times. Measured by weight. 1 adhaka (Indian) = 1/4 drona (E)	Indian
akaina (Greek, Attic)	–	length, distance	L	1 akaina (Greek, Attic) = 2.777040 m	Obsolete Greek unit of length employed in ancient times. 1 akaina (Attic) = 9 pous (E)	Attic
albert	Alb	radiation photosynthetic activity	$MT^{-3}N^{-1}$	(see note)	Unit of radiation photosynthetic activity proposed by Lewis in 1985. The unit is named after Albert Einstein. 1 Alb = 1 $\mu E.m^{-2}.s^{-1}$ (E)	
alm (Swedish)	–	length, distance	L	1 alm (Swedish) = 0.5938 m	Obsolete Swedish unit of length. 1 alm = 2 fot (E)	Swedish
almude (Portuguese)	–	capacity, volume	L^3	1 almude (Portuguese) $= 16.5 \times 10^{-3}$ m^3	Obsolete Portuguese unit of capacity used for liquid substances.	Portuguese
almude (Spanish, dry)	–	capacity, volume	L^3	1 almude (Spanish, dry) $= 4.6250833 \times 10^{-3}$ m^3	Obsolete Spanish unit of capacity used for dry substances. 1 almude (Spanish, dry) = 1/12 fanega (E)	Spanish
alqueira (Portuguese)	–	capacity, volume	L^3	1 alqueira (Portuguese) $= 2.75 \times 10^{-3}$ m^3	Obsolete Portuguese unit of capacity used for liquid substances. 1 alqueira (Portuguese) = 1/6 almude (E)	Portuguese

Unit (synonym, acronym)	Symbol	Physical quantity	Dimension	Conversion factor (SI equivalent unit)	Notes, definitions, other conversion factors	System
alqueira (Portuguese, dry)	–	capacity, volume	L^3	1 alqueira (Portuguese, dry) $= 13.50 \times 10^{-3}$ m^3	Obsolete Portuguese unit of capacity used for dry substances. 1 alqueira (Portuguese, dry) = 1/4 fanga (E)	Portuguese
amagat	amagat	molar density	NL^{-3}	1 amagat = 44.615906 $mol.m^{-3}$ (1 atm = 101 325 Pa) 1 amagat = 44.032476 $mol.m^{-3}$ (1 bar = 100 000 Pa)	It is the reciprocal of the molar volume. The unit is used sometimes to express molar volume and molar density. Its value changes according to the nature of gas. This variation results from the behaviour of gas versus the ideal gas in the standard T and P conditions. The unit is named after E.H. Amagat (1841–1915).	
amagat	amagat	molar volume	L^3N^{-1}	1 amagat $= 22.413531 \times 10^{-3} m^3.mol^{-1}$ (1 atm = 101 325 Pa) 1 amagat $= 22.710511 \times 10^{-3} m^3.mol^{-1}$ (1bar = 100000Pa)	It is the molar volume of an ideal gas in the standard T and P conditions. Its value changes according to the nature of gas. This variation results from the behaviour of gas versus the ideal gas in the standard T and P conditions. The unit is named after E.H. Amagat (1841–1915).	
American run	–	specific length	$M^{-1}L$	1 American run = 3225.45109489143785 $m.kg^{-1}$	Obsolete American unit employed in the textile industry. 1 American run = 100 yd/oz (US) (E)	US
amma (Greek, Attic) [Greek cord]	–	length, distance	L	1 amma (Greek, Attic) = 18.513600 m	Obsolete Greek unit of length employed in ancient times. 1 amma (Attic) = 60 pous (E)	Attic
ampere	A	electric current intensity	I	**SI base unit**	The ampere is that constant current which, if maintained in two straight parallel	

					conductors of infinite length, of negligible circular cross-section, and placed one metre apart in vacuum, would produce between these conductors a force equal to 2×10^{-7} newton per metre of length [9th CGPM (1948), Resolution 2 and 7]. The unit is named after A.M. Ampère (1775–1836).	SI, MKSA
ampere (int. mean)	–	electric current intensity	I	1 A (int. mean) = 0.99985 A	Obsolete (IEUS) unit of electric current intensity. It was defined in 1881. It is equal to current intensity which permits to deposit, in one second, by electrolysis from an aqueous silver nitrate solution, 0.00111800 grams of silver metal at the cathode.	IEUS, INT
ampere (int. US)	–	electric current intensity	I	1 A (int. US) = 0.999835 A	Obsolete IEUS unit	IEUS, US
ampere per metre	$A.m^{-1}$	magnetic field strength, lineic current density	IL^{-1}	**SI derived unit**	The ampere per metre is the magnetic field strength which is created tangentially at a distance of one metre from a straight conductor, of infinite length and with negligible cross section, by a circulating current of one ampere.	SI
ampere per square centimetre	$A.cm^{-2}$	electric current density	IL^{-2}	$1\ A.cm^{-2} = 10^4\ A.m^{-2}$ (E)	Unit of electric current density employed in electrochemical engineering.	
ampere per square decimetre	$A.dm^{-2}$	electric current density	IL^{-2}	$1\ A.dm^{-2} = 100\ A.m^{-2}$ (E)	Unit of electric current density employed in electrochemical engineering.	
ampere per square inch	$A.in^{-2}$ (A/sq.in)	electric current density	IL^{-2}	$1\ A.in^{-2} = 1.555000310 \times 10^3\ A.m^{-2}$	British and American unit of electric current density.	UK, US

Unit (synonym, acronym)	Symbol	Physical quantity	Dimension	Conversion factor (SI equivalent unit)	Notes, definitions, other conversion factors	System
ampere per square metre	$A.m^{-2}$	electric current density	IL^{-2}	**SI derived unit**	The ampere per square metre is the electric current density which is equal to an electric current of one ampere which circulates in a homogeneous conductor having a section area of one square metre.	SI
ampere-hour	Ah	quantity of electricity, electric charge	IT	1 Ah = 3600 C (E)	Non-SI unit used in electrical engineering in common use with the SI. It is a practical unit of electric charge equal to the charge flowing in one hour through a conductor passing one ampere.	
ampere-square metre	$A.m^2$	magnetic dipole moment	IL^2	**SI derived unit**	$1\ A.m^2 = 1\ J.T^{-1}$ (E)	SI
ampere-turn	A-tour	magnetomotive force	I	1 ampere-turn = 2864.77 A	Obsolete unit of magnetomotive force. The ampere-turn is equal to the magnetomotive-force produced when a current of one ampere flows through one turn of a magnetizing coil. 1 ampere-turn = 3600 Gb (E)	
amphora (Greek, Attic)	–	capacity, volume	L^3	1 amphora (Greek, Attic) $= 19.44 \times 10^{-3}\ m^3$	Obsolete Greek unit of volume employed in ancient times. It was used for capacity measurements of liquids.	Attic
amphora (large, Persian)	–	capacity, volume	L^3	1 amphora (large, Persian) $= 97.8 \times 10^{-3} m^3$ (of water)	Obsolete unit of capacity of the Assyrio-Chaldean-Persian system used in ancient times. Measured by weight. 1 amphora (large, Persian) = 3 amphora (E)	Persian

amphora (Persian)	–	capacity, volume	L^3	1 amphora (Persian) $= 32.6 \times 10^{-3} m^3$ (of water)	Obsolete unit of capacity of the Assyrio-Chaldean-Persian system used in ancient times. Measured by weight.	Persian
amphora (Roman)	–	capacity, volume	L^3	1 amphora (Roman) $= 26.364 \times 10^{-3}$ m^3	Obsolete Roman unit of volume employed in ancient times. It was used for capacity measurements of liquids. 1 amphora (Roman) = 48 sextarius (E)	Roman
ångström	Å	length, distance	L	1 Å $= 10^{-10}$ m (E)	Obsolete unit of wavelength employed in atomic and molecular physics for electromagnetic radiation measurements ranging from the UV to IR regions. The unit is named after the Swedish scientist A.J. Ångström (1814–1874). The unit is based on the wavelength of the emission spectra's red line of an atom of cadmium. The agreed numerical value is equal exactly to $\lambda = 6438.4696$ Å in dry air which contains at maximum a volume fraction of carbon dioxide of 30×10^{-6}, and measured under standard atmospheric pressure and at $T = 15°C$. The unit has the same order of magnitude as the radius of the atom.	
ångström star (WK_{α_1})	Å*	length, distance	L	1 Å* $= 1.0000148192 \times 10^{-10}$ m (E)	Obsolete unit of wavelength used in atomic physics and radiocrystallography. The definition of the unit is based on the wavelength of the spectral line K_{α_1} of an atom of tungsten which is precisely equal to 0.2090100 Å*.	

Unit (synonym, acronym)	Symbol	Physical quantity	Dimension	Conversion factor (SI equivalent unit)	Notes, definitions, other conversion factors	System
angula	–	length, distance	L	1 angula $= 1.904166667 \times 10^{-2}$ m	Obsolete Indian unit of length used in ancient times. 1 angula = 1/24 hasta (E)	Indian
ankar (Swedish)	–	capacity, volume	L^3	1 ankar (Swedish) $= 39.257430 \times 10^{-3}$ m^3	Obsolete Swedish unit of capacity for liquid substances. 1 ankar (Swedish) = 15 kanna (E)	Swedish
anker (Dutch)	–	capacity, volume	L^3	1 anker (Dutch) $= 38.400 \times 10^{-3}$ m^3	Obsolete Dutch unit of capacity used for liquid substances. 1 anker (Dutch) = 32 mingelen (E)	Dutch
anker (Prussian)	–	capacity, volume	L^3	1 anker (Prussian) $= 34.350900 \times 10^{-3}$ m^3	Obsolete German unit of capacity used for liquid substances. 1 anker (Prussian) = 30 quart (E)	German
anker (Russian)	–	capacity, volume	L^3	1 anker (Russian) $= 36.898230 \times 10^{-3}$ m^3	Obsolete Russian unit of capacity for liquid substances used before 1917. 1 anker (Russian) = 3 vedro (E)	Russian
API degree	API	specific gravity index of petroleum	nil	$^{\circ}\mathrm{API} = 141.5/d_{60^{\circ}\mathrm{F}}^{60^{\circ}\mathrm{F}} - 131.5$	Hydrometer unit adopted in 1952 by the American petroleum Institute (API). It serves to measure the density of raw petroleum and to avoid confusion in business transactions. It has a scale from 0°API (sp. gr. of 1.076) to 100°API (sp. gr. of 0.6112). Specific gravity is widely used in countries outside the US, and with the adoption of SI units API favours density at 15°C instead of degrees API.	US

apostilb	asb	luminous luminance	JL^{-2}	1 asb = 0.3183099 $cd.m^{-2}$	Obsolete German unit of luminous luminance. 1 asb = $[1/\pi]$ cd/m^2 (E) 1 asb = 10^{-4} lambert (E)	German
apt (Egyptian)	–	capacity, volume	L^3	1 apt (Egyptian) = $8.5 \times 10^{-3} m^3$ (of water)	Obsolete Egyptian unit of capacity used in ancient times. Measured by weight. 1 apt (Egyptian) = 1/4 khar (E)	Egyptian
aranzada (Spanish)	–	surface, area	L^2	1 aranzada (Spanish) = 4471.917881 m^2	Obsolete Spanish unit of area. 1 aranzada (Spanish) = 6400 sq. vara (E)	Spanish
arataba (large, Persian)	–	capacity, volume	L^3	1 arataba (large, Persian) = 65.2×10^{-3} m^3 (of water)	Obsolete unit of capacity of the Assyrio-Chaldean-Persian system used in ancient times. Measured by weight. 1 arataba (large, Persian) = 2 amphora (E)	Persian
arcmin (minute of arc)	–	length, distance	L	1 arcmin = $2.908882087 \times 10^{-4}$m ($R = 1$ m)	Unit of curvilinear abscissa. It is equal to the angle expressed in minutes multiplied by the circle radius. 1 arcmin (m) = $(\pi/10\,800)R(m)\theta(')$	
are	a	surface, area	L^2	1 are = 10^2 m^2 (E)	French metric unit of area used in surveyor's measurements. It was introduced after the French Revolution. The unit is still used in agriculture.	French
arienzo (Spanish)	–	mass	M	1 arienzo (Spanish) = $1.996931424 \times 10^{-4}$ kg	Obsolete Spanish unit of mass. 1 arienzo (Spanish) = 1/2304 libra (E)	Spanish
arpent (de Paris)	a	surface, area	L^2	1 arpent (de Paris) = 3418.868950 m^2	Obsolete French unit of area used in surveyor's measurements before the French Revolution (1789). 1 arpent (de Paris) = 32 400 sq. pieds (de Paris)	French

Unit (synonym, acronym)	Symbol	Physical quantity	Dimension	Conversion factor (SI equivalent unit)	Notes, definitions, other conversion factors	System
arpent (Eaux et Forêts)	a	surface, area	L^2	1 arpent (Eaux et Forêts) = 5107.199295 m^2	Obsolete French unit of area used in surveyor's measurements before the French Revolution (1789). 1 arpent (Eaux et Forêts) = 48 400 sq. pieds (de Paris)	French
arpent (ordinaire, metric)	a	surface, area	L^2	1 arpent (metric) = 4221 m^2	Obsolete French metric unit of area used in surveyor's measurements.	French
arpent (Québec)	–	length, distance	L	1 arpent (Québec) = 58.47131 m	Obsolete French unit of length sometimes still used in Québec (Canada).	CAN, French
arpent (Québec)	a	surface, area	L^2	1 arpent (Canada) = 3419.6 m^2	Obsolete French metric unit of area used in surveyor's measurements. Still used in Québec (Canada).	CAN, French
arratel (Portuguese)	–	mass	M	1 arratel (Portuguese) = 0.459 kg	Obsolete Portuguese unit of mass. 1 arratel (Portuguese) = 1 libra (E)	Portuguese
arroba (oil) (Spanish)	–	capacity, volume	L^3	1 arroba (oil) = 12.563×10^{-3} m^3	Obsolete Spanish unit of capacity used for liquid substances.	Spanish
arroba (Portuguese)	–	mass	M	1 arroba (Portuguese) = 14.688 kg	Obsolete Portuguese unit of mass. 1 arroba (Portuguese) = 32 libra (E)	Portuguese
arroba (Spanish)	–	mass	M	1 arroba (Spanish) = 11.502325 kg	Obsolete Spanish unit of mass. 1 arroba (Spanish) = 25 libra (E)	Spanish
arroba (water) (Spanish)	–	capacity, volume	L^3	1 arroba (water) = 15.643162×10^{-3} m^3	Obsolete Spanish unit of capacity used for liquid substances. It was equal to 34 libra of river water.	Spanish

arroba (wine) (Spanish)	–	capacity, volume	L^3	1 arroba (wine) = $16.133 \times 10^{-3} m^3$	Obsolete Spanish unit of capacity used for liquid substances.	Spanish
arshin (Russian)	–	length, distance	L	1 arshin (Russian) = 0.7112 m (E)	Obsolete Russian unit of length used before 1917. 1 arshin (Russian) = 7/3 foute (E)	Russian
artaba (short, Persian)	–	capacity, volume	L^3	1 artaba (short, Persian) = 48.9×10^{-3} m^3 (of water)	Obsolete unit of capacity of the Assyrio-Chaldean-Persian system used in ancient times. Measured by weight. 1 artaba (short, Persian) = 3/2 amphora (E)	Persian
artabe (Arabic) (amphora)	–	capacity, volume	L^3	1 artabe (Arabic) = $65.28 \times 10^{-3} m^3$ (of water)	Obsolete Arabic unit of capacity used in ancient times. Measured by weight. 1 artabe (Arabic) = 2 cafiz (E)	Arabic
artabe (Egyptian)	–	capacity, volume	L^3	1 artabe (Egyptian) = $51 \times 10^{-3}. m^3$ (of water)	Obsolete Egyptian unit of capacity used in ancient times. Measured by weight. 1 artabe (Egyptian) = 3/2 khar (E)	Egyptian
assbaa (Arabic) [Arabian finger]	–	length, distance	L	1 assbaa (Arabic) = 2×10^{-2} m	Obsolete Arabic unit of length used in ancient times. 1 assbaa (Arabic) = 1/16 feet (Arabic) (E)	Arabic
astronomical unit	AU, UA	length, distance	L	1 AU = $1.49597870 \times 10^{11}$ m	Unit of length employed in astronomy for describing planetary distance. One AU corresponds approximately to the mean distance between the Earth and the Sun. It was adopted by the International Astronomical Union in 1964.	
atmosphere (standard atmosphere)	atm, A_n	pressure, stress	$ML^{-1}T^{-2}$	1 atm = 101 325 Pa (E)	Obsolete pressure and stress unit which should be discontinued. Unit of pressure equal to the air pressure measured at mean sea level.	INT

Unit (synonym, acronym)	Symbol	Physical quantity	Dimension	Conversion factor (SI equivalent unit)	Notes, definitions, other conversion factors	System
atmosphere (technical)	at	pressure, stress	$ML^{-1}T^{-2}$	1 at = 9.80665×10^4 Pa (E)	Obsolete MKpS pressure and stress derived unit. Obsolete. 1 at = 1 kgf.cm^{-2} (E) 1 atm = 1.033 227 453 at	MKpS
atomic unit of mass [$^{12}C = 12.0000$]	u, uma, Da (^{12}C), AMU	mass	M	1 u (^{12}C) = $1.660540210 \times 10^{-27}$ kg	The atomic unit of mass u (^{12}C) is equal to the fraction 1/12 of the mass of the carbon 12 atom. 1 u (^{12}C) = 0.999682141 u (^{16}O)	
atomic unit of mass [$^{16}O = 16.0000$]	u, uma, Da (^{16}O) AMU	mass	M	1 u(^{16}O) = $1.660012432 \times 10^{-27}$ kg	The atomic unit of mass u (^{16}O) is equal to the fraction 1/16 of the mass of the oxygen 16 atom. 1 u (^{16}O) = 1.000317937 u (^{12}C)	
atomic unit of mass [$^{1}H = 1.0000$]	u, uma, Da (^{1}H) AMU	mass	M	1 u (^{1}H) = $1.673533995 \times 10^{-27}$ kg	The atomic unit of mass u (^{1}H) is equal to the fraction 1/1 of the mass of the hydrogen (^{1}H) atom.	
atour (Egyptian)	–	length, distance	L	1 atour (Egyptian) = 5.235×10^3 m	Obsolete Egyptian unit of length used in ancient times. 1 atour = 10 000 Royal cubit (E)	Egyptian
atour (Royal Egyptian)	–	length, distance	L	1 atour (Royal Egyptian) = 1.047×10^4 m	Obsolete Egyptian unit of length used in ancient times. 1 atour = 20 000 Royal cubit (E)	Egyptian
attogram	ag	mass	M	1 ag = 10^{-21} kg (E)	Submultiple of the SI base unit. 1 ag = 10^{-18} g (E)	SI
attometre	am	length, distance	L	1 am = 10^{-18} m (E)	Submultiple of the SI base unit.	SI

					in the textile industry for clothes measurements. Its was used in France during the Ancien Régime before the French Revolution (1789). 1 aune (de Paris) = 0.6024 toise (de Pérou)	
aurure (Egyptian)	–	surface, area	L^2	1 aurure (Egyptian) = 2740.5 m^2	Obsolete Egyptian unit of area used in ancient times. 1 aurure (Egyptian) = 100 pekeis (E)	Egyptian
Avogadro (Number)	N_A, L	number of entities per amount of substance	N^{-1}	1 $N_A = 6.0221367 \times 10^{23}$ mol^{-1}	Fundamental physical constant. The unit is named after A. Avogadro (1776–1856).	
azumbre (Spanish)	–	capacity, volume	L^3	1 azumbre (Spanish) = $1.955395250 \times 10^{-3}$ m^3	Obsolete Spanish unit of capacity used for liquid substances. 1 azumbre (Spanish) = 1/8 arroba (water) (E)	Spanish
B unit (pastille dose)	–	exposure	ITM^{-1}	1 pastille dose = 1.290×10^{-1} $C.kg^{-1}$ (E)	Obsolete unit of exposure of ionizing radiations. It was equal to the radiation dose required to change the variation colour of a basic barium platinocyanide pastille from apple green (tint A) to reddish brown (tint B). 1 B unit = 500 röntgens (E)	
bag (UK)	–	capacity, volume	L^3	1 bag (UK) = $109.1062080 \times 10^{-3}$ m^3	Obsolete British unit of capacity. 1 bag (UK) = 1 sack (E) 1 bag (UK) = 3 bushels (UK) (E) 1 bag (UK) = 6 buckets (UK) (E) 1 bag (UK) = 12 pecks (UK) (E) 1 bag (UK) = 24 gallons (UK) (E)	UK

Unit (synonym, acronym)	Symbol	Physical quantity	Dimension	Conversion factor (SI equivalent unit)	Notes, definitions, other conversion factors	System
bag (UK, cement)	–	mass	M	1 bag (UK, cement) = 42.63768278 kg (E)	Obsolete British unit of mass employed in civil engineering. 1 bag (UK, cement) = 94 lb (av.) of Portland cement (E)	UK
baha (Indian)	–	capacity, volume	L^3	1 baha (Indian) = 2.640 m^3 (of water)	Obsolete Indian unit of capacity used in ancient times. Measured by weight. 1 baha (Indian) = 200 drona (E)	Indian
Balling degree	–	specific gravity unit to express amount of sugar in syrups, juice	nil	$^{\circ}\text{Balling} = 200 - 200/d^{60^{\circ}F}_{60^{\circ}F}$	Obsolete hydrometer unit adopted in 1859.	
balmer	–	wavenumber	L^{-1}	1 balmer = 10^2 m^{-1} (E)	Obsolete wavenumber unit proposed in 1951 for wavenumber measurements in spectroscopy; it has not been adopted. The unit is named after J.J. Balmer (1825–1898). 1 balmer = 1 kayser 1 balmer = 1 cm^{-1}	
balthazar (belshazzar)	–	capacity, volume	L^3	1 balthazar = 12.122912×10^{-3} m^3	Obsolete British unit employed for expressing the capacity of wine containers. Still employed in oenology, especially in France. 1 balthazar = 8/3 gallons (UK) (E) 1 balthazar = 16 bouteilles (E)	UK, French
bar	bar	pressure, stress	$ML^{-1}T^{-2}$	1 bar = 10^5 Pa (E)	Pressure and stress unit employed in fluid mechanics. Its use is temporarily maintained with the SI. 1 atm = 1.01325 bar (E)	

bara (Indian)	–	mass	M	1 bara (Indian) = 94 kg	Obsolete Indian unit of mass used in ancient times. 1 bara (Indian) = 2 000 pala (E)	Indian
barad	barad	pressure, stress	$ML^{-1}T^{-2}$	1 barad = 10^{-1} Pa (E)	Pressure and stress unit proposed and named after the British Association in 1888. Now replaced by the barye. 1 barad = 1 dyne.cm^{-2} (E) 1 atm = 1 013 250 barad (E)	
barid (Arabic) [veredus]	–	length, distance	L	1 barid (Arabic) = 2.304×10^4 m	Obsolete Arabic unit of length used in ancient times. 1 barid (Arabic) = 72 000 feet (Arabic) (E)	Arabic
barile di olio (Italian)	–	capacity, volume	L^3	1 barile di olio (Italian) = 33.4×10^{-3} m^3	Obsolete Italian unit of capacity.	Italian
barile di vino (Italian)	–	capacity, volume	L^3	1 barile di vino (Italian) = 45.6×10^{-3} m^3	Obsolete Italian unit of capacity.	Italian
bark degree	°Bk	specific gravity of liquids, hydrometer index, hydrometer degree	nil	$°Bk = 1000 \times [d^{60°F}_{60°F} - 1]$	A barkometer is used for testing tanning liquor. It has a scale ranging 0°–80°Bk; the number to the right of the decimal point of the sp. gr. is the °Bk, thus 1.025 sp. gr. is 25°Bk.	
barleycorn (UK)	–	length, distance	L	1 barleycorn (UK) = 8.466667×10^{-3} m	Obsolete British unit of length. 1 barleycorn (UK) = 4 lines (UK) (E) 1 barleycorn (UK) = 1/3 inch (E) 1 barleycorn (UK) = 1/36 foot (E)	UK

Unit (synonym, acronym)	Symbol	Physical quantity	Dimension	Conversion factor (SI equivalent unit)	Notes, definitions, other conversion factors	System
barn	b	surface, area	L^2	1 barn = 10^{-28} m^2 (E)	Obsolete unit of area employed to express cross section of nuclides. It was introduced by H.G. Holloway and C.P. Baker in Chicago in 1942, to measure the probability of some phenomena interesting nuclear physics (fission, capture, diffusion, absorption, etc, of nuclides). 1 barn = 10^{-24} cm^2 (E) 1 barn = 100 fm^2 (E)	
baromil	–	length, distance	L	1 baromil = 7.500626793 × 10^{-4} m	Obsolete American unit used for graduating mercury barometers. It was introduced by the American Meteorological Society. It corresponds to the increasing height of mercury column, which is equal to an increasing atmospheric pressure of 1 mbar if measured at 0°C for a barometer at the sea level and at a latitude of N45°. 1 baromil ≈ 0.75 mm	US
barrel (UK, beer)	bbl (UK, beer)	capacity, volume	L^3	1 barrel (UK, beer) = 163.659312 × 10^{-3} m^3	Obsolete British unit of capacity used in the brewery industry. 1 barrel (UK, beer) = 36 gallons (UK) (E)	UK
barrel (UK, wine)	bbl (UK, wine)	capacity, volume	L^3	1 barrel (UK, wine) = 143.2018980 × 10^{-3} m^3	Obsolete British unit of capacity used for alcohols and spirits. 1 barrel (UK, wine) = 31.5 gal (UK) (E)	UK

barrel (UK) per long ton	bbl (UK). $(\text{lg t})^{-1}$	specific volume	L^3M^{-1}	1 barrel (UK).$(\text{lg t})^{-1}$ $= 1.409402428 \times 10^{-4}\ m^3.kg^{-1}$	Obsolete British unit of specific volume. 1 barrel (UK).$(\text{lg t})^{-1}$ = 31.5 gal (UK).$(2240\ \text{lb})^{-1}$ (E)	UK
barrel (UK, alcohol)	bbl (alcohol)	capacity, volume	L^3	1 barrel (UK, alcohol) $= 227.304600 \times 10^{-3}\ m^3$	Obsolete British unit of capacity. 1 barrel (UK, alcohol) = 50 gallons (UK) (E)	US
barrel (UK, cement)	–	mass	M	1 barrel (UK, cement) = 170.5507311 kg (E)	Obsolete British unit for mass measurements in civil engineering. 1 barrel (UK, cement) = 376 lb (av.) of Portland cement (E) 1 barrel (UK, cement) = 4 bags (cement) (E)	UK
barrel (UK, salt)	–	mass	M	1 barrel (UK, salt) = 127.00586360 kg (E)	Obsolete British unit for mass employed for measurements in the chemical process industry. 1 barrel (UK, salt) = 280 lb (av.) of sodium chloride (E)	UK
barrel (US, alcohol)	bbl (alcohol)	capacity, volume	L^3	1 barrel (US, alcohol) $= 189.2205892 \times 10^{-3}\ m^3$	Obsolete American unit of capacity. 1 barrel (US, alcohol) = 50 gallons (US, liq.) (E)	US
barrel (US, cranb.)	bbl (US, cranb.)	capacity, volume	L^3	1 barrel (US, cranb.) $= 95.4710 \times 10^{-3}\ m^3$	American unit of capacity.	US
barrel (US, dry)	bbl (US, dry)	capacity, volume	L^3	1 barrel (US, dry) $= 115.626205 \times 10^{-3}\ m^3$	Obsolete American unit of capacity employed for dry foodstuffs (flour, grain). 1 barrel (US, dry) = 26.25 gal (US, dry) (E)	US

Unit (synonym, acronym)	Symbol	Physical quantity	Dimension	Conversion factor (SI equivalent unit)	Notes, definitions, other conversion factors	System
barrel (US, liq.)	bbl (US, liq.)	capacity, volume	L^3	1 barrel (US, liq.) = $119.2404712 \times 10^{-3}$ m^3	Obsolete American unit of capacity employed for liquid foodstuffs. 1 barrel (US, liq.) = 31.5 gal (US, liq.) (E)	US
barrel (US, liq.) per foot	bbl (US, liq.).ft^{-1}	surface, area	L^2	1 barrel (US, liq.).ft^{-1} = $3.912088950 \times 10^{-1}$ m^2	American unit of area used in chemical engineering. 1 barrel (US, liq.).ft^{-1} = 31.5 gal (US, liq.).ft^{-1} (E) 1 barrel (US, liq.).ft^{-1} = 2.625 gal (US, liq.).in^{-1} (E)	US
barrel (US, liq.) per inch	bbl (US, liq.).in^{-1}	surface, area	L^2	1 barrel (US, liq.).in^{-1} = 4.694506740 m^2	American unit of area used in chemical engineering. 1 barrel (US, liq.).in^{-1} = 31.5 gal (US, liq.).in^{-1} (E) 1 barrel (US, liq.).in^{-1} = 12 barrels (US, liq.).ft^{-1} (E)	US
barrel (US, liq.) per short ton	bbl (US, liq.).(sh. t)$^{-1}$	specific volume	L^3M^{-1}	1 barrel (US, liq.).(sh. t)$^{-1}$ = $1.314401201 \times 10^{-4}$ m^3.kg^{-1}	Obsolete American unit of specific volume. 1 barrel (US, liq.).(sh. t)$^{-1}$ = 31.5 gal (US, liq.).(2000 lb)$^{-1}$ (E)	US
barrel (US, petrol)	bbl (US, petrol)	capacity, volume	L^3	1 barrel (US, petrol) = $158.9872956 \times 10^{-3}$ m^3	American unit of capacity usually employed in the petroleum industry and trading in fuels. 1 bbl (US, petrol) = 42 gallons (US, liq.) (E) 1 bbl (US, petrol) = 35 imperial gallons 1 bbl (US, petrol) ≈ 159 litres	US

barrel (US, petrol) per day-psi	bbl/(d-psi)	volume flow rate per pressure drop	$M^{-1}L^{4}T$	1 barrel (US, petrol) per day-psi $= 2.668883979 \times 10^{-10}$ $m^3.Pa^{-1}.s^{-1}$	British and American unit.	UK, US
barrel (US, petrol) per foot	bbl (US, petrol). ft^{-1}	surface, area	L^2	1 barrel (US, petrol).ft^{-1} $= 5.216118599 \times 10^{-1}$ m^2	American unit of area used in chemical engineering. 1 barrel (US, petrol).ft^{-1} = 42 gal (US, liq.).ft^{-1}. (E) 1 barrel (US, petrol).ft^{-1} = 3.5 gal (US, liq.).in^{-1} (E)	US
barrel (US, petrol) per inch	bbl (US, petrol). in^{-1}	surface, area	L^2	1 barrel (US, petrol).in^{-1} = 6.259342319 m^2	American unit of area used in chemical engineering. 1 barrel (US, petrol).in^{-1} = 42 gal (US, liq.).in^{-1} (E) 1 barrel (US, petrol).in^{-1} = 12 barrel (US, petrol).ft^{-1} (E)	US
barrel (US, petrol)-mile (statute)	bbl (US, petrol)-mi (stat.)	quadratic moment of a plane area	L^4	1 barrel (US, petrol)-mile (stat.) $= 2.558652491 \times 10^{2}$ m^4	Obsolete American unit.	US
barrel oil equivalent	bboe	energy, work, heat	ML^2T^{-2}	1 bboe $= 6.12 \times 10^{9}$ J	Usual American large unit of energy employed in oil industry and economics to express energy balances. 1 bboe = 6.12 GJ	US
barrers	barrers	permeability coefficient (volume flow rate)	$M^{-1}L^3T$	1 barrers $= 7.500616827 \times 10^{-17}$ $m^2.s^{-1}.Pa^{-1}$	Obsolete American unit employed in gas separation by membrane processes. It was used when the flow rate of the gas through the membrane was expressed in cubic centimetre per second. The membrane thickness and its surface area was expressed in centimetre and square *(continued overleaf)*	US

Unit (synonym, acronym)	Symbol	Physical quantity	Dimension	Conversion factor (SI equivalent unit)	Notes, definitions, other conversion factors	System
barrers *(continued)*					centimetres respectively. Finally, the pressure drop across the membrane was expressed in mmHg (0°C). 1 barrers $= 10^{-10}$ cm^3(STP).cm.(cm^2.s.cmHg)$^{-1}$ (E)	US
barril (Spanish)	–	mass	M	1 barril (Spanish) = 23.004650 kg	Obsolete Spanish unit of mass. 1 barril (Spanish) = 50 libra (E)	Spanish
barye (barrie, dynes per square centimetre, microbar)	–	pressure, stress	$ML^{-1}T^{-2}$	1 barye = 10^{-1} Pa (E)	Obsolete cgs derived unit of pressure and stress with a special name. 1 barye = 1 dyn.cm^{-2} (E) 1 atm = 1 013 250 barye (E)	cgs
base box (basis box)	–	surface mass density	ML^{-2}	1 base box $= 2.241931058 \times 10^{-2}$ kg.m^{-2}	Obsolete American and British unit of surface mass density used in metallurgy. It serves to describe the amount of tin coating obtained by dip coating. It corresponds to the mass of coating, expressed in pounds, needed to cover an area of 31 360 square inches which corresponds to 112 sheets 14 inches high long by 10 inches wide. base box (US). basis box (UK).	UK, US
Bates degree	°Bates	specific gravity of liquids, hydrometer index, hydrometer degree	nil	°Bates $= 1000 \times \{[d(60°F) - 1]/2.78\}$	Hydrometer scale introduced in 1918	
bath (Hebrew, new)	–	capacity, volume	L^3	1 bath (Hebrew, new) $= 21.420 \times 10^{-3}$ m^3 (of water)	Obsolete Hebrew unit of capacity used in ancient times. Measured by weight.	Hebrew

obsolete)				$= 29.376 \times 10^{-3}$ m^3 (of water)	ancient times. Measured by weight.	Hebrew
baud	bd	rate of data transmission, quantity of information sent per time unit, telegraph signalling speed	T^{-1}	1 baud = 1 bit.s^{-1} (E)	Unit of rate of data transmission employed in computer science and telecommunications. It corresponds to one impulse per second. In computer science, it is equal to one bit per second. It was introduced in 1927 in Berlin. The unit is named after J.M.E. Baudot (1845–1903).	
Baumé degree (Holland, obsolete)	°Bé (Holl.)	specific gravity of liquids, hydrometer index, hydrometer degree	nil	• liquids lighter than water °Bé (Holl.) = $144/d_{60°F}^{60°F} - 144$ • liquids heavier than water °Bé (Holl.) = $144 - 144/d_{60°F}^{60°F}$		
Baumé degree (rational)	°Bé (Rat.)	specific gravity of liquids, hydrometer index, hydrometer degree	nil	• liquids lighter than water °Bé (Rat) = $144.30/d_{15°C}^{15°C} - 144.30$ • liquids heavier than water °Bé (Rat) = $144.30 - 144.30/d_{15°C}^{15°C}$	The rational Baumé index was an old specific gravity index. It is now obsolete. The rational Baumé scale was used for liquids heavier and lighter than water. For liquids heavier than water the zero index corresponded to the graduation when the hydrometer was immersed in pure water at 15°C and the 10°Bé index was equal to the graduation when the hydrometer was immersed in a aqueous sodium chloride solution containing 10% wt of pure salt. This range was divided into ten equal parts and the graduation was continued to the 67°Bé index. For liquids lighter than water, the zero graduation was equal to the depth *(continued overleaf)*	

Unit (synonym, acronym)	Symbol	Physical quantity	Dimension	Conversion factor (SI equivalent unit)	Notes, definitions, other conversion factors	System
Baumé degree (rational) *(continued)*					of the hydrometer in an aqueous sodium chloride solution containing 10% wt of pure salt measured at 12.5°C and the 10°Bé index was measured in pure water.	
Baumé degree (American)	°Bé (US)	specific gravity of liquids, hydrometer index, hydrometer degree	nil	• liquids lighter than water $°Bé = 140/d_{60°F}^{60°F} - 130$ • liquids heavier than water $°Bé\,(Rat) = 145 - 145/d_{60°F}^{60°F}$		
Baumé degree (Gerlach)	°Bé (Gerlach)	specific gravity of liquids, hydrometer index, hydrometer degree	nil	• liquids lighter than water $°Bé\,(Gerlach) = 146.3/d_{15°C}^{15°C} - 146.3$ • liquids heavier than water $°Bé\,(Gerlach) = 146.3 - 146.3/d_{15°C}^{15°C}$		
Baumé degree (NIST)	°Bé (NIST)	specific gravity of liquids, hydrometer index, hydrometer degree	nil	• liquids lighter than water $°Bé\,(NIST) = 140/d_{60°F}^{60°F} - 130$ • liquids heavier than water $°Bé\,(NIST) = 145 - 145/d_{60°F}^{60°F}$		
Baumé degree (obsolete scale)	°Bé (obsolete)	specific gravity of liquids, hydrometer index, hydrometer degree	nil	• liquids lighter than water $°Bé\,(old) = 146.78/d_{17.5°C}^{17.5°C} - 146.78$ • liquids heavier than water $°Bé\,(old) = 146.78 - 146.78/d_{17.5°C}^{17.5°C}$		
bboe (barrel oil equivalent)	bboe	energy, work, heat	ML^2T^{-2}	$1\ bboe = 6.12 \times 10^9\ J$	American large unit of energy employed in oil industry and economics to express energy balances. 1 bboe = 6.12 GJ	US

becher (Austrian, dry)	–	capacity, volume	L^3	1 becher (Austrian, dry) = $4.80382813 \times 10^{-4}$ m^3	Obsolete Austrian unit of capacity used for dry substances. 1 becher (Austrian, dry) = 1/128 metzel (E)	Austrian
Beck degree	% Beck, °Beck	specific gravity unit to express amount of sugar in syrups, juice	nil	$°Beck = 170 - (170/d^{12.5°C}_{12.5°C})$	Hydrometer unit introduced in 1830. The Beck hydrometer has 0°Beck corresponding to sp. gr. 1000 and 30°Beck to sp. gr. 0.850; equal divisions on the scale are continued as far as required in both directions.	UK, US
becquerel	Bq	radioactivity	T^{-1}	**SI derived unit** 1 Bq = 1 s^{-1} (E)	The becquerel is the SI derived unit, with a special name, for radioactivity. It describes a radioactivity of an amount of radionuclide decaying at the rate, on average, of one spontaneous nuclear transitions per second. This excepted the nature of the emitted particles (15th CGPM, May 1975). The unit is named after the French scientist A.H. Becquerel (1852–1908). 1 Bq = 2703×10^{-11} Ci	SI
becquerel per cubic metre	$Bq.m^{-3}$	radioactivity concentration	$L^{-3}T^{-1}$	**SI derived unit** 1 $Bq.m^{-3}$ = 1 $s^{-1}.m^{-3}$ (E)	The $Bq.m^{-3}$ is the SI derived unit. The becquerel per cubic metre is the ratio of one becquerel of radioactivity of the contained radionucleide to a volume of one cubic metre of the material.	SI
becquerel per kilogram	$Bq.kg^{-1}$	specific radioactivity	$M^{-1}T^{-1}$	**SI derived unit** 1 $Bq.kg^{-1}$ = 1 $s^{-1}.kg^{-1}$ (E)		SI

Unit (synonym, acronym)	Symbol	Physical quantity	Dimension	Conversion factor (SI equivalent unit)	Notes, definitions, other conversion factors	System
bekah (Hebrew) [sacred system]	–	mass	M	1 bekah (Hebrew) $= 7.083333 \times 10^{-3}$ kg	Obsolete Hebrew unit of mass used in ancient times. Sacred system. 1 bekah (Hebrew) = 1/120 mina (E)	Hebrew
bel	B	logarithm (Briggsian) of a ratio of two sound powers	nil	1 B = 10 dB (E) (see note)	The bel expresses the ratio of two sound powers or intensity levels P_1 and P_2 as a Briggs logarithm difference according to the equation $N(\mathrm{B}) = \log_{10} P_1/P_2$. For sound pressure the number of decibels is $N(\mathrm{B}) = 2 \times \log_{10} P_1/P_2$. The unit is named after A.G. Bell (1847–1922). 1 B = 10 dB (E) $1\ \mathrm{B} = 1/2\ \ln_{10} N_P(\mathrm{E}) \approx 1.151293 N_P$	
bema (Greek, Attic) [Greek pace]	–	length, distance	L	1 bema (Greek, Attic) $= 7.714 \times 10^{-1}$ m	Obsolete Greek base unit of length employed in ancient times. 1 bema (Attic) = 5/2 pous (E)	Attic
benz	Bz, $\mathrm{m.s^{-1}}$	velocity, speed	$\mathrm{LT^{-1}}$	1 Bz = 1 $\mathrm{m.s^{-1}}$ (E)	Name proposed and refuted for the SI derived unit of velocity. The unit was named after the German Karl Benz (1844–1929). Anecdotal.	
berkovets (Russian)	–	mass	M	1 berkovets = 163.8068717 kg	Obsolete Russian unit of mass used before 1917 for general purposes. 1 berkovets = 400 funts (E)	Russian
bes (Roman)	–	mass	M	1 bes (Roman) $= 218 \times 10^{-3}$ kg	Obsolete Roman unit of mass used in ancient times. 1 bes (Roman) = 8 unciae (Roman) (E)	Roman

BeV (billion eV)	BeV	energy, work, heat	ML^2T^{-2}	1 BeV = 1.60217733 × 10^{-10} J	Obsolete American unit of energy employed in nuclear physics. The name of the unit derived from the English acronym: **B**illion **e**lectron **V**olt 1 BeV = 1 GeV (E)	US
bicron (micromicron)	μμ	length, distance	L	1 μμ = 10^{-12} m (E)	Obsolete unit of length used in atomic spectroscopy. The unit is named after the acronym of dou**b**le m**icron**. 1 μμ = 1 pm (E)	
biot	–	optical rotatory power		(see note)	Obsolete unit used in spectroscopy. It serves to express rotatory power of matter which has circular dichroism. $R = \frac{3 \times 10^3 hc}{8\pi^3 N \log_{10} e} \int \frac{\Delta\varepsilon d\lambda}{\lambda}$ The unit is named after J.B. Biot (1774–1862). 1 biot = 10^{-40} cgs	
biot (deca-ampere)	Bi	electric current intensity	I	1 Bi = 10 A (E)	Obsolete cgs unit of electric current in the emu subsystem. The biot is that constant current which, if maintained in two straight parallel conductors of infinite length, of negligible circular cross-section, and placed one centimetre apart in vacuo, would produce between these conductors a force equal to 2 dynes per centimetre of length (1961). The unit is named after J.B. Biot (1774–1862). 1 Bi = 1 abampere (E)	cgs

Unit (synonym, acronym)	Symbol	Physical quantity	Dimension	Conversion factor (SI equivalent unit)	Notes, definitions, other conversion factors	System
bisquare foot	ft^4	quadratic moment of a plane area	L^4	1 ft^4 = 8630.974841 m^4	American and British unit of quadratic moment.	UK, US
bisquare inch	in^4	quadratic moment of a plane area	L^4	1 $in^4 = 4.162314256 \times 10^{-7}$ m (E)	American and British unit of quadratic moment.	UK, US
bisquare metre	m^4	quadratic moment of a plane area	L^4	**SI derived unit**		SI
bit	bit	quantity of information	nil	1 bit = 0 or 1	A bit is the information described by a symbol with two values. The name is a contraction of **bi**nary digi**t**.	
blink	–	time, period, duration	T	1 blink = 0.86400 s (E)	Anecdotal unit 1 blink = 10^{-5} days (E)	
blondel	–	luminous luminance	JL^{-2}	1 Blondel = 0.318309886 $cd.m^{-2}$	Obsolete French unit of luminous luminance adopted in 1942. The unit is named after A.E. Blondel (1863–1938). 1 blondel = $1/\pi$ $cd.m^{-2}$ (E) 1 blondel = 10^{-4} lambert (E)	French
board foot measure	fbm, B.M.	capacity, volume	L^3	1 board foot = $2.359737216 \times 10^{-3}$ m^3	Obsolete British and American unit of capacity which is equal to the product of one foot of length per one foot width per one inch thickness. 1 board foot = 1/12 ft^3 (E) 1 board foot = 144 in^3 (E)	UK, US
bohr (a.u. of length, 1st Bohr radius)	a_0, b	length, distance	L	1 bohr = $5.29177249 \times 10^{-11}$ m	Fundamental physical constant. Base unit of a.u. system. The unit is named after N. Bohr (1885–1962). 1 bohr = $4\pi\varepsilon_0\hbar^2/m_0e^2$	a.u.

boisseau (de Paris, dry)	–	capacity, volume	L^3	1 boisseau (de Paris) $= 1.86278 \times 10^{-3}$ m^3	Obsolete French unit of volume employed before the French Revolution. It was used for capacity measurements of dry substances. 1 boisseau (de Paris) = 1/144 muid (de Paris, dry) (E)	French
bole	g.cm.s^{-1}	linear momentum, momentum	MLT^{-1}	1 bole $= 10^{-5}$ kg.m.s^{-1} (E)	Name suggested for the cgs linear momentum unit. Anecdotal. 1 bole = 1 g.cm.s^{-1} (E)	cgs
bolt (US cloth)	–	length, distance	L	1 bolt (US) = 36.576 m (E)	Obsolete American unit of length. 1 bolt (US) = 120 feet (E)	US
bota (Spanish)	–	capacity, volume	L^3	1 bota (Spanish) $= 469.294860 \times 10^{-3}$ m^3	Obsolete Spanish unit of capacity used for liquid substances. 1 bota (Spanish) = 30 arroba (water) (E)	Spanish
botchka (Russian)	–	capacity, volume	L^3	1 botchka (Russian) $= 491.976400 \times 10^{-3}$ m^3	Obsolete Russian unit of capacity for liquid substances, used before 1917. 1 botchka (Russian) = 40 vedro (E)	Russian
bottle (vodka) (Russian)	–	capacity, volume	L^3	1 bottle (vodka) (Russian) $= 6.149705 \times 10^{-4}$ m^3	Obsolete Russian unit of capacity for liquid substances, used before 1917. 1 bottle (vodka) (Russian) = 5 tcharka (E)	Russian
bottle (wine) (Russian)	–	capacity, volume	L^3	1 bottle (wine) (Russian) $= 7.68713125 \times 10^{-4}$ m^3	Obsolete Russian unit of capacity for liquid substances, used before 1917. 1 bottle (wine) (Russian) = 25/4 tcharka (E)	Russian
bougie décimale	bd	luminous intensity	J	1 bougie décimale = 1.02 cd (E)	Obsolete unit of luminous intensity. It was equal to 1/20 of the Violle intensity. It was used during the period 1889–1919.	
bougie internationale	bi	luminous intensity	J	1 bougie internationale = 1 cd (E)	Precursor of the SI candela.	INT

Unit (synonym, acronym)	Symbol	Physical quantity	Dimension	Conversion factor (SI equivalent unit)	Notes, definitions, other conversion factors	System
bougie nouvelle	bn	luminous intensity	J	1 bougie nouvelle = 1 cd (E)	Obsolete unit of luminous intensity. It was equal to 1/60 of the intensity of one square centimetre of a blackbody radiator at the temperature of solidification of platinum (2046 K) [1948]. It was a precursor of the SI candela.	INT
bouteille champenoise (reputed quart)	–	capacity, volume	L^3	1 bouteille champenoise $= 0.757682 \times 10^{-3}$ m^3	Obsolete British unit employed for expressed the capacity of wine containers. Still employed in oenology, especially in France. 1 bout. champ. = 1/6 gallon (UK) (E)	UK, French
brasse (French fathom)	–	length, distance	L	1 brasse = 1.8288 m (E)	Obsolete French unit of length used in navigation. It expressed the depth of sea water.	French
brewster	B	photoelastic work	$M^{-1}LT^2$	1 brewster = 10^{-12} $m^2.N^{-1}$ (E)	The unit is named after Sir David Brewster (1781–1868).	
British thermal unit (39°F; 4°C)	Btu (39°F)	energy, work, heat	ML^2T^{-2}	1 Btu (39°F) = 1059.67 J	Obsolete British unit of energy.	UK
British thermal unit (60°F; 15.56°C)	Btu (60°F)	energy, work, heat	ML^2T^{-2}	1 Btu (60°F) = 1054.678 J	Obsolete British unit of energy. The Btu (60°F) is equal to the heat needed to raise the temperature of one pound of air-free water by 1°F (from 59.5 to 60.5°F) at the constant pressure of one standard atmosphere (101 325 Pa).	UK

(Gas Inspection Act Regulations – 60.5°F)	Insp. Act)					
British thermal unit (ISO/TC 12)	Btu (ISO)	energy, work, heat	ML^2T^{-2}	1 Btu (ISO) = 1055.06 J	Obsolete British unit of energy.	UK
British thermal unit (IT, International Steam Table)	Btu (IT)	energy, work, heat	ML^2T^{-2}	1 Btu (IT) = 1055.05585262 J	Obsolete British unit of energy. The Btu (IT) is the unit of energy whose magnitude is such that one Btu per pound equals 2326 joules per kilogram.	UK
British thermal unit (IT) per (pound-Fahrenheit degree)	Btu (IT).(lb-°F)$^{-1}$	specific heat capacity	$L^2T^{-2}\Theta^{-1}$	1 Btu (IT).(lb-°F)$^{-1}$ = 4 186.800 585 J.kg^{-1}.K^{-1}	Obsolete British energy density unit. 1 Btu (IT).(lb-°F)$^{-1}$ = 1 cal (IT).g^{-1}.°C^{-1} (E)	UK
British thermal unit (IT) per barrel (US, petrol)	Btu (IT)bbl (US petrol)$^{-1}$	energy density	$ML^{-1}T^{-2}$	1 Btu (IT)bbl (US, petrol)$^{-1}$ = 6.636101662 × 10^3 J.m^{-3}	Obsolete American energy density unit.	US
British thermal unit (IT) per cubic foot	Btu (IT).ft^{-3}	energy density	$ML^{-1}T^{-2}$	1 Btu (IT).ft^{-3} = 37.258945790 × 10^3 J.m^{-3}	Obsolete British and American energy density unit.	UK, US
British thermal unit (IT) per Fahrenheit degree	Btu (IT).°F^{-1}	entropy	$ML^2T^{-2}\Theta^{-1}$	1 Btu (IT).°F^{-1} = 1.899100534 × 10^3 J.K^{-1}	Obsolete British and American entropy unit.	UK, US
British thermal unit (IT) per foot per hour per Fahrenheit degree	Btu (IT) .h^{-1}.ft^{-1} °F^{-1}	thermal conductivity	$MLT^{-3}\Theta^{-1}$	1 Btu (IT).h^{-1}.ft^{-1}.°F^{-1} = 1.730734665 W.m^{-1}.K^{-1}	Obsolete British and American unit of thermal conductivity used in chemical engineering and heat transfer technology.	UK, US
British thermal unit (IT) per gallon (UK)	Btu (IT). gal (UK)$^{-1}$	energy density	$ML^{-1}T^{-2}$	1 Btu (IT).gal (UK)$^{-1}$ = 2.320797406 × 10^5 J.m^{-3}	Obsolete British energy density unit.	UK

Unit (synonym, acronym)	Symbol	Physical quantity	Dimension	Conversion factor (SI equivalent unit)	Notes, definitions, other conversion factors	System
British thermal unit (IT) per gallon (US)	Btu (IT).gal (US)	energy density	$ML^{-1}T^{-2}$	1 Btu (IT).gal (US)$^{-1}$ = 2.787162698 × 10^5 J.m^{-3}	Obsolete American energy density unit.	US
British thermal unit (IT) per hour	Btu (IT).h^{-1}	power	ML^2T^{-3}	1 Btu (IT).h^{-1} = 0.293071070 W	Obsolete British and American power unit.	UK, US
British thermal unit (IT) per hour per square foot	Btu (IT).h^{-1}.ft^{-2}	energy flux, heat flux	MT^{-3}	1 Btu (IT).h^{-1}.ft^{-2} = 3.154590743 W.m^{-2}	Obsolete British and American unit.	UK, US
British thermal unit (IT) per minute	Btu (IT).min^{-1}	power	ML^2T^{-3}	1 Btu (IT).min^{-1} = 17.58426420 W	Obsolete British and American unit.	UK, US
British thermal unit (IT) per pound	Btu (IT).lb^{-1}	specific heat	L^2T^{-2}	1 Btu (IT).lb^{-1} = 2326.000324 J.kg^{-1}	Obsolete British unit.	UK
British thermal unit (IT) per second	Btu (IT).s^{-1}	power	ML^2T^{-3}	1 Btu (IT).s^{-1} = 1055.05585262 W	Obsolete British and American unit.	UK, US
British thermal unit (IT) per square foot per hour per Fahrenheit degree	Btu (IT).h^{-1}.ft^{-2}.°F^{-1}	coefficient of heat transfer	$MT^{-1}\Theta^{-1}$	1 Btu (IT).h^{-1}.ft^{-2}.°F^{-1} = 5.678264135 W.m^{-2}.K^{-1}	Obsolete British and American unit.	UK, US
British thermal unit (IT)-inch per square foot per hour per Fahrenheit degree	Btu (IT).in.h^{-1}.ft^{-2}.°F^{-1}	thermal conductivity	$MLT^{-3}\Theta^{-1}$	1 Btu (IT).in.h^{-1}.ft^{-2}.°F^{-1} = 1.4422788809 W.m^{-1}.K^{-1}	Obsolete British and American unit of thermal conductivity used in chemical engineering and heat transfer technology.	UK, US
British thermal unit (IT)-inch per square foot per second per Fahrenheit degree	Btu (IT).in.s^{-1}.ft^{-2}.°F^{-1}	thermal conductivity	$MLT^{-3}\Theta^{-1}$	1 Btu (IT).in.s^{-1}.ft^{-2}.°F^{-1} = 519.220399 W.m^{-1}.K^{-1}	Obsolete British and American unit of thermal conductivity used in chemical engineering and heat transfer technology.	UK, US

British thermal unit (mean)	Btu (mean), B	energy, work, heat	ML^2T^{-2}	1 Btu (mean) = 1055.87 J	Obsolete British unit of energy. The Btu (mean) is equal to 1/180 of the heat needed to raise the temperature of one pound of air-free water from 32°F (0°C) to 212°F (100°C) at the constant pressure of one standard atmosphere (101 325 Pa).	UK, FPS
British thermal unit (thermochemical)	Btu (therm.)	energy, work, heat	ML^2T^{-2}	1 Btu (therm.) = 1054.350 J	Obsolete British unit of energy.	UK
British thermal unit (therm.) per hour	Btu (therm.).h^{-1}	power	ML^2T^{-3}	1 Btu (therm.).h^{-1} = 0.292875000 W	Obsolete British and American unit.	UK, US
British thermal unit (therm.) per foot per hour per Fahrenheit degree	Btu (therm.).$ft^{-1}.h^{-1}.°F^{-1}$	thermal conductivity	$MLT^{-3}\Theta^{-1}$	1 Btu (therm.).$ft^{-1}.h^{-1}.°F^{-1}$ = 1.729 576 772 $W.m^{-1}.K^{-1}$	Obsolete British and American unit of thermal conductivity used in chemical engineering and heat transfer technology.	UK, US
British thermal unit (therm.) per hour per square foot	Btu (therm.).$h^{-1}. ft^{-2}$	energy flux, heat flux	MT^{-3}	1 Btu (therm.).$h^{-1}.ft^{-2}$ = 3.152480263 $W.m^{-2}$	Obsolete British and American unit.	UK, US
British thermal unit (therm.) per minute	Btu (therm.) .min^{-1}	power	ML^2T^{-3}	1 Btu (therm.).min^{-1} = 17.57250000 W	Obsolete British and American unit.	UK, US
British thermal unit (therm.) per pound	Btu (therm.).lb^{-1}	specific heat	L^2T^{-2}	1 Btu (therm.).lb^{-1} = 2324.443861 $J.kg^{-1}$	Obsolete British unit.	UK
British thermal unit (therm.) per (pound-fahrenheit degree)	Btu (therm.).$(lb\text{-}°F)^{-1}$	specific heat capacity	$L^2T^{-2}\Theta^{-1}$	1 Btu (therm.).$(lb\text{-}°F)^{-1}$ = 4183.998950 $J.kg^{-1}.K^{-1}$	Obsolete British unit. 1 Btu (therm.).$(lb\text{-}°F)^{-1}$ = 1 cal (therm.).$g^{-1}.°C^{-1}$ (E)	UK

Unit (synonym, acronym)	Symbol	Physical quantity	Dimension	Conversion factor (SI equivalent unit)	Notes, definitions, other conversion factors	System
British thermal unit (therm.) per second	Btu (therm.). s^{-1}	power	ML^2T^{-3}	1 Btu (therm.).s^{-1} = 1054.350 W	Obsolete British and American unit.	UK, US
British thermal unit (therm.) per square foot per hour per Fahrenheit degree	Btu (therm.). h^{-1}. ft^{-2}. $°F^{-1}$	coefficient of heat transfer	$MT^{-1}\Theta^{-1}$	1 Btu (therm.).h^{-1}.ft^{-2}.$°F^{-1}$ = 5.674464475 W.m^{-2}.K^{-1}	Obsolete British and American unit.	UK, US
British thermal unit (therm.)-inch per square foot per hour per Fahrenheit degree	Btu (therm.). in. h^{-1}.ft^{-2}. $°F^{-1}$	thermal conductivity	$MLT^{-3}\Theta^{-1}$	1 Btu (therm.).in.h^{-1}. ft^{-2}.$°F^{-1}$ = 1.441313976 W.m^{-1}.K^{-1}	Obsolete British and American unit of thermal conductivity used in chemical engineering and heat transfer technology.	UK, US
British thermal unit (therm.)-inch per square foot per second per Fahrenheit degree	Btu (therm.). in. s^{-1}. ft^{-2}. $°F^{-1}$	thermal conductivity	$MLT^{-3}\Theta^{-1}$	1 Btu (therm.).in.s^{-1}.ft^{-2}.$°F^{-1}$ = 518.8730315 W.m^{-1}.K^{-1}	Obsolete British and American unit of thermal conductivity used in chemical engineering and heat transfer technology.	UK, US
British thermal unit (UK gas industry)	Btu (UK Gas industry)	energy, work, heat	ML^2T^{-2}	1 Btu (UK Gas industry) = 1054.76 J	Obsolete British unit of energy.	UK
Brix degree (°Fischer)	% Brix, °Brix	specific gravity unit to express amount of sugar in syrups, juice	nil	°Brix = $400/d_{60°F}^{60°F} - 400$	Saccharometer scale introduced in 1892. 10°Brix corresponds to an aqueous solution which contains 10 g.l^{-1} of sugar and has the specific gravity 1.0386.	
bu (Japanese)	–	length, distance	L	1 bu (Japanese) = 3.030303030 × 10^{-3} m	Obsolete Japanese unit of length. 1 bu = 1/100 shaku (E)	Japanese

bucket (UK)	bk (UK)	capacity, volume	L	1 bucket (UK) = 18.18436800 × 10^{-3} m^3	Obsolete British unit used for capacity measurements for all merchandise (solids, liquids, foodstuffs, etc). 1 bucket (UK) = 4 gallons (UK) (E) 1 bucket (UK) = 2 pecks (UK) (E)	UK
bushel (UK)	bu (UK)	capacity, volume	L^3	1 bushel (UK) = 36.36873600 × 10^{-3} m^3	Obsolete British unit used for capacity measurements for all merchandise (solids, liquids, foodstuffs, etc). It contains 80 pounds distilled water at 62°F. 1 bushel (UK) = 4 pecks (UK) (E) 1 bushel (UK) = 8 gallons (UK) (E) 1 bushel (UK) = 32 quarts (UK) (E) 1 bushel (UK) = 64 pints (UK) (E) 1 bushel (UK) = 256 gills (UK) (E) 1 bushel (UK) = 1.032067144 bushels (US, dry)	UK
bushel (US, dry) (Winchester bushel)	bu (US, dry)	capacity, volume	L^3	1 bushel (US, dry) = 35.239072 × 10^{-3} m^3	Obsolete American unit of capacity used to measure the volume of powdered or divided solid materials (flour, sand, cement, ores, etc). It contains 77.601 pounds distilled water at 62°F. 1 bushel (US, dry) = 4 pecks (US, dry) (E) 1 bushel (US, dry) = 32 quarts (US, dry) (E) 1 bushel (US, dry) = 64 pints (US, dry) (E) 1 bushel (US, dry) = 0.9689385959 bushels (UK)	US

Unit (synonym, acronym)	Symbol	Physical quantity	Dimension	Conversion factor (SI equivalent unit)	Notes, definitions, other conversion factors	System
butt (UK)	bt (UK)	capacity, volume	L^3	1 butt (UK) = $490.977936 \times 10^{-3}$ m^3	Obsolete British unit used for capacity measurements for all merchandise (solids, liquids, foodstuffs, etc). 1 butt (UK) = 108 gallons (UK) (E) 1 butt (UK) = 54 pecks (UK) (E) 1 butt (UK) = 27 buckets (UK) (E)	UK
button (UK) (UK line)	line (UK)	length, distance	L	1 line (UK) = 2.1166667×10^{-3} m	Obsolete British unit of length. 1 line (UK) = 1/4 barleycorn (UK) (E) 1 line (UK) = 1/12 inch (E) 1 line (UK) = 1/144 foot (E) 1 line (UK) = 1/432 yard (E) 1 line (UK) $\approx$ 2.117 mm	UK
button (US) (US line)	line (US)	length, distance	L	1 line (US) = 6.35×10^{-4} m (E)	Obsolete American unit of length. It was used in botany to describe flower measurements. 1 line (US) = 1/40 inch (E) 1 line (US) = 1/480 foot (E) 1 line (US) = 1/1440 yard (E) 1 line (US) = 0.635 mm (E)	US
byte (octet)	o, B	quantity of information	nil	1 byte = 8 bits (E)	Unit employed in computer science. It is equal to a sequence of adjacent binary digits operated upon as a unit in a computer and usually shorter than a word. Multiple of 8 bits. 1 KB = 1024 B ($K = 2^{10}$) 1 MB = 1 048 576 B ($M = 2^{20}$)	

		capacity, volume	L^3	1 cab (Hebrew) $= 1.190 \times 10^{-3}$ m^3 (of water)	Obsolete Hebrew unit of capacity used in ancient times. Measured by weight. 1 cab (Hebrew) = 1/18 ephah (new)	Hebrew
cabda (Arabic) [Arabic palm]	–	length, distance	L	1 cabda (Arabic) $= 8 \times 10^{-2}$ m	Obsolete Arabic unit of length used in ancient times. 1 cabda (Arabic) = 1/4 feet (Arabic) (E)	Arabic
cable length (int.)	–	length, distance	L	1 cable length (int.) = 185.200 m	Obsolete international unit of length used in navigation. It was equal to one tenth of the international mile. 1 cable length (UK) = 1/10 mile (int.) (E)	INT
cable length (UK)	–	length, distance	L	1 cable length (UK) = 185.3184 m (E)	Obsolete British unit of length used in navigation. It was equal to one tenth of a mile (UK, naut.). 1 cable length (UK) = 7296 inches (E) 1 cable length (UK) = 608 feet (E) 1 cable length (UK) = 1/10 mile (UK, naut.) (E)	UK
cable length (US)	–	length, distance	L	1 cable length (US) = 219.456 m	Obsolete American unit of length used in navigation. It was equal to 9/76 of a mile (US, naut.). 1 cable length (US) = 8640 inches (E) 1 cable length (US) = 720 feet (E) 1 cable length (US) = 240 yards (E) 1 cable length (US) = 120 fathoms (E) 1 cable length (US) = 9/76 mile (US, naut.) (E)	US
cados (Persian)	–	capacity, volume	L^3	1 cados (Persian) $= 1.018750 \times 10^{-3}$ m^3 (of water)	Obsolete unit of capacity of the Assyrio-Chaldean-Persian system used in ancient times. Measured by weight. 1 cados (Persian) = 1/32 amphora (E)	Persian

Unit (synonym, acronym)	Symbol	Physical quantity	Dimension	Conversion factor (SI equivalent unit)	Notes, definitions, other conversion factors	System
cafiz (Arabic)	–	capacity, volume	L^3	1 cafiz (Arabic) $= 32.64 \times 10^{-3}$ m^3 (of water)	Obsolete Arabic unit of capacity used in ancient times. Measured by weight.	Arabic
cafiz (Arabic)	–	surface, area	L^2	1 cafiz (Arabic) $= 147.500$ m^2	Obsolete Arabic unit of area used in ancient times. 1 cafiz (Arabic) = 1/40 feddan (E)	Arabic
cahiz (Spanish, dry)	–	capacity, volume	L^3	1 cahiz (Spanish, dry) $= 666.012 \times 10^{-3}$ m^3	Obsolete Spanish unit of capacity used for dry substances. 1 cahiz (Spanish, dry) = 12 fanega (E)	Spanish
calemin (Spanish)	–	surface, area	L^2	1 calemin (Spanish) $= 536.630\,1457$ m^2	Obsolete Spanish unit of area. 1 calemin (Spanish) = 768 sq. vara (E)	Spanish
calemin (Spanish, dry)	–	capacity, volume	L^3	1 calemin (Spanish, dry) $= 4.6250833 \times 10^{-3}$ m^3	Obsolete Spanish unit of capacity used for dry substances. 1 calemin (Spanish, dry) = 768 square vara (E)	Spanish
calibre (centiinch)	calibre, cin	length, distance	L	1 calibre $= 2.54 \times 10^{-4}$ m (E)	Obsolete American and British unit of length used in for calibre measurements of weapons. 1 calibre = 10 mils (E) 1 calibre = 10 thous (E) 1 calibre $= 10^{-2}$ inch (E)	UK, US
calorie (15°C)	cal_{15}	energy, work, heat	ML^2T^{-2}	1 $cal_{15} = 4.185$ J	The calorie (15°C) is an old heat unit. It was equal to the heat needed to raise the temperature of one gram of air-free water from 14.5°C to 15.5°C at the constant pressure of one standard atmosphere (101 325 Pa). The use of the calorie should have ceased from 31st December, 1977.	

calorie (diet kilocalorie)	Cal, kcal	energy, work, heat	ML^2T^{-2}	1 Cal = 4180 J (E)	Obsolete unit of thermal energy employed usually in dietetics. It should be discontinued to avoid confusion between 'small and large' calorie. Now abolished. 1 Cal = 1kcal (E)	
calorie (IT) (International Steam Table)	cal (IT)	energy, work, heat	ML^2T^{-2}	1 cal (IT) = 4.18674 J	Obsolete unit of energy. It was used in steam data tables of Keenan and Keyes. 1 cal (IT) = (1/860) W (int.).h (E)	INT
calorie (IT) per centimetre per second per degree Celsius	cal (IT).cm^{-1}.$°C^{-1}$	thermal conductivity	$MLT^{-3}\Theta^{-1}$	1 cal (IT).cm^{-1}.$°C^{-1}$ = 418.674 W.m^{-1}.K^{-1} (E)	Obsolete unit of thermal conductivity employed in heat transfer measurements.	
calorie (IT) per gram	cal (IT).g^{-1}	specific heat	L^2T^{-2}	1 cal (IT).g^{-1} = 4186.74 J.Kg^{-1} (E)	Obsolete unit.	INT
calorie (IT) per hour	cal (IT).h^{-1}	power	ML^2T^{-3}	1 cal (IT).h^{-1} = 1.162983333 × 10^{-3} W	Obsolete unit.	
calorie (IT) per minute	cal (IT).min^{-1}	power	ML^2T^{-3}	1 cal (IT).min^{-1} = 6.977899999 × 10^{-2} W	Obsolete unit.	
calorie (IT) per second	cal (IT).s^{-1}	power	ML^2T^{-3}	1 cal (IT).s^{-1} = 4.18674 W	Obsolete unit.	
calorie (mean)	cal_{mean}	energy, work, heat	ML^2T^{-2}	1 cal_{mean} = 4.19002 J	Obsolete unit of energy. The calorie (mean) is equal to 1/100 of the heat needed to raise the temperature of one gram of air-free water from 0°C to 100°C at the constant pressure of one standard atmosphere (101 325 Pa). The use of the calorie should have ceased from 31st December, 1977.	

Unit (synonym, acronym)	Symbol	Physical quantity	Dimension	Conversion factor (SI equivalent unit)	Notes, definitions, other conversion factors	System
calorie (thermochemical)	cal (therm.)	energy, work, heat	ML^2T^{-2}	1 cal (therm.) = 4.1840 J (E)	Obsolete unit of energy. It was defined by the National Bureau of Standards (NBS) in 1953.	US
calorie (therm.) per centimetre per second and per degree Celsius	cal (therm.). $cm^{-1}.°C^{-1}$	thermal conductivity	$MLT^{-3}\Theta^{-1}$	1 cal (therm.).$cm^{-1}.°C^{-1}$ = 418.4 $W.m^{-1}.K^{-1}$ (E)	Obsolete unit of thermal conductivity employed in heat transfer measurements.	
calorie (therm.) per gram	cal (therm.). g^{-1}	specific heat	$LT^{-3}\Theta^{-1}$	1 cal (therm.).g^{-1} = 4184.00 $J.Kg^{-1}$ (E)	Obsolete unit.	
calorie (therm.) per hour	cal (therm.). h^{-1}	power	ML^2T^{-3}	1 cal (therm.).h^{-1} = $1.622222222 \times 10^{-3}$ W	Obsolete unit.	
calorie (therm.) per minute	cal (therm.). min^{-1}	power	ML^2T^{-3}	1 cal (therm.).min^{-1} = $6.973333333 \times 10^{-2}$ W	Obsolete unit.	
calorie (therm.) per second	cal (therm.) .s^{-1}	power	ML^2T^{-3}	1 cal (therm.).s^{-1} = 4.1840 W	Obsolete unit.	
canada (Portuguese)	–	capacity, volume	L^3	1 canada (Portuguese) = 1.375×10^{-3} m^3	Obsolete Portuguese unit of capacity used for liquid substances. 1 canada (Portuguese) = 1/12 almude (E)	Portuguese
candela	cd	luminous intensity	J	**SI base unit**	The candela is the luminous intensity, in a given direction, of a source that emits monochromatic radiation of frequency 540×10^{12} Hz and that has a radiant	SI

					intensity in that direction of 1/683 watt per steradian. [16th CGPM (1979), Resolution 3].	
candela per square metre	$cd.m^{-2}$	luminous luminance	$J.L^{-2}$	**SI derived unit**		SI
candle (int.)	c	luminous intensity	J	1 candle (int.) = 1.019 367 992 cd	Obsolete British unit of luminous intensity introduced in 1860 by the *Metropolitain Gas Act*. One candle (int.) was specified as a spermaceti candle weighing six to the pound and burning at the rate of 120 grain (av.) an hour. The US unit is a specified fraction of the average horizontal candlepower of a group of 45 carbon-filament lamps preserved at the Bureau of Standards (US).	UK, INT
candle (new unit)	–	luminous intensity	J	1 candle (new unit) = 1 cd (E)	Obsolete unit of luminous intensity. It was equal to 1/60 of the intensity of one square centimetre of a blackbody radiator at the temperature of solidification of platinum (2046 K) [1948]. It was a precursor of the SI candela.	INT
candle (pentane)	–	luminous intensity	J	1 candle (pentane) = 1.019 367 992 cd	Obsolete British unit of luminous intensity. 1 candle (pentane) = 1 candle (int.)	UK
candlepower (spherical)	–	luminous flux	JΩ	1 candlepower (spherical) = 12.566370 lm	Obsolete American and British unit of luminous flux intensity. The candlepower (spherical) of a lamp is the average candlepower of a lamp in all directions in space. It is equal to the total luminous flux of the lamp in lumens divided per 4π.	UK, US

Unit (synonym, acronym)	Symbol	Physical quantity	Dimension	Conversion factor (SI equivalent unit)	Notes, definitions, other conversion factors	System
canna (Italian)	–	length, distance	L	1 canna (Italian) = 2.055080 m	Obsolete national Italian unit of length. 1 canna (Italian) = 4 piedi liprando (E)	Italian
canne (Egyptian)	–	length, distance	L	1 canne (Egyptian) = 4.071666667 m	Obsolete Egyptian unit of length used in ancient times. 1 canne = 70/9 Royal cubit (E)	Egyptian
cantara (Spanish)	–	capacity, volume	L^3	1 cantara (Spanish) $= 15.643162 \times 10^{-3}$ m^3	Obsolete Spanish unit of capacity used for liquid substances. 1 cantara (Spanish) = 1 arroba (water) (E)	Spanish
cantaro (Italian)	–	mass	M	1 cantaro (Italian) = 460.5 kg	Obsolete Italian unit of mass. 1 cantaro (Italian) = 150 libbra (E)	Italian
caphite (Arabic) (kiladja)	–	capacity, volume	L^3	1 caphite (Arabic) $= 1.36 \times 10^{-3}$ m^3 (of water)	Obsolete Arabic unit of capacity used in ancient times. Measured by weight. 1 caphite (Arabic) = 1/24 cafiz (E)	Arabic
carat	ct, Kt	fraction, relative values, yields, efficiencies, abundance	nil	1 carat $= 41.666666667 \times 10^{-3}$	It corresponds to the mass fraction of gold in precious alloys used in jewellery. It is commonly used in precious metal business transactions. 1 carat = 41.67‰ (E) 24 carats = 1000‰ (E) 24 carats = 100% (E)	INT
carat (metric)	ct.	mass	M	1 carat (metric) $= 2 \times 10^{-4}$ kg (E)	Metric unit of mass introduced since 1932 and employed in jewellery for the weighing of precious metals, precious stones and gems (e.g. diamond, ruby, sapphire). The name of the unit is derived from the Sanskrit word *quirrat* meaning the small and very uniform seeds of the carob tree	INT

					which in antiquity were used to weigh precious metals and stones. 1 carat (metric) = 200 mg (E)	
carat (obsolete)	ct (obsolete)	mass	M	1 carat (obsolete) $= 2.05 \times 10^{-4}$ kg (E)	Obsolete unit of mass employed in jewellery for the weighing of precious metals, precious stones and gems (e.g. diamond, ruby, sapphire). It has now been replaced by the metric carat. The name of the unit is derived from the Sanskrit word *quirrat* meaning the small and very uniform seeds of the carob tree which in antiquity were used to weigh precious metals and stones. 1 carat (obsolete) = 205 mg (E)	INT
carcel	–	luminous intensity	J	1 carcel = 10 cd (E)	Obsolete French luminous intensity unit.	French
Carcel unit	–	luminous intensity	J	1 Carcel unit = 9.796126403 cd	Obsolete luminous intensity unit. The Carcel unit is the horizontal intensity of the Carcel lamp, burning 42 grams of colza oil per hour. It was used in the UK during the period 1880–1884. 1 Carcel unit = 9.61 candle (int.) (E)	UK
carnot	–	entropy	$ML^2T^{-2}\Theta^{-1}$	1 carnot = 1 $J.K^{-1}$ (E)	Obsolete unit of entropy.	
Cartier degree	°Cartier	specific gravity of liquids, hydrometer index, hydrometer degree	nil	°Cartier $= 126.1 - (136.8/d)$ $d = d(12.5°C)$	Obsolete hydrometer unit introduced in 1800. The Cartier hydrometer floats in water at the 10° scale division and at 30° corresponding to 32°Bé.	
Cé	–	time, period, duration	T	1 Cé = 86.400 s (E)	Obsolete decimal unit of time suggested in 1900 by the *Congrès International de Chronométrie*. 1 Cé = 1/1000 day (E)	

Unit (synonym, acronym)	Symbol	Physical quantity	Dimension	Conversion factor (SI equivalent unit)	Notes, definitions, other conversion factors	System
celo	$ft.s^{-2}$	acceleration	LT^{-2}	1 Celo = 0.3048 $m.s^{-2}$ (E)	Name suggested for the unit of acceleration in the FPS system. Anecdotal.	FPS
Celsius degree	°C	temperature	Θ	T (K) = T(°C) − 273.15	Usual temperature scale employed in most countries. 0°C (melting ice) 100°C (boiling water)	INT
Celsius-heat unit (centigrade-heat unit)	Chu, chu	energy, work, heat	ML^2T^{-2}	1 chu = 1899.18 J	The chu is an obsolete British and American heat unit. It was equal to the heat needed to raise by 1°C the temperature of one pound of air-free water at the constant pressure of one standard atmosphere (101 325 Pa).	UK, US
cent	–	logarithmic musical interval	nil	$I = 1200 \log_2(f_1/f_2)$ (see note)	The cent is the interval between two musical sounds having as a basic frequency ratio the 1200th root of two. The number of cents between frequencies f_1 and f_2 is: $I = 1200 \log_2(f_1/f_2) \approx 3986314 \log_{10}(f_1/f_2)$ 1 cent = 1/1200 octave (E) 1 cent = 301/1200 savart (E)	
cental (kintal, centner, hundredweight)	cH, cwt	mass	M	1 cental = 45.359237 kg (E)	Obsolete British and American unit of mass used in business transactions especially in agriculture. It serves to measure the weight of foods and grain. 1 cental = 1 cwt (E) 1 cental = 100 lb av. (E) 1 cental = 1 sh. cwt (US) (E)	UK, US

(Celsius-heat unit) (15°C)	chu (15°C)				American heat unit. It was equal to the heat needed to raise the temperature of one pound of air-free water from 14.5°C to 15.5°C at the constant pressure of one standard atmosphere (101 325 Pa). 1 Chu ≈ 455 cal 1 Chu ≈ 1.8 Btu	
centigram	cg	mass	M	1 cg = 10^{-5} kg (E)	Submultiple of the SI base unit. 1 cg = 10^{-2} g (E)	SI
centiHg	–	pressure, stress	$ML^{-1}T^{-2}$	1 centiHg (0°C) = 1333.223684 Pa	Obsolete unit of pressure. Abbreviation of **centi**metre of mercury (**Hg**). It was defined as the pressure exerted by a column of mercury 1 cm high, measured at 0°C (32°F). 1 centiHg = 1 cmHg (0°C) (E) 76 centiHg (0°C) = 101 325 (E) 76 centiHg (0°C) = 1 atm (E). 1 centiHg (0°C) = 101 325/76 Pa (E)	
centilitre	cl, cL	capacity, volume	L^3	1 cl = 10^{-5} m^3 (E)	Submultiple of the litre.	
centimetre	cm	length, distance	L	1 cm = 10^{-2} m (E)	Submultiple of the SI base unit, cgs base unit	cgs, SI
centimetre (electromagnetic)	'cm'	electric inductance	$ML^2T^{-2}I^{-2}$	1 'cm' = 10^{-9} H (E)	Obsolete unit of electric inductance employed in electrical engineering and electronics. One 'cm' is equal to an e.m.f. of one emu cgs of induced voltage in a circuit by a variation of current of one emu cgs unit per second.	

Unit (synonym, acronym)	Symbol	Physical quantity	Dimension	Conversion factor (SI equivalent unit)	Notes, definitions, other conversion factors	System
centimetre (electrostatic)	'cm'	electric capacitance	$M^{-1}L^{-2}T^4I^2$	1 'cm' = $1/9 \times 10^{-11}$ F (E)	Obsolete unit of electric capacitance employed in electrical engineering and electronics. One 'cm' describes an electric capacitance which is equal to one esu cgs of electric potential during the increase of one esu cgs of charge.	UK, US
centimetre of mercury (0°C)	cmHg (0°C)	pressure, stress	$ML^{-1}T^{-2}$	1 cmHg (0°C) = 1333.223684 Pa	Obsolete unit of pressure employed in physics to measure small pressures. It was defined as the pressure exerted by a column of mercury 1 cm high, measured at 0°C (32°F). 76 cmHg (0°C) = 101 325 (E) 76 cmHg (0°C) = 1 atm (E). 1 cmHg (0°C) = 101 325/76 Pa (E)	
centimetre of mercury (32°F)	cmHg (32°F)	pressure, stress	$ML^{-1}T^{-2}$	1 cmHg (32°F) = 1333.223684 Pa	Obsolete unit of pressure employed in physics to measure small pressures. It was defined as the pressure exerted by a column of mercury 1 cm high, measured at 0°C (32°F). 76 cmHg (0°C) = 101 325 (E) 76 cmHg (0°C) = 1 atm (E). 1 cmHg (32°F) = 101 325/76 Pa (E)	UK, US
centimetre of water (15.56°C)	cmH_2O (15.56°C)	pressure, stress	$ML^{-1}T^{-2}$	1 cmH_2O (15.56°C) = 97.97059096 Pa	Obsolete unit of pressure employed in physics to measure small pressures. It is equal to the pressure exerted by a column of water 1 cm high measured at 15.56°C (60°F). 1 atm = 1034.241008 cmH_2O (60°F)	

(39.2°F)	(39.2°F)				physics to measure small pressures. It is equal to the pressure exerted by a column of water 1 cm high measured at 4°C (39.2°F). 1 atm = 1033.256383 cmH_2O (4°C)	
centimetre of water (4°C)	cmH_2O (4°C)	pressure, stress	$ML^{-1}T^{-2}$	1 cmH_2O (4°C) = 98.06375414 Pa	Obsolete unit of pressure employed in physics to measure small pressures. It is equal to the pressure exerted by a column of water 1 cm high measured at 4°C (39.2°F). 1 atm = 1033.256383 cmH_2O (4°C)	
centimetre of water (60°F)	cmH_2O (60°F)	pressure, stress	$ML^{-1}T^{-2}$	1 cmH_2O (60°F) = 97.97059096 Pa	Obsolete unit of pressure employed in physics to measure small pressures. It is equal to the pressure exerted by a column of water 1 cm high measured at 15.56°C (60°F). 1 atm = 1034.241008 cmH_2O (60°F)	UK, US
centimetre per second	$cm.s^{-1}$	velocity, speed	LT^{-1}	1 $cm.s^{-1} = 10^{-2}$ $m.s^{-1}$	Obsolete cgs unit of velocity.	SI
centioctave	–	logarithmic musical interval	nil	(see note)	The centioctave is the interval between two musical sounds having as a basic frequency ratio the 1st root of two. The number of cents between frequencies f_1 and f_2 is: $I = 100 \times \log_2(f_1/f_2)$ 1 centioctave = 0.301 savart (E) 1 centioctave = 12 cent (E)	

Unit (synonym, acronym)	Symbol	Physical quantity	Dimension	Conversion factor (SI equivalent unit)	Notes, definitions, other conversion factors	System
centipoise	cP, cPo	dynamic viscosity, absolute viscosity	$ML^{-1}T^{-1}$	1 cP = 10^{-3} Pa.s (E)	Unit of dynamic viscosity or absolute viscosity in the cgs system. In the UK the written symbol is cPo. 1 cP = 1 mPa.s (E)	cgs
centistokes	cSt	kinematic viscosity	L^2T^{-1}	1 cSt = 10^{-6} $m^2.s^{-1}$ (E)	Submultiple of the old cgs unit of kinematic viscosity. 1 cSt = 1 $mm^2.s^{-1}$ (E)	cgs
centner (kintal, cental, hundredweight)	cH, cwt	mass	M	1 centner = 45.359 237 kg (E)	Obsolete British unit of mass used in business transactions, especially in agriculture. It serves to measure food grain. 1 cental = 1 cwt (E) 1 cental = 100 lb (av.) (E) 1 cental = 1 sh. cwt (US) (E)	UK, US
centuria (Roman)	–	surface, area	L^2	1 centuria (Roman) = 5.053503744×10^5 m^2	Obsolete Roman unit of area employed in ancient times. 1 centuria (Roman) = 5 760 000 quadratus pes (E)	Roman
century	–	time, period, duration	T	1 century = 100 years (E)		
chad	chad	neutron fluence rate	$L^{-2}T^{-1}$	1 chad = 10^4 $n.m^{-2}s^{-1}$ (E)	Obsolete unit of neutron fluence rate used in nuclear physics. The unit is named after Sir J. Chadwick (1891–1974). 1 chad = 1 $n.cm^{-2}s^{-1}$ (E) Sometimes: 1 Chad = 10^{16} $n.cm^{-2}s^{-1}$ (E)	

chain (engineer's)	ch	length, distance	L	1 chain (engineer's) = 30.48 m (E)	Obsolete American surveyor's unit of length. 1 chain (engineer's) = 1/6.6 furlong (US) (E) 1 chain (engineer's) = 6 rods (E) 1 chain (engineer's) = 100 feet (E) 1 chain (engineer's) = 100 links (US) (E)	US
chain (Gunter's)	ch	length, distance	L	1 chain (Gunter's) = 20.1168 m (E)	Obsolete British surveyor's unit of length. 1 chain (Gunter's) = 1/10 furlong (UK) (E) 1 chain (Gunter's) = 4 rods (E) 1 chain (Gunter's) = 66 feet (E) 1 chain (Gunter's) = 100 links (UK) (E)	UK
chain (Ramsden's)	ch	length, distance	L	1 chain (Ramsden's) = 30.48 m (E)	Obsolete American surveyor's unit of length. 1 chain (Ramsden's) = 1/6.6 furlong (US) (E) 1 chain (Ramsden's) = 6 rods (E) 1 chain (Ramsden's) = 100 feet (E) 1 chain (Ramsden's) = 100 links (US) (E)	US
chain (surveyor's)	ch	length, distance	L	1 chain (surveyor's) = 20.1168 m (E)	Obsolete British surveyor's unit of length. 1 chain (surveyor's) = 1/10 furlong (UK) (E) 1 chain (surveyor's) = 4 rods (E) 1 chain (surveyor's) = 66 feet (E) 1 chain (surveyor's) = 100 links (E)	UK

Unit (synonym, acronym)	Symbol	Physical quantity	Dimension	Conversion factor (SI equivalent unit)	Notes, definitions, other conversion factors	System
chaldron (UK)	chal (UK)	capacity, volume	L^3	1 chaldron (UK) = 1.309274496 m^3	Obsolete British unit used for capacity measurements for all merchandise (solids, liquids, foodstuffs, etc). 1 chaldron (UK) = 12 sacks (UK) (E) 1 chaldron (UK) = 36 bushels (UK) (E) 1 chaldron (UK) = 72 buckets (UK) (E) 1 chaldron (UK) = 144 pecks (UK) (E) 1 chaldron (UK) = 288 gallons (UK) (E)	UK
chalque (Greek, Attic)	–	mass	M	1 chalque (Greek, Attic) = 5.854167×10^{-5} kg	Obsolete Greek unit of mass used in ancient times. 1 obol (Greek, Attic) = 1/48 drachma (Greek, Attic)	Attic
chao (Chinese)	–	capacity, volume	L^3	1 chao (Chinese) = 10.3544×10^{-6} m^3	Obsolete Chinese unit of capacity used in ancient times. 1 chao (Chinese) = 1/100 tcheng (E)	Chinese
chebel (Persian)	–	length, distance	L	1 chebel (Persian) = 25.600 m	Obsolete unit of length in the Assyrio-Chaldean-Persian system used in ancient times. 1 chebel = 80 zereths (E)	Persian
chei (Chinese)	–	capacity, volume	L^3	1 chei (Chinese) = 103.544×10^{-3} m^3	Obsolete Chinese unit of capacity used in ancient times. 1 chei (Chinese) = 100 tcheng (E)	Chinese
cheng (Chinese)	–	capacity, volume	L^3	1 cheng (Chinese) = 1.03544×10^{-3} m^3	Obsolete Chinese unit of capacity used in ancient times.	Chinese
chenix (Greek, Attic) [χοινιξ]	–	capacity, volume	L^3	1 chenix (Greek, Attic) = 1.08×10^{-3} m^3	Obsolete Greek unit of volume employed in ancient times. It was	Attic

					substances.	
cheval-vapeur (horsepower)	cv, HP	power	ML^2T^{-3}	1 HP = 735.498750 W	Obsolete unit of power introduced by James Watt in 1784 to allow to describe the power of steam machinery. It was equal to the work effort of a horse needed to raise vertically 528 cubic feet of water to one metre high in one minute. In 1850, the following conversion factor was adopted: 1 HP = 75 $kgf.m.s^{-1}$ (E)	UK, INT
chi (Chinese)	–	length, distance	L	1 chi (Chinese) = 0.23 m	Obsolete Chinese unit of length used in ancient times. Base unit of the Chinese system of length.	Chinese
ching (Chinese)	–	surface, area	L^2	1 ching (Chinese) = 61 440 m^2	Obsolete Chinese unit of area used in ancient times. 1 ching (Chinese) = 100 meou (E)	Chinese
chkalik (Russian)	–	capacity, volume	L^3	1 chkalik (Russian) = 6.149705×10^{-5} m^3	Obsolete Russian unit of capacity for liquid substances used before 1917. 1 chkalik (Russian) = 1/2 tcharka (E)	Russian
chô (Japanese)	–	surface, area	L^2	1 chô = 9917355372 m^2	Obsolete Japanese unit of area. 1 chô = 3000 bu (E)	Japanese
chô (Japanese)	–	length, distance	L	1 chô (Japanese) = 109.8 m (E)	Obsolete Japanese unit of length. 1 chô = 360 shaku (E)	Japanese
chopine (de Paris)	–	capacity, volume	L^3	1 chopine (de Paris) = $0.465694500 \times 10^{-3}$ m^3	Obsolete French unit of capacity employed before the French Revolution. It serves to express the capacity of liquids and grains. It varies according to the location and merchandise. 1 chopine (de Paris) = 1/2 pinte (de Paris)	French

Unit (synonym, acronym)	Symbol	Physical quantity	Dimension	Conversion factor (SI equivalent unit)	Notes, definitions, other conversion factors	System
chopine (UK)	–	capacity, volume	L^3	1 chopine (UK) $= 0.568261500 \times 10^{-3}$ m^3	Obsolete British unit of capacity. 1 chopine (UK) = 1/8 gallons (UK) (E) 1 chopine (UK) = 4 gills (UK) (E)	UK
chopine (US, dry)	–	capacity, volume	L^3	1 chopine (US, dry) $= 0.550610500 \times 10^{-3}$ m^3	Obsolete American unit of capacity. 1 chopine (US, dry) = 1/8 gallons (US, dry) (E)	US
chopine (US, liq.)	–	capacity, volume	L^3	1 chopine (US, liq.) $= 0.473176473 \times 10^{-3}$ m3	Obsolete American unit of capacity. 1 chopine (US, liq.) = 1/8 gallons (US, liq.) (E) 1 chopine (US, liq.) = 4 gills (US, liq.) (E)	US
choryos (Egyptian) [Egyptian palm]	–	length, distance	L	1 choryos (Egyptian) $= 8.725 \times 10^{-2}$ m	Obsolete Egyptian unit of length used in ancient times. 1 choryos = 1/6 Royal cubit (E)	Egyptian
chronon (tempon)	–	time, period, duration	T	1 chronon $= 10^{-23}$ s (E)	Obsolete unit of time employed in atomic physics. It corresponds to the time needed by light to cover a distance equal to the electron radius.	
cicéro	–	length, distance	L	1 cicéro $= 4.511658334 \times 10^{-3}$ m	Obsolete French unit employed in typography. 1 cicéro = 12 points Didot (E)	French
circular inch	cin, cir. in	surface, area	L^2	1 circular inch $= 5.067074791 \times 10^{-4}$ m^2	Obsolete British and American unit of area. It was used for the measurement of small circular areas, such as the cross section of a wire. It was equal to the area of	UK, US

					a disc with a diameter of one inch. 1 circular inch = 10^6 circular mils (E)	
circular mil	cmil, cir. mil	surface, area	L^2	1 circular mil = $5.067074791 \times 10^{-10}$ m^2	Obsolete British and American unit of area. It was used for the measurement of small circular areas, such as the cross section of a wire. It was equal to the area of a disc with a diameter of one mil. 1 circular mil = $\pi/4 \times 10^{-6}$ in^2 (E)	UK, US
circular mile (int. naut.)	cmi, cir. mi (int. naut.)	surface, area	L^2	1 circular mile (int. naut.) = 2.693840302×10^6 m^2	Obsolete international unit of area. It was used for the measurement of circular areas. It was equal to the area of a disc with a diameter of one mile (int. naut.).	INT
circular mile (int.)	cmi, cir. mi (int.)	surface, area	L^2	1 circular mile (int.) = 2.034179489×10^6 m^2	Obsolete international unit of area. It was used for the measurement of circular areas. It was equal to the area of a disc with a diameter of one mile (int.).	INT
circular mile (statute)	cmi, cir. mi (stat.)	surface, area	L^2	1 circular mile (stat.) = 2.034171905×10^6 m^2	Obsolete British and American unit of area. It was used for the measurement of circular areas. It was equal to the area of a disc with a diameter of one mile (statute).	UK, US
circular mile (US, naut.)	cmi, cir. mi (US, naut.)	surface, area	L^2	1 circular mile (US, naut.) = 2.697285795×10^6 m^2	Obsolete British and American unit of area. It was used for the measurement of circular areas. It was equal to the area of a disc with a diameter of one mile (US, naut.).	UK, US
circular mile (US, survey)	cmi, cir. mi (US, survey)	surface, area	L^2	1 circular mile (US, survey) = 2.034418004×10^6 m^2	Obsolete American unit of area. It was used for the measurement of circular areas. It was equal to the area of a disc with a diameter of one mile (US, survey).	US

Unit (synonym, acronym)	Symbol	Physical quantity	Dimension	Conversion factor (SI equivalent unit)	Notes, definitions, other conversion factors	System
circular millimetre	cmm, cir. mm	surface, area	L^2	1 circular millimetre $= 7.853981634 \times 10^{-7}$ m^2	Obsolete British and American unit of area. It was used for the measurement of small circular areas, such as the cross section of a wire. It was equal to the area of a disc with a diameter of one millimetre. 1 circular mm $= \pi/4 \times 10^{-6}$ m2 (E)	UK
circumference	–	plane angle	α	1 circumference $= 2\pi$ rad (E)	1 circumference = 360 degrees (E) 1 circumference = 400 gons (E)	INT
clarke	–	mass fraction, relative abundance	nil	1 clarke $= 10^{-6}$ m.m^{-1} (E)	Obsolete index used in geology and geochemistry. It described the average mass fraction of an element in the Earth's crust. In metallogeny, the Clarke index serves to express the ratio of the abundance of an element in an ore versus its average abundance in the Earth's crust.	
Clarke degree	–	percentage of alcohol in wines and spirits	nil		Unit introduced in 1730	
clausius (rank)	Cl	entropy	$ML^2T^{-2}\Theta^{-1}$	1 clausius = 4.184 J.K^{-1} (E)	Obsolete unit of entropy. The unit is named after R.J.L. clausius (1822-1888). 1 clausius = 1 cal$_{th}$.K^{-1} 1 clausius = 4.184 carnot (E)	
clima (Roman)	–	surface, area	L^2	1 clima (Roman) = 315.843984 m^2	Obsolete Roman unit of area employed in ancient times. 1 clima (Roman) = 3 600 quadratus pes (E)	Roman

		(thermal resistance multiplied by area)			employed in the British textile industry. One tog corresponds to the heat insulation coefficient of clothing which conserves a temperature difference of 0.1°F between its surfaces when the heat flux is equal to 1 $kcal_{15}.h^{-1}.m^{-2}$. 1 clo = 0.180 $(kcal_{15}.h^{-1})^{-1}\ m^2.°C$ (E) 1 clo = 0.648 $(cal_{15}.s^{-1})^{-1}.m^2.°C$ (E) 1 clo = 0.879109939 $(Btu_{IT}.h^{-1})^{-1}.ft^2.°F$	
clove (UK)	–	mass	M	1 clove (UK) = 3.628738960 kg (E)	Obsolete British unit of mass. 1 clove (UK) = 8 lb (av.) (E)	UK
clusec	clusec	power	ML^2T^{-3}	1 clusec = 1.3×10^{-3} W	Obsolete unit of power employed in vacuum technology.	
ço (Chinese)	–	capacity, volume	L^3	1 ço (Chinese) = $1.03544 \times 10^{-6}\ m^3$	Obsolete Chinese unit of capacity used in ancient times. 1 ço (Chinese) = 1/1000 tcheng (E)	Chinese
coal skip	–	mass	M	1 coal skip = 45.359237 kg (E)	Obsolete British unit employed in coal mining industry. 1 coal skip = 100 lb (av.) (E) 1 coal skip = 1 cwt (E)	UK
codos (Spanish)	–	length, distance	L	1 codos (Spanish) = 4.179525×10^{-1} m	Obsolete Spanish unit of length. 1 codos (Spanish) = 1/2 vara (E)	Spanish
condylos (Greek, Attic)	–	length, distance	L	1 condylos (Greek, Attic) = 3.857×10^{-2} m	Obsolete Greek unit of length employed in ancient times. 1 condylos (Attic) = 1/8 pous (E)	Attic

Unit (synonym, acronym)	Symbol	Physical quantity	Dimension	Conversion factor (SI equivalent unit)	Notes, definitions, other conversion factors	System
congius (Roman gallon)	–	capacity, volume	L^3	1 congius (Roman gallon) $= 3.283 \times 10^{-3}$ m^3	Obsolete Roman unit of volume employed in ancient times. It was used for capacity measurements of liquids. 1 congius (Roman gallon) = 6 sextarius (E)	Roman
coomb (UK)	–	capacity, volume	L^3	1 coomb (UK) $= 145.474944 \times 10^{-3}$ m^3	Obsolete British unit of volume. 1 coomb (UK) = 4 bushels (UK) (E) 1 coomb (UK) = 32 gallons (UK) (E)	UK
copas (Spanish)	–	capacity, volume	L^3	1 copas (Spanish) $= 0.122212203 \times 10^{-3}$ m^3	Obsolete Spanish unit of capacity used for liquid substances. 1 copas (Spanish) = 1/128 arroba (water) (E)	Spanish
cor (Hebrew)	–	capacity, volume	L^3	1 cor (Hebrew) $= 214.20 \times 10^{-3}$ m^3 (of water)	Obsolete Hebrew unit of capacity used in ancient times. Measured by weight. 1 cor (Hebrew) = 10 ephah (new)	Hebrew
cord (UK, wood)	cd (UK)	capacity, volume	L^3	1 cord (UK, wood) $= 3.624556364$ m^3	Obsolete British unit of capacity which was employed for stacked logs of wood. It was equal to a pile of wood cut 4 feet long, piled 4 feet high and 8 feet wide. 1 cord (UK) = 8 cord-ft (E) 1 cord (UK) = 128 ft^3 (E) 1 cord (UK) = 1536 board foot measure (E)	UK
cord-foot	cord-ft	capacity, volume	L^3	1 cord-foot $= 453.0695455 \times 10^{-3}$ m^3	Obsolete British unit of capacity. 1 cord-foot = 1/8 cord (E) 1 cord-foot = 16 ft^3 (E)	UK

cotton (hand)	–	specific length	$M^{-1}L$	1 cotton = 1693.361817 m.kg^{-1}	Obsolete American and British unit employed in the textile industry. 1 cotton = 840 yd.lb^{-1} (E)	UK, US
cotyle (Greek, Attic) [χοτυλη]	–	capacity, volume	L^3	1 cotyle (Greek, Attic) $= 0.27 \times 10^{-3}$ m^3	Obsolete Greek unit of volume employed in ancient times. It was used for capacity measurements of dry substances. 1 cotyle (Greek, Attic) = 1/4 chenix (E)	Attic
coudée (de Paris) (French cubit)	–	length, distance	L	1 coudée (de Paris) = 0.50 m (E)	Obsolete French unit of length employed before the French Revolution (1789). It was equal to the distance between the elbow and the tip of the middle finger.	French
coulomb	C	quantity of electricity, electric charge	IT	**SI derived unit** 1 C = 1 A.s (E)	The coulomb is the electric charge transported in one second by a current of one ampere. The unit is named after C.A. Coulomb (1736–1806).	SI
coulomb (int.)	C (int.)	quantity of electricity, electric charge	IT	coulomb (int.) = 0.99985 C	Obsolete American unit.	IEUS
coulomb metre per joule	C.m.J^{-1}	1st hyper-susceptibility	$M^{-1}L^{-1}T^3I$	**SI derived unit** 1 C.m.J^{-1} = 1 kg^{-1}.m^{-1}.s^3.A (E)		SI
coulomb metre per square joule	C.m.J^{-2}	2nd hyper-susceptibility	$M^{-2}L^{-3}T^5I$	**SI derived unit** 1 C.m.J^{-2} = 1 kg^{-2}.m^{-3}.s^5 .A (E)		SI
coulomb per cubic metre	C.m^{-3}	electric charge density	IT.L^{-3}	**SI derived unit**		SI

Unit (synonym, acronym)	Symbol	Physical quantity	Dimension	Conversion factor (SI equivalent unit)	Notes, definitions, other conversion factors	System
coulomb per kilogram	$C.kg^{-1}$	exposure	ITM^{-1}	**SI derived unit.**	1 $C.kg^{-1}$ = 3876 R	
coulomb per square metre	$C.m^{-2}$	surface density of charge, electric flux density, electric displacement	ITL^{-2}	**SI derived unit.**	The coulomb per square metre is the unit of electric displacement in a capacitor where electrodes are two parallel planes, of infinite area, separated by a vacuum and which has an electric surface charge density of one coulomb per square metre.	SI
covada (Portuguese)	–	length, distance	L	1 covada (Portuguese) = 6.570×10^{-1} m	Obsolete Portuguese unit of length. 1 covada (Portuguese) = 2 pe (E)	Portuguese
cran (mease)	cran	capacity, volume	L^3	1 cran = $170.478450 \times 10^{-3}$ m^3	Obsolete unit of capacity employed in the fishing industry. It describes the amount of herrings and it is equal to the number of herrings which can be packed into a standard box of volume of 37.5 gal (UK). 1 cran ≈ 750 herrings	UK
crinal	crinal	force, weight	MLT^{-2}	1 crinal = 10^{-1} N (E)	Obsolete unit of force. 1 crinal = 1 $kg.dm.s^{-2}$ (E)	
crith	crith	mass	M	1 crith = $8.994298964 \times 10^{-5}$ kg (1 atm, 273.15 K), 1 crith = $8.876682914 \times 10^{-5}$ kg (1 bar, 273.15 K)	Obsolete unit of weight employed to measure mass of gas. It was equal to the mass of one litre of hydrogen in the standard state. Old standard (P = 101 325 Pa, T = 273.15 K New standard $P = 10^5$ Pa, T = 273.15 K	

crocodile	-	electric potential, electric potential difference, electromotive force			employed in nuclear physics. 1 crocodile = 1 MV (E)	
cron	cron	time, period, duration	T	1 cron = 10^6 years (E)	Unit of time employed in geology. It was suggested by J.S. Huxley in 1957. 1 cron = 1 My (E)	
crosa	-	length, distance	L	1 crosa = 3656 m	Obsolete Indian unit of length used in ancient times. 1 crosa = 8000 hasta (E)	Indian
cuartilla (Spanish)	-	capacity, volume	L^3	1 cuartilla (Spanish) = $3.910790500 \times 10^{-3}$ m^3	Obsolete Spanish unit of capacity used for liquid substances. 1 cuartilla (Spanish) = 1/4 arroba (water) (E)	Spanish
cuartilla (Spanish)	-	surface, area	L^2	1 cuartilla (Spanish) = 17.46842922 m^2	Obsolete Spanish unit of area. 1 cuartilla (Spanish) = 25 sq. vara (E)	Spanish
cuartilla (Spanish, dry)	-	capacity, volume	L^3	1 cuartilla (Spanish, dry) = 13.875250×10^{-3} m^3	Obsolete Spanish unit of capacity used for dry substances. 1 cuartilla (Spanish, dry) = 1/4 arroba (E)	Spanish
cuartillo (Spanish)	-	capacity, volume	L^3	1 cuartillo (Spanish) = 0.4888488×10^{-3} m^3	Obsolete Spanish unit of capacity used for liquid substances. 1 cuartillo (Spanish) = 1/32 arroba (water) (E)	Spanish
cuartillo (Spanish, dry)	-	capacity, volume	L^3	1 cuartillo (Spanish, dry) = $1.15627083 \times 10^{-3}$ m^3	Obsolete Spanish unit of capacity used for dry substances. 1 cuartillo (Spanish, dry) = 1/48 fanega (E)	Spanish

Unit (synonym, acronym)	Symbol	Physical quantity	Dimension	Conversion factor (SI equivalent unit)	Notes, definitions, other conversion factors	System
cubem (cubic int. stat. mile)	cubem, cu. mi (int. stat.)	capacity, volume	L^3	1 cubem = 4.168205135×10^9 m^3	Obsolete American and British unit of volume. 1 cubem = 1 (int. stat. mile)^3	UK, US
cubic attometre	am^3	capacity, volume	L^3	1 $am^3 = 10^{-54}$ m^3 (E)	Submultiple of the SI derived unit.	SI
cubic centimetre	cm^3	capacity, volume	L^3	1 $cm^3 = 10^{-6}$ m^3(E)	Submultiple of the SI derived unit.	SI
cubic centimetre (Mohr cubic centimetre)	cc	capacity, volume	L^3	1 cc = 1.00238×10^{-6} m^3(E)	Obsolete unit of capacity which was employed in pharmacy. It is equal to the volume occupied by one gram of pure water measured at 17.5°C. The unit is named after C.F. Mohr (1806–1879).	
cubic Coulomb cubic metre per square joule	$C^3.m^3.J^{-2}$	1st hyper-polarizability	$M^{-2}L^{-1}T^7I^3$	**SI derived unit** 1 $C^3.m^3.J^{-2} = 1$ $kg^{-2}.m^{-1}.s^7.A^3$ (E)		SI
cubic decametre	dam^3	capacity, volume	L^3	1 $dam^3 = 10^3$ m^3 (E)	Multiple of the SI derived unit.	SI
cubic decimetre	dm^3	capacity, volume	L^3	1 $dm^3 = 10^{-3}$ m^3 (E)	Submultiple of the SI derived unit.	SI
cubic exametre	Em^3	capacity, volume	L^3	1 $Em^3 = 10^{54}$ m^3 (E)	Multiple of the SI derived unit.	SI
cubic femtometre	fm^3	capacity, volume	L^3	1 $fm^3 = 10^{-45}$ m^3 (E)	Submultiple of the SI derived unit.	SI
cubic foot	ft^3, cu. ft	capacity, volume	L^3	1 $ft^3 = 2.8316846592 \times 10^{-2}$ m^3 (E)	British and American legal unit of capacity. 1 $ft^3 = 1728$ in^3 (E)	UK, US, FPS
cubic foot per minute	cfm, $ft^3.min^{-1}$	volume flow rate	L^3T^{-1}	1 $ft^3.min^{-1}$ = $4.719474432 \times 10^{-4}$ $m^3.s^{-1}$	1 $ft^3.min^{-1} = 0.4719474432$ $l.s^{-1}$	UK, US

cubic foot per pound	$ft^3.lb^{-1}$	specific volume	L^3M^{-1}	$1\ ft^3.lb^{-1} = 6.242796057 \times 10^{-2}\ m^3.kg^{-1}$	British and American unit of specific volume.	UK, US, FPS
cubic foot per second	cfs, $ft^3.s^{-1}$	volume flow rate	L^3T^{-1}	$1\ ft^3.s^{-1} = 28.316846590 \times 10^{-3}\ m^3.s^{-1}$	$1\ ft^3.s^{-1} = 28.316846659\ l.s^{-1}$	UK, US, FPS
cubic foot per UK ton	$ft^3.ton^{-1}$	specific volume	L^3M^{-1}	$ft^3.(UK)\ ton^{-1} = 2.786962526 \times 10^{-5}\ m^3.kg^{-1}$	Obsolete British unit of specific volume.	UK
cubic gigametre	Gm^3	capacity, volume	L^3	$1\ Gm^3 = 10^{27}\ m^3$ (E)	Multiple of the SI base unit.	SI
cubic hectometre	hm^3	capacity, volume	L^3	$1\ hm^3 = 10^6\ m^3$ (E)	Multiple of the SI derived unit.	SI
cubic inch	in^3, cu.in	capacity, volume	L^3	$1\ in^3 = 1.6387064 \times 10^{-5}\ m^3$ (E)	British and American leagal unit of capacity.	UK, US
cubic inch per pound	$in^3.lb^{-1}$	specific volume	L^3M^{-1}	$1\ in^3.lb^{-1} = 3.612729200 \times 10^{-5}\ m^3.kg^{-1}$	British and American unit of specific volume.	UK, US
cubic kilometre	km^3	capacity, volume	L^3	$1\ km^3 = 10^9\ m^3$ (E)	Multiple of the SI derived unit.	SI
cubic megametre	Mm^3	capacity, volume	L^3	$1\ Mm^3 = 10^{18}\ m^3$ (E)	Multiple of the SI derived unit.	SI
cubic metre	m^3	capacity, volume	L^3	**SI derived unit**	The cubic metre is the unit of capacity which is equal to the volume of a metre cube.	SI
cubic metre per coulomb	$m^3.C^{-1}$	Hall coefficient	$L^3I^{-1}T^{-1}$	**SI derived unit**		SI
cubic metre per hour	$m^3.h^{-1}$	volume flow rate	L^3T^{-1}	$1\ m^3.h^{-1} = 2.7777777 \times 10^{-4}\ m^3.s^{-1}$	$1\ m^3.h^{-1} = [1/3600]\ m^3.s^{-1}$ (E)	
cubic metre per kilogram	$m^3.kg^{-1}$	specific volume	L^3M^{-1}	**SI derived unit**		SI

Unit (synonym, acronym)	Symbol	Physical quantity	Dimension	Conversion factor (SI equivalent unit)	Notes, definitions, other conversion factors	System
cubic metre per mole	$m^3.mol^{-1}$	molar refraction	L^3N^{-1}	**SI derived unit**		SI
cubic metre per mole	$m^3.mol^{-1}$	second virial coefficient	L^3N^{-1}	**SI derived unit**		SI
cubic metre per second	$m^3.s^{-1}$	volume flow rate	L^3T^{-1}	**SI derived unit**	The cubic metre per second is the SI derived unit of flow rate. It is equal to one cubic metre of a homogenous fluid which flows uniformly in one second.	SI
cubic micrometre	μm^3	capacity, volume	L^3	$1\ \mu m^3 = 10^{-18}\ m^3$ (E)	Submultiple of the SI derived unit.	SI
cubic millimetre	mm^3	capacity, volume	L^3	$1\ mm^3 = 10^{-9}\ m^3$ (E)	Submultiple of the SI derived unit.	SI
cubic nanometre	nm^3	capacity, volume	L^3	$1\ nm^3 = 10^{-27}\ m^3$ (E)	Submultiple of the SI derived unit.	SI
cubic petametre	Pm^3	capacity, volume	L^3	$1\ Pm^3 = 10^{45}\ m^3$ (E)	Multiple of the SI derived unit.	SI
cubic picometre	pm^3	capacity, volume	L^3	$1\ pm^3 = 10^{-36}\ m^3$ (E)	Submultiple of the SI derived unit.	SI
cubic terametre	Tm^3	capacity, volume	L^3	$1\ Tm^3 = 10^{36}\ m^3$ (E)	Multiple of the SI derived unit.	SI
cubic yard	yd^3, cu.yd	capacity, volume	L^3	$1\ yd^3 = 764.554858 \times 10^{-3}\ m^3$(E)	British and American official unit of capacity. $1\ yd^3 = 27\ ft^3$ (E) $1\ yd^3 = 46\,656\ in^3$ (E)	UK, US
cubic yoctometre	ym^3	capacity, volume	L^3	$1\ ym^3 = 10^{-72}\ m^3$ (E)	Submultiple of the SI derived unit.	SI
cubic yottametre	Ym^3	capacity, volume	L^3	$1\ Ym^3 = 10^{72}\ m^3$ (E)	Multiple of the SI derived unit.	SI
cubic zeptometre	zm^3	capacity, volume	L^3	$1\ zm^3 = 10^{-63}\ m^3$ (E)	Submultiple of the SI derived unit.	SI
cubic zettametre	Zm^3	capacity, volume	L^3	$1\ Zm^3 = 10^{63}\ m^3$ (E)	Multiple of the SI derived unit.	SI

					ancient times. 1 cubit (Arabic) = 3/2 feet (Arabic) (E)	
cubit (Greek, Attic)	–	length, distance	L	1 cubit (Greek, Attic) = 6.171200×10^{-1} m	Obsolete Greek unit of length employed in ancient times. 1 cubit (Attic) = 2 pous (E)	Attic
cubit (Hachemic, Arabic)	–	length, distance	L	1 cubit (Hachemic, Arabic) = 6.40×10^{-1} m	Obsolete Arabic unit of length used in ancient times. 1 cubit (Hachemic, Arabic) = 2 feet (Arabic) (E)	Arabic
cubit (Hebrew)	–	length, distance	L	1 cubit (Hebrew) = 0.555 m	Obsolete Hebrew unit of length used in ancient times. Base unit in the system of length.	Hebrew
cubit (new, Arabic)	–	length, distance	L	1 cubit (new, Arabic) = 4.80×10^{-1} m	Obsolete Arabic unit of length used in ancient times. 1 cubit (new, Arabic) = 3/2 feet (Arabic) (E)	Arabic
cubit (Persian)	–	length, distance	L	1 cubit (Persian) = 0.640 m	Obsolete unit of length in the Assyrio-Chaldean-Persian system used in ancient times. 1 cubit = 2 zereth (E)	Persian
cubit (UK)	–	length, distance	L	1 cubit (UK) = 0.4572 m (E)	Obsolete British unit of length. It was equal to the distance between the elbow and the tip of the middle finger. 1 cubit (UK) = 1.5 foot (E) 1 cubit (UK) = 4.5 hands (UK) (E) 1 cubit (UK) = 6 palms (UK) (E) 1 cubit (UK) = 18 inches (UK) (E) 1 cubit (UK) = 216 lines (UK) (E)	UK

Unit (synonym, acronym)	Symbol	Physical quantity	Dimension	Conversion factor (SI equivalent unit)	Notes, definitions, other conversion factors	System
cubitus (Roman) [Roman cubit]	–	length, distance	L	1 cubitus (Roman) = 0.4416 m	Obsolete Roman unit of length employed in ancient times. 1 cubitus (Roman) = 3/2 pes (E)	Roman
cudava (Indian)	–	capacity, volume	L^3	1 cudava (Indian) = 4.125×10^{-4} m^3 (of water)	Obsolete Indian unit of capacity used in ancient times. Measured by weight. 1 cudava (Indian) = 1/32 drona (E)	Indian
culleus (Roman hogshead) [dolium]	–	capacity, volume	L^3	1 culleus (Roman hogshead) = 527.28×10^{-3} m^3	Obsolete Roman unit of volume employed in ancient times. It was used for capacity measurements of liquids. 1 culleus (Roman hogshead) = 960 sextarius (E)	Roman
cumbha (Indian)	–	capacity, volume	L^3	1 cumbha (Indian) = 264×10^{-3} m^3(of water)	Obsolete Indian unit of capacity used in ancient times. Measured by weight. 1 cumbha (Indian) = 20 drona (E)	Indian
cun (Chinese) [tsouen]	–	length, distance	L	1 cun (Chinese) = 2.3×10^{-2} m	Obsolete Chinese unit of length used in ancient times. 1 cun = 1/10 tchi (E) 1 cun = 2.3 cm (E)	Chinese
cup (metric)	–	length, distance	L	1 cup (metric) = 0.2 m (E)	Obsolete American metric unit of length.	US
cup (US, length)	–	length, distance	L	1 cup (US) = 0.2365882 m	Obsolete American unit of length.	US
cup (US, liq.)	cup (US, liq.)	capacity, volume	L^3	1 cup (US, liq.) = $2.365882365 \times 10^{-4}$ m^3	Obsolete American unit of capacity. 1 cup (US, liq.) = 1/16 gal (US, liq.) 1 cup (US, liq.) ≈ 236.59 cm^3	US

curie	Ci	radioactivity	T[illegible]	1 Ci = 3.7 × 10[illegible] Bq (E)	Obsolete unit of radioactivity. One Curie is equal exactly to the radioactivity of a source which has the same radioactivity as one gram of the radium isotope $^{226}_{88}$Ra in secular equilibrium with its radon derivative, $^{222}_{86}$Rn. The unit is named after the two famous French scientists Pierre and Marie Curie. The Ci is still in use.	
Curie per litre	Ci/l	radioactivity concentration	$L^{-3}T^{-1}$	1 Ci/l = 3.7×10^{13} Bq.m^{-3}	Obsolete unit.	
cusec	–	volume flow rate	L^3T^{-1}	1 cusec = $2.831684659 \times 10^{-2}$ m^3.s^{-1}	Obsolete British unit of flow rate. It was used in vacuum pump technology. The name of the unit derived from the acronym of cubic foot per second. 1 cusec = 1 ft^3.s^{-1} (E)	UK
cyanthos (Greek, Attic)	–	capacity, volume	L^3	1 cyanthos (Greek, Attic) = 4.5×10^{-5} m^3	Obsolete Greek unit of volume employed in ancient times. It was used for capacity measurements of dry substances. 1 cyanthos (Greek, Attic) = 1/24 chenix (E)	Attic
cyathus (Roman)	–	capacity, volume	L^3	1 cyathus (Roman) = 4.558333×10^{-3} m^3	Obsolete Roman unit of volume employed in ancient times. It was used for capacity measurements of liquids. 1 cyathus (Roman gallon) = 1/12 sextarius (E)	Roman
cycle per second (cycle)	cps, c.s^{-1}, cy, c	frequency	T^{-1}	1 cycle.s^{-1} = 1 s^{-1} (E)	Old frequency unit employed in broadcasting. Precursor of the hertz. 1 cycle.s^{-1} = 1 Hz (E)	UK, INT
D unit	–	exposure	ITM^{-1}	1 D unit = 259.08×10^{-4} C.kg^{-1}	Obsolete unit of exposure to X-rays. It was introduced in 1925. 1 D unit = 100 röntgens (E)	

Unit (synonym, acronym)	Symbol	Physical quantity	Dimension	Conversion factor (SI equivalent unit)	Notes, definitions, other conversion factors	System
daktylos (Greek, Attic) [Greek finger]	–	length, distance	L	1 daktylos (Greek, Attic) $= 1.928500 \times 10^{-2}$ m	Obsolete Greek unit of length employed in ancient times. 1 daktylos (Attic) = 1/16 pous (E)	Attic
dalton (atomic unit of mass)	u, uma, Da	mass	M	1 u $= 1.660540210 \times 10^{-27}$ kg	The atomic unit of mass $u(^{12}C)$ is equal to the fraction 1/12 of the mass of the carbon 12 atom. The unit is named after J. Dalton (1766–1844).	
dan (Persian)	–	surface, area	L^2	1 dan (Persian) = 1 474.56 m^2	Obsolete Persian unit of area used in ancient times. 1 dan (Persian) = 100 gar (E)	Persian
dan (shih)	–	mass	M	1 dan = 30 kg	Obsolete Chinese unit of mass. 1 dan = 120 jin (E)	Chinese
daneq (Arabic)	–	surface, area	L^2	1 daneq (Arabic) = 983.333333 m^2	Obsolete Arabic unit of area used in ancient times. 1 daneq (Arabic) = 1/6 feddan (E)	Arabic
daraf	–	electric elastance	$ML^2T^{-4}I^{-2}$	1 daraf = 1 F^{-1} (E)	Obsolete unit employed to express the reciprocal electric capacitance. The name is derived by writing farad backwards.	US
darce	–	hydrodynamic permeability	L^2	1 darce $= 10^{-12}$ m^2 (E)	Obsolete permeability unit employed in hydrology and civil engineering.	
darcy	–	hydrodynamic permeability	L^2	1 darcy $= 9.869233 \times 10^{-13}$ m^2	Obsolete cgs unit of permeability. It was employed in hydrology and civil engineering to measure the permeability of porous matter. One darcy corresponds to the volume of liquid, having a dynamic viscosity of one centipoise, which flows	cgs

					through an area of one square centimetre of a porous medium in one second when it undergoes a pressure gradient of one atm per cm of length. The unit is named after H. Darcy (1803–1858).	
darwin	–	evolutionary rate of change	T^{-1}	(see note)	Unit of evolutionary rate of change. The unit is named after C. Darwin (1809–1882). It was proposed for evolutionary rate of change measurements of a biological species (animal or botanical). One darwin corresponds to an evolution (disappearance, increase) which changes with a factor $e \approx 2.78$ in one million years.	
day (anomalistic)	d (anom.)	time, period, duration	T	1 day (anomalistic) = 86 402.28063 s		
day (Bessel)	d (Bessel)	time, period, duration	T	1 day (Bessel) = 86 400.75452 s		
day (calendar)	d	time, period, duration	T	1 day (calendar) = 86 400 s (E)		
day (Gaussian)	d (Gauss.)	time, period, duration	T	1 day (Gaussian) = 86 402.10483 s		
day (sidereal)	d (sider.)	time, period, duration	T	1 day (sidereal) = 86 164 s		
day (solar mean)	d (solar mean)	time, period, duration	T	1 day (solar mean) = 86 400 s (E)		
day (tropical)	d (tropical)	time, period, duration	T	1 day (tropical) = 86 398.16556 s		

Unit (synonym, acronym)	Symbol	Physical quantity	Dimension	Conversion factor (SI equivalent unit)	Notes, definitions, other conversion factors	System
deben	–	mass	M	1 deben = 13.65×10^{-3} kg	Obsolete Egyptian unit of mass used in ancient times.	Egyptian
debye	D	electric dipole moment	$ML^2T^{-2}I^{-1}$	1 D = 0.358×10^{-30} C.m (E)	Obsolete unit of electric dipole moment employed in molecular physics and in physical chemistry. The unit is named after P.J.W. Debye (1884–1966). 1 D = 10^{-18} esu (cgs)	
decagram	dag	mass	M	1 dag = 10^{-2} kg (E)	Submultiple of the SI base unit. 1 dag = 10 g (E)	SI
decametre	dam	length, distance	L	1 dam = 10 m (E)	Multiple of the SI base unit.	SI
decastère	dast	capacity, volume	L^3	1 dast = 10 m^3 (E)	Multiple of MTS unit.	MTS
decempeda (Roman) [Roman perch]	–	length, distance	L	1 decempeda (Roman) = 2.944 m	Obsolete Roman unit of length employed in ancient times. 1 decempeda (Roman) = 10 pes (E)	Roman
decempeda quadrata (Roman)	–	surface, area	L^2	1 decempeda quadrata (Roman) = 8.773440 m^2	Obsolete Roman unit of area employed in ancient times. 1 decempeda quadrata (Roman) = 100 quadratus pes (E)	Roman
decibel	dB	logarithm (Briggsian) of a ratio of two sound powers	nil	1 dB = 0.1 B (E) (see note)	The decibel submultiple of the bel expresses the ratio of two sound powers or intensity levels P_1 and P_2 as a Briggs logarithm difference according to the equation $N(\text{db}) = 10 \times \log_{10} P_1/P_2$. For sound pressure the number of decibels is $N(\text{dB}) = 20 \times \log_{10} P_1/P_2$. The unit is named after A.G. Bell (1847–1922).	

					$1\ dB = 1/20\ ln(10) N_p(E) \approx 0.1151293 Np$	
decigram	dg	mass	M	$1\ dg = 10^{-4}\ kg$ (E)	Submultiple of the SI base unit $1\ dg = 10^{-1}\ g$ (E)	SI
decimetre	dm	length, distance	L	$1\ dm = 10^{-1}\ m$ (E)	Submultiple of the SI base unit.	SI
decistère	dst	capacity, volume	L^3	$1\ dst = 10^{-1}\ m^3$ (E)	Submultiple of MTS unit	MTS
decitonne (quintal)	dt, q	mass	M	1 dt = 100 kg (E)	Obsolete French metric unit of mass still used in a few countries. Submultiple of the MTS base unit.	MTS, French
degree	°	plane angle	α	$1° = 1.7453 \times 10^{-2}$ rad (E)	$1° = \pi/180$ rad (E) $1° = 400/360$ grade (E)	INT
degree hydrotimètrique français	°THF, °f	hardness of water, concentration of calcium and magnesium in water	nil	1 °THF = 10 mg of $CaCO_3$ per 1000 cm^3 of water	French unit employed to express hardness of water.	French
degree per second	$°.s^{-1}$	angular velocity, angular frequency, circular frequency	αT^{-1}	$1\ °.s^{-1} = 0.0174533\ rad.s^{-1}$		
demal	D	normality	NL^{-3}	$1\ demal = 10^3\ eq.m^{-3}$ (E)	Obsolete unit of normality employed in chemistry. $1\ demal = 1\ eq.l^{-1}$ (E)	
demiard	–	capacity, volume	L^3	$1\ demiard = 2.8413 \times 10^{-4}\ m^3$		
denaro (Italian)	–	mass	M	1 denaro (Italian) $= 1.0659722 \times 10^{-3}$ kg	Obsolete Italian unit of mass. 1 denaro (Italian) = 1/288 libbra (E)	Italian

Unit (synonym, acronym)	Symbol	Physical quantity	Dimension	Conversion factor (SI equivalent unit)	Notes, definitions, other conversion factors	System
denier	den, denier	linear mass density	ML^{-1}	1 denier $= 1.111111111 \times 10^{-7}$ kg.m^{-1}	Obsolete French unit of linear mass density employed in the textile industry. It served to express the thickness of thread, yarn and textile fibres. The denier is equal to the weight expressed in grams, of a thread of 9000 m length. 1 denier = 1/9 tex (E)	French
denier (de Paris) [French scruple]	–	mass	M	1 denier (de Paris) $= 1.274754818 \times 10^{-3}$ kg	Obsolete French unit of mass employed before the French Revolution (1789). 1 denier (de Paris) = 1/384 livre (de Paris) (E)	French
derah (Egyptian) [long cubit]	–	length, distance	L	1 derah (Egyptian) = 0.449580 m	Obsolete Egyptian unit of length used in ancient times.	Egyptian
derah (Egyptian) (Royal cubit)	–	length, distance	L	1 derah (Royal, Egyptian) = 0.524 m	Obsolete Egyptian unit of length used in ancient times.	Egyptian
dessatine	–	surface, area	L^2	1 dessatine = 10.925 m^2	Obsolete Russian unit of area 1 dessatine = 2400 square saashens (E)	Russian
deunx (Roman)	–	mass	M	1 deunx (Roman) $= 299.75 \times 10^{-3}$ kg	Obsolete Roman unit of mass used in ancient times. 1 deunx (Roman) = 11 unciae (Roman) (E)	Roman
dextans (Roman)	–	mass	M	1 dextans (Roman) $= 272.50 \times 10^{-3}$ kg	Obsolete Roman unit of mass used in ancient times. 1 dextans (Roman) = 10 unciae (Roman) (E)	Roman

dhanush [orgyla]		length, distance	L	1 dhanush = [illegible] m	Obsolete Indian unit of length used in ancient times. 1 dhanush = 4 hasta (E)	Indian
dharana (Indian)	–	mass	M	1 dharana (Indian) = 4.70×10^{-3} kg	Obsolete Indian unit of mass used in ancient times. 1 dharana (Indian) = 10 pala (E)	Indian
dichas (Egyptian)	–	length, distance	L	1 dichas (Egyptian) = 1.745×10^{-1} m	Obsolete Egyptian unit of length used in ancient times. 1 dichas = 1/3 Royal cubit (E)	Egyptian
dichas (Greek, Attic)	–	length, distance	L	1 dichas (Greek, Attic) = 1.542800×10^{-1} m	Obsolete Greek unit of length employed in ancient times. 1 dichas (Attic) = 1/2 pous (E)	Attic
diedo (Spanish)	–	length, distance	L	1 diedo (Spanish) = $1.741468750 \times 10^{-2}$ m	Obsolete Spanish unit of length. 1 diedo (Spanish) = 1/48 vara (E)	Spanish
digit (Egyptian)	–	length, distance	L	1 digit (Egyptian) = $9.357142857 \times 10^{-3}$ m	Obsolete Egyptian unit of length used in ancient times. 1 digit = 1/56 Royal cubit (E)	Egyptian
digitus (Roman) [Roman finger]	–	length, distance	L	1 digitus (Roman) = 1.84×10^{-2} m	Obsolete Roman unit of length employed in ancient times. 1 digitus (Roman) = 1/16 pes (E)	Roman
dihrem (Arabic)	–	mass	M	1 dihrem (Arabic) = 2.833333×10^{-3} kg	Obsolete Arabic unit of mass used in ancient times. System of the Prophet. 1 dihrem (Arabic) = 1/120 rotl (E)	Arabic
dinero (Spanish)	–	mass	M	1 dinero (Spanish) = $1.198158854 \times 10^{-3}$ kg	Obsolete Spanish unit of mass. 1 dinero (Spanish) = 1/384 libra (E)	Spanish

Unit (synonym, acronym)	Symbol	Physical quantity	Dimension	Conversion factor (SI equivalent unit)	Notes, definitions, other conversion factors	System
diobol (Greek, Attic)	–	mass	M	1 diobol (Greek, Attic) $= 1.44 \times 10^{-3}$ kg	Obsolete Greek unit of mass used in ancient times. 1 obol (Greek, Attic) = 1/3 drachma (Greek, Attic)	Attic
dioptre	δ	refractive power	L^{-1}	**SI derived unit** 1 dioptre = 1 m^{-1} (E)	The dioptre describes the refractive power of an optical system which has a focal length of 1 metre in a medium having a refractive index of one.	SI
djarib (Arabic)	–	surface, area	L^2	1 djarib (Arabic) = 1475 m^2	Obsolete Arabic unit of area used in ancient times. 1 djarib (Arabic) = 1/4 feddan (E)	Arabic
dobson (Dobson unit)	DU	concentration of ozone in the air	$ML^{-1}T^{-2}$	1 DU = 10^{-3} mmHg	Unit employed in meteorology which is equal to the amount of ozone contained in atmospheric air. One dobson is equal to the height of the column expressed in mmHg (0°C) which corresponds to the amount of ozone contained in air. The unit is named after G.M.B. Dobson (1889–1970).	
dodrans (Roman)	–	mass	M	1 dodrans (Roman) $= 245.25 \times 10^{-3}$ kg	Obsolete Roman unit of mass used in ancient times. 1 dodrans (Roman) = 9 unciae (Roman) (E)	Roman
doli [Russian grain]	–	mass	M	1 doli $= 4.443545782 \times 10^{-5}$ kg	Obsolete Russian unit of mass used before 1917 for general uses. 1 doli = 1/9216 funt (E)	Russian

doli (Russian)	–	mass	M	1 doli (Russian) $= 4.443494030 \times 10^{-5}$ kg	Obsolete Russian unit of mass used before 1917 in pharmacy.	Russian
dolium (Roman hogshead) [culleus]	–	capacity, volume	L^3	1 culleus (Roman hogshead) $= 527.28 \times 10^{-3}$ m^3	Obsolete Roman unit of volume employed in ancient times. It was used for capacity measurements of liquids. 1 culleus (Roman hogshead) = 960 sextarius (E)	Roman
dollar	dollar	reactivity of nuclear power reactor	nil	1 dollar $= 10^{-5}$ (E)	Obsolete American unit of nuclear reactivity employed in nuclear engineering.	US
donkey	–	power	ML^2T^{-3}	1 donkey = 250 W (E)	Obsolete British unit of power.	UK
doppelzentner (Prussian)	–	mass	M	1 doppelzentner = 102.896420 kg (E)	Obsolete German unit of mass. 1 doppelzentner = 220 pfund (E)	German
dots (points) per inch	dpi, ppi	graphic resolution	nil	1 dpi = 39.37007874 $point.m^{-1}$	Unit employed in computer science to describe graphic resolution.	
double remen (Egyptian)	–	length, distance	L	1 double remen (Egyptian) = 0.741 m	Obsolete Egyptian unit of length used in ancient times. It was equal to the length of the diagonal of a square having its sides equal to one royal cubit (0.524 m).	Egyptian
drachm (UK, fluid)	fl dr (UK)	capacity, volume	L^3	1 fl dr (UK) $= 3.551634375 \times 10^{-6}$ m^3	British unit used for capacity measurements for all merchandise (solids, liquids, foodstuffs, etc). 1 fl dr (UK) = 1/40 gill (UK) (E) 1 fl dr (UK) = 1/8 fl oz (UK) (E) 1 fl dr (UK) = 60 minims (UK) (E) 1 (UK) fl dr = 0.960760363 fl dr (US, liq.)	UK

Unit (synonym, acronym)	Symbol	Physical quantity	Dimension	Conversion factor (SI equivalent unit)	Notes, definitions, other conversion factors	System
drachm (US, liq.)	fl dr (US)	capacity, volume	L^3	1 fluid drachm (US) $= 3.696691198 \times 10^{-6}$ m^3	American unit used for capacity measurements of liquids. 1 fl dr (US) = 1/32 gills (US, liq.) (E) 1 fl dr (US) = 1/8 fl oz(US) (E) 1 fl dr (US) = 1.040840 fl dr (UK)	US
drachma (Greek, Attic)	–	mass	M	1 drachma (Greek, Attic) $= 4.32 \times 10^{-3}$ kg	Obsolete Greek unit of mass used in ancient times. 1 drachma (Greek, Attic) = 1/6000 Greek talent (E)	Attic
drachme (Austrian, apothecary)	–	mass	M	1 drachme (Austrian, apothecary) $= 4.375104166 \times 10^{-3}$ kg	Obsolete Austrian unit of mass used in pharmacy. 1 drachme (Austrian, apothecary) = 1/96 pfund (Austrian, apothecary) (E)	Austrian
drachme (Dutch)	–	mass	M	1 drachme (Dutch) $= 3.845060313 \times 10^{-3}$ kg	Obsolete Dutch unit of mass. 1 drachme (Dutch) = 1/128 pond (Dutch) (E)	Dutch
drachme (Russian)	–	mass	M	1 drachme (Russian) $= 3.643665105 \times 10^{-3}$ kg	Obsolete Russian unit of mass used before 1917 in pharmacy. 1 drachme (Russian) = 84 doli (E)	Russian
dram or drachm (apothecary)	dr (ap.), dr (apoth.), ʒ	mass	M	1 dr (apoth.) $= 3.887934580 \times 10^{-3}$ kg	Obsolete British and American unit of weight employed in pharmacy. It was used as a weight unit for drugs and medicine preparations (lotions, potions, ointments, plant extracts, etc). 1 dr (apoth.) = 1/96 lb (apoth.) (E) 1 dr (apoth.) = 1/8 oz (apoth.) (E) 1 dr (apoth.) = 3 scruples (apoth.) (E)	UK, US

					1 dr (apoth.) = 60 grains (apoth.) (E) 1 dr (apoth.) = 3.387934580 g	
dram or drachm (avoirdupois)	dr (avdp), dr (av.)	mass	M	1 dr (av.) $= 1.771845195 \times 10^{-3}$ kg (E)	Legal British and American unit of weight used since the WMA of 1963. 1 dr (av.) = 1/256 lb (av.) (E) 1 dr (av.) = 1/16 oz (av.) (E) 1 dr (av.) = 1.771845195 g	UK, US
dram or drachm (troy)	dr (troy)	mass	M	1 dr (troy) $= 3.887934580 \times 10^{-3}$ kg	Obsolete British unit of mass employed for the weighing of precious metals, precious stones and gems (diamond, ruby, sapphire) in the United Kingdom. Now obsolete in the UK but remains in common use in the USA. 1 dr (troy) = 1/96 lb (troy) (E) 1 dr (troy) = 1/8 oz (troy) (E) 1 dr (troy) = 3 scruples (troy) (E) 1 dr (troy) = 60 grains (troy) (E) 1 dr (troy) = 3.387934580 g	UK, US
dreiling (Austrian)	–	capacity, volume	L^3	1 dreiling (Austrian) = 1698.120 m^3	Obsolete Austrian unit of capacity used for liquid substances. 1 dreiling (Austrian) = 1200 mass (E)	Austrian
drex	drex	linear mass density	ML^{-1}	1 drex $= 10^{-7}$ $kg.m^{-1}$ (E)	Obsolete American unit of linear mass density employed in the textile industry. 1 drex = 1 g.(10 000 m)$^{-1}$ (E)	US, CAN
drona (Indian)	–	capacity, volume	L^3	1 drona (Indian) $= 13.2 \times 10^{-3}$ m^3 (of water)	Obsolete Indian unit of capacity used in ancient times. Measured by weight.	Indian

Unit (synonym, acronym)	Symbol	Physical quantity	Dimension	Conversion factor (SI equivalent unit)	Notes, definitions, other conversion factors	System
dry barrel (US)	bbl (US, dry)	capacity, volume	L^3	1 dry barrel (US) $= 115.6271 \times 10^{-3}$ m^3	Obsolete American unit of capacity used to measure the volume of powdered or divided solid materials (flour, sand, cement, ores, etc).	US
dry bushel (US) (Winchester bushel)	bu. (US, dry)	capacity, volume	L^3	1 dry bushel (US) $= 35.239072 \times 10^{-3}$ m^3	Obsolete American unit of capacity used to measure the volume of powdered or divided solid materials (flour, sand, cement, ores, etc). It holds 77.601 pounds distilled water at 62°F. 1 bushel (US, dry) = 4 pecks (US, dry) (E) 1 bushel (US, dry) = 8 gallons (US, dry) (E) 1 bushel (US, dry) = 32 quarts (US, dry) (E) 1 bushel (US, dry) = 64 pints (US, dry) (E) 1 bushel (US, dry) = 0.9689385959 bushel (UK)	US
dry peck (US)	pk (US, dry)	capacity, volume	L^3	1 dry peck (US) $= 8.809767500 \times 10^{-3}$ m^3	Obsolete American unit of capacity used to measure the volume of powdered or divided solid materials (flour, sand, cement, ores, etc). 1 peck (US, dry) = 16 pints (US, dry) (E) 1 peck (US, dry) = 8 quarts (US, dry) (E) 1 peck (US, dry) = 0.9689385959 peck (UK)	US

dry pint (US)	pt (US, dry)	capacity, volume	L^3	1 dry pint (US) = $0.550610469 \times 10^{-3}$ m^3	Obsolete American unit of capacity used to measure the volume of powdered or divided solid materials (flour, sand, cement, ores, etc). 1 dry pint (US) = 1/8 gallons (US, dry) (E)	US
dry quart (US)	qt (US, dry)	capacity, volume	L^3	1 dry quart (US) = $1.101220938 \times 10^{-3}$ m^3	Obsolete American unit of capacity used to measure the volume of powdered or divided solid materials (flour, sand, cement, ores, etc). 1 quart (US, dry) = 2 pints (US, dry) (E) 1 dry quart (US) = 1/4 gallons (US, dry) (E)	US
duime (Dutch)	–	length, distance	L	1 duime (Dutch) = 25.733 m	Obsolete unit of length used in Amsterdam (Netherlands).	Dutch
duïme (Russian inch)	–	length, distance	L	1 duïme (Russian) = 2.54×10^{-2} m	Obsolete Russian unit of length used before 1917. 1 duïme (Russian) = 1/12 foute (E)	Russian
duty (foot-pound-force)	ft-lbf	energy, work, heat	ML^2T^{-2}	1 ft-lbf = 1.355817948 J	Obsolete British unit of energy. It was the work needed to displace horizontally without friction a weight of one pound-force by one foot length.	UK, FPS
dycas degree	–	percentage of alcohol in wines and spirits	nil		Obsolete unit introduced in 1790.	
dyne	dyn	force, weight	MLT^{-2}	1 dyn = 10^{-5} N (E)	Obsolete cgs unit of force.	cgs
dyne per centimetre	dyn.cm^{-1}	surface tension, interfacial tension	MT^{-2}	1 dyn.cm^{-1} = 10^{-3} N.m^{-1} (E)	Obsolete cgs unit of interfacial and surface tension.	cgs

Unit (synonym, acronym)	Symbol	Physical quantity	Dimension	Conversion factor (SI equivalent unit)	Notes, definitions, other conversion factors	System
dyne per square centimetre (barye, microbar, barrie)	dyn. cm^{-2}	pressure, stress	$ML^{-1}T^{-2}$	1 $dyn.cm^{-2} = 10^{-1}$ Pa (E)	Obsolete cgs derived unit of pressure and stress. 1 $dyn.cm^{-2}$ = 1 barye (E) 1 atm = 1 013 250 $dyn.cm^{-2}$ (E)	cgs
dyne-centimetre	dyn.cm	moment of a force, torque	ML^2T^{-2}	1 dyn.cm = 10^{-7} J.m (E)	Obsolete cgs unit of torque.	cgs
E unit	–	rate of radiation exposure, intensity of X-rays	IM^{-1}	1 E unit = 2.54×10^{-4} $C.kg.s^{-1}$	Obsolete unit of rate of exposure for X-ray radiation employed in radiochemistry. It was introduced by W. Daune in 1914. 1 E unit = 1 $R.s^{-1}$ (E)	
eimer (Austrian)	–	capacity, volume	L^3	1 eimer (Austrian) = 56.604×10^{-3} m^3	Obsolete Austrian unit of capacity used for liquid substances. 1 eimer (Austrian) = 40 mass (E)	Austrian
eimer (Prussian)	–	capacity, volume	L^3	1 eimer (Prussian) = 68.701800×10^{-3} m^3	Obsolete German unit of capacity used for liquid substances. 1 eimer (Prussian) = 60 quart (E)	German
eimer (Swedish)	–	capacity, volume	L^3	1 eimer (Swedish) = 78.514860×10^{-3} m^3	Obsolete Swedish unit of capacity for liquid substances. 1 eimer (Swedish) = 30 khanna(E)	Swedish
einstein	–	molar energy of electromagnetic radiation	$ML^{-2}T^{-2}N^{-1}$	1 einstein = 3.990313×10^{-1} ν(Hz)	Obsolete unit of molar energy of an electromagnetic radiation used in photochemistry. It was equal to the energy of one mole of photons having a frequency ν expressed in hertz. The unit is named after A. Einstein (1879–1955).	

radius					$r_e = \alpha^2/a_0$	
electron rest mass (a.u. of mass)	m_0, m_e	mass	M	1 $m_0 = 9.1093897 \times 10^{-31}$ kg	Fundamental physical constant. Base unit of the a.u. system.	a.u.
electronvolt	eV	energy, work, heat	$ML^{-2}T^{-2}$	1 eV $= 1.60217733 \times 10^{-19}$ C (E)	Unit of energy employed in nuclear and atomic physics. It is equal to the work done to an electron in moving it through a potential difference of one volt. The name is derived from the English acronym: **e**quivalent **V**olt.	
ell	–	length, distance	L	1 ell = 1.143 m (E)	Obsolete American and British unit of length. 1 ell = 45 inches (E)	UK, US
elle (Dutch)	–	length, distance	L	1 elle (Dutch) = 0.687813 m	Obsolete unit of length used in Amsterdam (Netherlands).	Dutch
elle (Prussian)	–	length, distance	L	1 elle (Prussian) $= 6.881961249 \times 10^{-1}$ m	Obsolete German unit of length. 1 elle = 17/8 fuss (E)	German
eman	–	radioactivity concentration	$T^{-1}L^{-3}$	1 eman = 3700 $Bq.m^{-3}$	Obsolete unit of radioactivity concentration. It was defined in 1930 by The International Radium Standards Committee. It was employed in radiotherapy. The eman is a unit used in balneology to express the radioactivity concentration of hot springs containing the radon isotope $^{222}_{86}Rn$ (**eman**ation). 1 eman $= 10^{-10}$ $Ci.dm^{-3}$ 3.64 eman = 1 Mache unit	
emu of charge (abcoulomb)	aC	quantity of electricity, electric charge	IT	1 abcoulomb = 10 C (E)	Obsolete cgs emu of electric charge.	cgs

Unit (synonym, acronym)	Symbol	Physical quantity	Dimension	Conversion factor (SI equivalent unit)	Notes, definitions, other conversion factors	System
emu of conductance (abmho)	aS, $(a\Omega)^{-1}$	electrical conductance	$M^{-1}L^{-2}T^3I^2$	1 abmho = 10^9 S (E)	Obsolete cgs emu of electric conductance.	cgs
emu of current (abampere)	aA, Bi	electric current intensity	I	1 abampere = 10 A (E)	Obsolete cgs emu of electric current.	cgs
emu of electric capacitance (abfarad)	aF	electric capacitance	$M^{-1}L^{-2}T^4I^2$	1 abfarad = 10^9 F (E)	Obsolete cgs emu of electric capacitance.	cgs
emu of electric potential (abvolt)	aV	electric potential, electric potential difference, electromotive force	$ML^2T^{-3}I^{-1}$	1 abvolt = 10^{-8} V (E)	Obsolete cgs emu of electric potential.	cgs
emu of inductance (abhenry)	aH	electric inductance	$ML^2T^{-2}I^{-2}$	1 abhenry = 10^{-9} H (E)	Obsolete cgs emu of electric inductance.	cgs
emu of power (abwatt)	aW	power	ML^2T^{-3}	1 abwatt = 10^{-8} W (E)	Obsolete cgs emu of power.	cgs
emu of resistance (abohm)	aΩ	electric resistance	$ML^2T^{-3}I^{-2}$	1 abohm = $10^{-9}\Omega$ (E)	Obsolete cgs emu of electric resistance.	cgs
encablure (French cable length)	–	length, distance	L	1 encablure = 220 m (E)	Obsolete French unit of length employed in navigation to describe short distances. 1 encablure = 120 brasses (E).	French
engel (Dutch)	–	mass	M	1 engel (Dutch) = $1.538024125 \times 10^{-3}$ kg	Obsolete Dutch unit of mass. 1 engel (Dutch) = 1/320 pond (Dutch) (E)	Dutch
engler (degree, second)	–	index of kinematic viscosity	nil	(see note)	Obsolete British empirical index of kinematic viscosity. The engler degree corresponds to the flow time expressed in seconds, of 200 cm^3 of liquid compared	UK

					with the flow time of 200 cm^3 of pure water measured at 20°C in an engler viscosity meter.	
entropy unit (unit of entropy)	ue, eu	molar entropy	$ML^2T^{-2}\Theta^{-1}N^{-1}$	1 ue = 1 $cal_{th}.K^{-1}.mol^{-1}$ (E)	Obsolete unit of entropy.	
eon	eon	time, period, duration	T	1 eon = 10^9 years (E)	Obsolete unit of time employed in Earth science. 1 eon = 1 Ga (E)	
eotvös	Eo, E	acceleration gradient	T^{-2}	1 eotvös = 10^{-9} s^{-2} (E)	Obsolete unit of acceleration gradient employed in geophysics (gravimetry) and mine prospecting. The unit is named after the Hungarian physicist R. Eotvös (1848–1919). 1 eotvös = 10^{-9} $galcm^{-1}$ (E)	
ephah (Hebrew, new)	–	capacity, volume	L^3	1 ephah (Hebrew, new) = 21.420 × 10^{-3} m^3 (of water)	Obsolete Hebrew unit of capacity used in ancient times. Measured by weight.	Hebrew
ephah (Hebrew, obsolete)	–	capacity, volume	L^3	1 ephah (Hebrew, obsolete) = 29.376 × 10^{-3} m^3 (of water)	Obsolete Hebrew unit of capacity used in ancient times. Measured by weight.	Hebrew
erg	erg	energy, work, heat	ML^2T^{-2}	1 erg = 10^{-7} J (E)	Obsolete cgs unit of energy.	cgs
erg per second	$erg.s^{-1}$	power	ML^2T^{-3}	1 $erg.s^{-1}$ = 10^{-7} W (E)	Obsolete cgs unit of power.	cgs
erg per square centimetre	$erg.cm^{-2}$	surface tension, interfacial tension	MT^{-2}	1 $erg.cm^{-2}$ = 10^{-3} $J.m^{-2}$ (E)	Obsolete cgs unit of interfacial and surface tension.	cgs
erlang (traffic unit)	–	telephone traffic unit	nil	(see note)	Obsolete unit employed in telegraphy. One erlang was equal to the product $C \times T$ where C is the number of calls per hour and T is the mean duration of call. The unit is named after A.K. Erlang (1878–1929).	

Unit (synonym, acronym)	Symbol	Physical quantity	Dimension	Conversion factor (SI equivalent unit)	Notes, definitions, other conversion factors	System
escrupolo (Portuguese)	–	mass	M	1 escrupolo (Portuguese) $= 1.195312500 \times 10^{-3}$ kg	Obsolete Portuguese unit of mass. 1 escrupolo (Portuguese) = 1/384 libra (E)	Portuguese
escrupulo (Spanish)	–	mass	M	1 escrupulo (Spanish) $= 21.566859370 \times 10^{-3}$ kg	Obsolete Spanish unit of mass. 1 escrupulo (Spanish) = 3/64 libra (E)	Spanish
estadal (Spanish)	–	length, distance	L	1 estadal (Spanish) = 3.343620 m	Obsolete Spanish unit of length. 1 estadal (Spanish) = 4 vara (E)	Spanish
estadio (Portuguese)	–	length, distance	L	1 estadio (Portuguese) = 258 m	Obsolete Portuguese unit of length. 1 estadio (Portuguese) = 785 pe (E)	Portuguese
estado (Spanish)	–	length, distance	L	1 estado (Spanish) = 1.671810 m	Obsolete Spanish unit of length. 1 estado (Spanish) = 2 vara (E)	Spanish
esu of charge (statcoulomb, franklin)	statC	quantity of electricity, electric charge	IT	1 statcoulomb $= 3.355641 \times 10^{-10}$ C	Obsolete cgs esu of electric charge. 1 statcoulomb = 1 Fr (E)	cgs
esu of conductance (statmho)	statmho	electrical conductance	$M^{-1}L^{-2}T^{3}I^{2}$	1 statmho $= 1.112650 \times 10^{-12}$ S	Obsolete cgs esu of electric conductance.	cgs
esu of current (statampere)	statA	electric current intensity	I	1 statampere $= 3.355641 \times 10^{-10}$ A	Obsolete cgs esu of electric current.	cgs
esu of electric capacitance (statfarad)	statF	electric capacitance	$M^{-1}L^{-2}T^{4}I^{4}$	1 statfarad $= 1.112650 \times 10^{-12}$ F	Obsolete cgs esu of electric capacitance.	cgs
esu of electric potential (statvolt)	statV	electric potential, electric potential difference, electromotive force	$ML^{2}T^{-3}I^{-1}$	1 statvolt $= 2.99792458 \times 10^{2}$ V	Obsolete cgs esu of electric potential.	cgs

esu of inductance (stathenry)	statH	electric inductance	$ML^2T^{-2}I^{-2}$	1 stathenry = 8.987552 × 10^{11} H	Obsolete cgs esu of electric inductance.	cgs
esu of resistance (statohm)	statohm	electric resistance	$ML^2T^{-3}I^{-2}$	1 statohm = 8.987554 × 10^{11} Ω	Obsolete cgs esu of resistance.	cgs
exagram	Eg	mass	M	1 Eg = 10^{15} kg (E)	Multiple of the SI base unit 1 Eg = 10^{18} g (E)	SI
exametre	Em	length, distance	L	1 Em = 10^{18} m (E)	Multiple of the SI base unit	SI
Fahrenheit degree	°F	temperature	Θ	$T(\text{K}) = 5/9[T(°\text{F}) - 459.67]$	Usual temperature scale in English-speaking countries. The scale is no longer in scientific use. It was invented in 1714 by the German scientist G.D. Fahrenheit (1686–1736), who set the zero at the lowest temperature he could obtain in the laboratory (by mixing ice and common salt) and took his own body temperature as 96°F. 32°F = 0°C (melting ice) 212°F = 100°C (boiling water) $T(°\text{C}) = 5/9[T(°\text{F}) - 32]$ $T(°\text{F}) = 9/5T(°\text{C}) + 32$	UK, US
famm (Swedish)	–	length, distance	L	1 famm (Swedish) = 1.78140 m	Obsolete Swedish unit of length.	Swedish
fanega (Spanish, dry)	–	capacity, volume	L^3	1 fanega (Spanish, dry) = 55.501 × 10^{-3} m^3	Obsolete Spanish unit of capacity used for dry substances.	Spanish
fanegada (Spanish)	–	surface, area	L^2	1 fanegada (Spanish) = 6 439.561749 m^2	Obsolete Spanish unit of area. 1 fanegada (Spanish) = 9 216 sq. vara (E)	Spanish
fang (Chinese)	–	capacity, volume	L^3	1 fang (Chinese) = 3.276800 m^3	Obsolete Chinese unit of capacity used in ancient times. 1 fang (Chinese) = 100 cubic tchi (E)	Chinese

Unit (synonym, acronym)	Symbol	Physical quantity	Dimension	Conversion factor (SI equivalent unit)	Notes, definitions, other conversion factors	System
fanga (Portuguese, dry)	–	capacity, volume	L^3	1 fanga (Portuguese, dry) $= 54 \times 10^{-3}$ m^3	Obsolete Portuguese unit of capacity used for dry substances.	Portuguese
farad	F	electric capacitance	$M^{-1}L^{-2}T^4I^2$	**SI derived unit** 1 F $= 1 C.V^{-1} = 1\ kg^{-1}m^{-2}s^4\ A^2$ (E)	The farad is the capacitance of an electric capacitor between the two plates of which there appears a difference of electric potential of one volt when it is charged by a quantity of electricity equal to one coulomb. The unit is named after M. Faraday (1791–1867).	SI
farad (int.)	F	electric capacitance	$M^{-1}L^{-2}T^4I^2$	1 F (int.) = 0.99951 F	Obsolete international unit of capacitance.	IEUS
faraday (based on ^{12}C)	F	molar electric charge	ITN^{-1}	1 F = 96 485.309 $C.mol^{-1}$	$F = N_a \times e$ 1 F ≈ 27.6 Ah	
faraday (chemical)	F	molar electric charge	ITN^{-1}	1 F = 96 495.7 $C.mol^{-1}$	$F = N_a \times e$ 1 F ≈ 27.6 Ah	
faraday (physical)	F	molar electric charge	ITN^{-1}	1 F = 96 512.9 $C.mol^{-1}$	$F = N_a \times e$ 1 F ≈ 27.6 Ah	
fass (Austrian)	–	capacity, volume	L^3	1 fass (Austrian) $= 566.040 \times 10^{-3}$ m^3	Obsolete Austrian unit of capacity used for liquid substances. 1 fass (Austrian) = 400 mass (E)	Austrian
fathom	fath	length, distance	L	1 fathom = 1.8288 m (E)	Obsolete British and American unit of length employed in marine applications. It describes the depth of water. 1 fathom = 72 inches (E) 1 fathom = 6 feet (E) 1 fathom = 2 yards (E)	UK, US

feddan (Arabic)	–	surface, area	L^2	1 feddan (Arabic) = 5900 m^2	Obsolete Arabic unit of area used in ancient times.	Arabic
femtogram	fg	mass	M	1 fg = 10^{-18} kg (E)	Submultiple of the SI base unit. 1 fg = 10^{-15} g (E)	SI
femtometre	fm	length, distance	L	1 fm = 10^{-15} m (E)	Submultiple of the SI base unit.	SI
fen (Chinese) [gros]	–	length, distance	L	1 fen (Chinese) = 27.6 m	Obsolete Chinese unit of length used in ancient times. 1 fen = 120 tchi (E)	Chinese
fen (Chinese)	–	length, distance	L	1 fen (Chinese) = 2.3×10^{-3} m	Obsolete Chinese unit of length used in ancient times. 1 fen = 1/100 tchi (E)	Chinese
fen (Chinese)	–	surface, area	L^2	1 fen (Chinese) = 61.440 m^2	Obsolete Chinese unit of area used in ancient times. 1 fen (Chinese) = 1/10 meou (E)	Chinese
ferk (Arabic)	–	capacity, volume	L^3	1 ferk (Arabic) = 8.16×10^{-3} m^3 (of water)	Obsolete Arabic unit of capacity used in ancient times. Measured by weight. 1 ferk (Arabic) = 1/4 cafiz (E)	Arabic
fermi	F, f	length, distance	L	1 F = 10^{-15} m (E)	Obsolete unit of length used in nuclear physics to describe nuclide dimensions. The unit is named after the Italo-American scientist E. Fermi (1901–1954). It has the same order of magnitude as nuclides' radii. 1 F = 1 fm (E)	
ferrado (Portuguese)	–	surface, area	L^2	1 ferrado (Portuguese) = 725.410125 m^2	Obsolete Portuguese unit of area. 1 ferrado (Portuguese) = 4840 square vara (E)	Portuguese

Unit (synonym, acronym)	Symbol	Physical quantity	Dimension	Conversion factor (SI equivalent unit)	Notes, definitions, other conversion factors	System
feuillette (de Paris)	–	capacity, volume	L^3	1 feuillette (de Paris) $= 134.120016 \times 10^{-3}$ m^3	Obsolete French unit of capacity employed before the French Revolution. It served to express the capacity of liquids and grains. It varies according to the location and merchandise. 1 feuillette (de Paris) = 144 pintes (de Paris)	French
finger (Greek, Attic)	–	length, distance	L	1 finger (Greek, Attic) $= 3.857 \times 10^{-2}$ m	Obsolete Greek unit of length employed in ancient times. 1 finger (Attic) = 1/16 pous (E)	Attic
finger (Hebrew)	–	length, distance	L	1 finger (Hebrew) $= 2.3125 \times 10^{-2}$ m	Obsolete Hebrew unit of length used in ancient times. 1 finger = 1/24 cubit (E)	Hebrew
finger (Persian)	–	length, distance	L	1 finger (Persian) $= 2 \times 10^{-2}$ m	Obsolete unit of length in the Assyrio-Chaldean-Persian system used in ancient times. 1 finger = 1/16 zereth (E)	Persian
Finsen unit	FU	irradiance, radiant flux received, energy flux	MT^{-3}	1 FU $= 10^{-5}$ W.m^{-2} (E)	Obsolete unit of irradiance employed for ultraviolet radiation measurements. UV radiation of $\lambda = 296.7$ nm has an irradiance of one Finsen unit when the energy density is equal to 10 μW.m^{-2}. The unit is named after N.R. Finsen (1860–1904).	
firkin (UK)	fir (UK)	capacity, volume	L^3	1 firkin (UK) $= 40.914828 \times 10^{-3}$ m^3	British unit used for capacity measurements for all merchandise (solids, liquids, foodstuffs, etc). 1 firkin (UK) = 9 gallons (UK) (E)	UK

	liq.)			$= 34.06870606 \times 10^{-3}$ m^3	measurements of liquids. 1 firkin (US) = 9 gallons (US, liq.) (E)	
first Bohr radius (Bohr, a.u. of length)	a_0, b, a.u.	length, distance	L	$a_0 = 5.29177249 \times 10^{-11}$ m	Physical fundamental constant. Base unit of a.u. system. The unit is named after N. Bohr (1885-1962). $a_0 = 4\pi\varepsilon_0\hbar^2.m\text{-}1_0e^2$	a.u.
Fischer degree	–	specific gravity of liquids, hydrometer index, hydrometer degree	nil	°Fischer = 400 – (400/d) d = d (12°Ré) = d (15.625°C)		
fjerdingar (Swedish, dry)	–	capacity, volume	L^3	1 fjerdingar (Swedish, dry) $= 18.319 \times 10^{-3}$ m^3	Obsolete Swedish unit of capacity for dry substances. 1 fjerdingar (Swedish, dry) = 7 kanna (E)	Swedish
FLOP	FLOP	number of computations made per unit of time	T^{-1}	1 FLOP = 1 operation.s^{-1} (E)	Unit used in computer science. The unit is named after the English acronym: **Fl**oating point **Op**erations per second.	
fluid dram (UK)	fl dr (UK)	capacity, volume	L^3	1 fluid dram (UK) $= 3.551634375 \times 10^{-6}$ m^3	British unit used for capacity measurements for all merchandise (solids, liquids, foodstuffs, etc). 1 fl dr (UK) = 1/1280 gal (UK) (E) 1 fl dr (UK) = 1/40 gill (UK) (E) 1 fl dr (UK) = 1/8 fl oz (UK) (E) 1 fl dr (UK) = 60 minims (UK) (E) 1 (UK) fl dr = 0.960759 fl dr (US)	UK

Unit (synonym, acronym)	Symbol	Physical quantity	Dimension	Conversion factor (SI equivalent unit)	Notes, definitions, other conversion factors	System
fluid dram (US)	fl dr (US)	capacity, volume	L^3	1 fluid drachm (US) $= 3.696691198 \times 10^{-6}$ m^3	American unit used for capacity measurements of liquids. 1 fl dr (US) = 1/1024 gal (US, liq.) (E) 1 fl dr (US) = 1/32 gill (US, liq.) (E) 1 fl dr (US) = 1/8 fl oz (US) (E) 1 fl dr (US) = 1.040840 fl dr (UK)	US
fluid ounce (UK)	fl oz (UK)	capacity, volume	L^3	1 fluid ounce (UK) $= 28.41307500 \times 10^{-6}$ m^3	British unit used for capacity measurements for all merchandise (solids, liquids, foodstuffs, etc). 1 fl oz (UK) = 1/160 gal (UK) (E) 1 fl oz (UK) = 1/40 quart (UK) (E) 1 fl oz (UK) = 1/20 pint (UK) (E) 1 fl oz (UK) = 1/4 gill (UK) (E) 1 fl oz (UK) = 8 fl dr (UK) (E) 1 fl oz (UK) = 480 minims (UK) (E) 1 fl oz (UK) = 0.960759 fl oz (US)	UK
fluid ounce (US)	fl oz (US)	capacity, volume	L^3	1 fluid ounce (US) $= 29.57352958 \times 10^{-6}$ m^3	American unit used for capacity measurements of liquids. 1 fl oz (US) = 1/128 gal (US, liq.) (E) 1 fl oz (US) = 1/32 quart (US, liq.) (E) 1 fl oz (US) = 1/16 pint (US, liq.) (E) 1 fl oz (US) = 1/4 gill (US, liq.) (E) 1 fl oz (US) = 8 fl dr (US) (E) 1 fl oz (US) = 480 minims (US, liq.) (E) 1 fl oz (US) = 1.040840 fl oz (UK)	US
foot	ft (′)	length, distance	L	1 ft = 0.3048 m (E)	Base unit of FPS system. Legal unit in the UK system (since the WMA, 1963) and US system (since the USMB, 1959).	UK, US, FPS

					1 foot = 144 lines (UK) (E) 1 foot = 480 lines (US) (E) 1 foot = 12 inches (E) 1 foot = 1/3 yards (E)	
foot (Arabic)	–	length, distance	L	1 foot (Arabic) = 0.320 m	Obsolete Arabic unit of length used in ancient times.	Arabic
foot (old)	ft (′)	length, distance	L	1 ft (old) = 0.3047997 m	Obsolete British unit of length. It was used in the UK before the WMA of 1963. 1 ft (old) = 12 inches (old)	UK
foot (Prussian)	–	length, distance	L	1 fuss (Prussian) = 0.313857 m	Obsolete German unit of length.	German
foot (US Survey)	ft (US Survey)	length, distance	L	1 ft (US Survey) = 0.30480060960 m (E)	Obsolete American unit of length used in the USA for geodetic measurements. Discontinued since the USMB, 1959. 1 ft (US Survey) = (1200/3937) m (E)	US
foot board measure (board foot measure)	fbm, B.M.	capacity, volume	L^3	1 fbm = $2.359737216 \times 10^{-3}$ m^3	Measure of capacity which is equal to the product of 1 foot of length per 1 foot width per 1 inch thickness. 1 bfm = 1/12 ft^3 (E) 1 bfm = 144 in^3 (E)	UK, US
foot of water (39.2°F) [conventional]	ft H_2O (39.2°F)	pressure, stress	$ML^{-1}T^{-2}$	1 ft H_2O (39.2°F) = 2.988983226×10^3 Pa	British and American unit of pressure employed in physics to measure small pressures. It is equal to the pressure exerted by a column of water one foot high measured at 4°C (39.2°F). 1 atm ≈ 33.9 ft H_2O (4°C)	UK, US
foot of water (15.56°C)	ft H_2O (15.5°C)	pressure, stress	$ML^{-1}T^{-2}$	1 ft H_2O (15.56°C) = 2.986143612×10^3 Pa	British and American unit of pressure employed in physics to measure small pressures. It is equal to the pressure *(continued overleaf)*	UK, US

Unit (synonym, acronym)	Symbol	Physical quantity	Dimension	Conversion factor (SI equivalent unit)	Notes, definitions, other conversion factors	System
foot of water (15.56°C) *(continued)*					exerted by a column of water one foot high measured at 15.56°C (60°F). 1 atm ≈ 33.93 ft H_2O (60°F).	
foot of water (4°C) [conventional]	ft H_2O (4°C)	pressure, stress	$ML^{-1}T^{-2}$	1 ft H_2O (4°C) = 2.988983226×10^3 Pa	British and American unit of pressure employed in physics to measure small pressures. It is equal to the pressure exerted by a column of water one foot high measured at 4°C (39.2°F). 1 atm ≈ 33.90 ft H_2O (4°C).	UK, US
foot of water (60°F)	ft H_2O (60°F)	pressure, stress	$ML^{-1}T^{-2}$	1 ft H_2O (60°F) = 2.986143612×10^3 Pa	British and American unit of pressure employed in physics to measure small pressures. It is equal to the pressure exerted by a column of water one foot high measured at 15.56°C (60°F). 1 atm ≈ 33.93 ft H_2O (60°F).	UK, US
foot per hour	$ft.h^{-1}$, fph	velocity, speed	LT^{-1}	1 fph = 8.466667×10^{-5} $m.s^{-1}$	British and American unit of speed. 1 fph = 3.048×10^{-4} $km.h^{-1}$ (E)	UK, US
foot per minute	$ft.min^{-1}$, fpm	velocity, speed	LT^{-1}	1 fpm = 5.08×10^{-3} $m.s^{-1}$ (E)	British and American unit of speed. 1 fpm = 1.82880×10^{-2} $km.h^{-1}$ (E)	UK, US
foot per second	$ft.s^{-1}$, fps	velocity, speed	LT^{-1}	1 $ft.s^{-1}$ = 3.048×10^{-1} $m.s^{-1}$ (E)	British and American unit of speed. 1 fps = 1.097280 $km.h^{-1}$ (E)	UK, US, FPS
foot per square second	$ft.s^{-2}$	acceleration	LT^{-2}	1 $ft.s^{-2}$ = 0.3048 $m.s^{-2}$ (E)	Unit of acceleration in the FPS system.	FPS
foot-candle	ft.C, ft.cd, fc	illuminance	$JL^{-2}\Omega$	1 ft.C = 10.76391 lx	Obsolete British unit of illuminance. The foot-candle is equal to one lumen	UK

					incident per square foot. 1 ft.C = 1.076391 milli-phot	
foot-lambert	ft.L	luminous luminance	JL^{-2}	1ft.L = 3.426259 $cd.m^{-2}$	Obsolete American unit of luminous luminance. 1ft.L = $[1/\pi]$ $cd.ft^{-2}$ (E) 1ft.L = $[1/144\pi]$ $cd.in^{-2}$ (E) 1ft.L = 1.0764 mL (E)	US
foot-pound-force (duty)	ft lbf	energy, work, heat	ML^2T^{-2}	1 ft-lbf = 1.355817948 J	Obsolete British unit of energy. It was the work needed to displace horizontally without friction a weight of one pound-force on one foot length.	UK, FPS
foot-pound-force per hour	ft-lbf.h^{-1}	power	ML^2T^{-3}	1 ft-lbf.h^{-1} = 3.766160967 $\times 10^{-4}$ W	British and American unit of power.	UK, US, FPS
foot-pound-force per minute	ft-lbf.min^{-1}	power	ML^2T^{-3}	1 ft-lbf.min^{-1} = 2.259696580 $\times 10^{-2}$ W	British and American unit of power.	UK, US, FPS
foot-pound-force per second	ft-lbf.s^{-1}	power	ML^2T^{-3}	1 ft-lbf.s^{-1} = 1.355817948 W	British and American unit of power.	UK, US, FPS
foot-poundal	ft-pdl	energy, work, heat	ML^2T^{-2}	1 ft-pdl = 4.2140111 $\times 10^{-2}$ J	Obsolete British unit of energy. It was the work needed to displace horizontally without friction a weight of one poundal a distance of one foot.	UK, FPS
foot-poundal per second	ft-pdl.s^{-1}	power	ML^2T^{-3}	1 ft-pdl.s^{-1} = 4.214011007 $\times 10^{-2}$ W	Obsolete British unit of power.	UK, FPS
forty foot equivalent unit	FEU, FEQ	capacity, volume	L^3	1 FEU = 72.491127280 m^3 (E)	Unit of capacity employed in marine applications. 1 FEU = 40 $\times$ 8 $\times$ 8 ft^3	UK, US
fot (Swedish) [Swedish foot]	–	length, distance	L	1 fot (Swedish) = 0.296900 m	Obsolete Swedish unit of length.	Swedish

Unit (synonym, acronym)	Symbol	Physical quantity	Dimension	Conversion factor (SI equivalent unit)	Notes, definitions, other conversion factors	System
fourier	$W.cm^{-1}.°C^{-1}$	thermal conductivity	$MLT^{-3}\Theta^{-1}$	1 $W.cm^{-1}.K^{-1} = 100\ W.m^{-1}.K^{-1}$	Obsolete cgs unit of thermal conductivity. The name was proposed but not adopted. The unit is named after J. Fourier (1768–1830).	cgs
foute (Russian foot)	–	length, distance	L	1 foute (Russian) = 0.3048 m (E)	Obsolete Russian unit of length used before 1917.	Russian
francoeur degree	–	percentage of alcohol in wines and spirits	nil		Obsolete unit introduced in 1842.	
franklin	Fr	quantity of electricity, electric charge	IT	1 Fr = $3.335640952 \times 10^{-10}$ C	Obsolete cgs unit of electric charge in the esu subsystem. The franklin is that charge which exerts on an equal charge at a distance of one centimetre in vacuo a force of one dyne (1941). The unit is named after B. Franklin (1706–1790). 1 franklin = $(1/10 \times c)$ C (E)	cgs
fraunhofer	–	wavelength resolution for spectral lines	nil	1 fraunhofer = $10^6 \times (\delta\lambda/\lambda)$	Obsolete unit of used to express the wavelength resolution for spectral lines in atomic and molecular spectroscopy. The unit is named after J. von Fraunhofer (1787–1826).	
fresnel	f	frequency	T^{-1}	1 fresnel = 10^{12} Hz (E)	Obsolete unit of frequency employed in spectroscopy. The unit is named after A.J. Fresnel (1788–1827).	
frigorie	fg	energy, work, heat	ML^2T^{-2}	1 fg = −4185.5 J	Obsolete heat unit employed in refrigeration and cryogenics.	

					liquid substances. 1 fuder (Austrian) = 1280 mass (E)	
fuder (Prussian)	–	capacity, volume	L^3	1 fuder (Prussian) $= 824.421600 \times 10^{-3}$ m^3	Obsolete German unit of capacity used for liquid substances. 1 fuder (Prussian) = 720 quart (E)	German
fuder (Swedish)	–	capacity, volume	L^3	1 fuder (Swedish) $= 942.178320 \times 10^{-3}$ m^3	Obsolete Swedish unit of capacity for liquid substances. 1 fuder (Swedish) = 360 kanna (E)	Swedish
funal (sthène)	sth	force, weight	MLT^{-2}	1 sth = 10^3 N (E)	Obsolete MTS unit of force.	MTS
funt [Russian pound]	–	mass	M	1 funt = 0.4095171792 kg	Obsolete Russian unit of mass used before 1917 for general use. 1 kg = 2.4419 funts (E)	Russian
furlong (UK)	fur	length, distance	L	1 furlong (UK) $= 2.01168 \times 10^2$ m (E)	Obsolete British unit of length. 1 furlong (UK) = 1 000 links (UK) (E) 1 furlong (UK) = 660 feet (E) 1 furlong (UK) = 220 yards (E) 1 furlong (UK) = 10 chains (UK) (E)	UK
furlong (US)	fur	length, distance	L	1 furlong (US) $= 2.01168 \times 10^2$ m (E)	Obsolete American unit of length. 1 furlong (US) = 1000 links (US) (E) 1 furlong (UK) = 660 feet (E) 1 furlong (UK) = 220 yards (E) 1 furlong (US) = 6.6 chains (US) (E)	US
fuss (Austrian) [Austrian foot]	–	length, distance	L	1 fuss (Austrian) = 0.316080 m	Obsolete Austrian unit of length.	Austrian
fuss (German) [Prussian foot]	–	length, distance	L	1 fuss (Prussian) = 0.313857 m	Obsolete German unit of length.	German

Unit (synonym, acronym)	Symbol	Physical quantity	Dimension	Conversion factor (SI equivalent unit)	Notes, definitions, other conversion factors	System
futtermassel (Austrian, dry)	–	capacity, volume	L^3	1 futtermassel (Austrian, dry) $= 1.921531250 \times 10^{-3}$ m^3	Obsolete Austrian unit of capacity used for dry substances. 1 futtermassel (Austrian, dry) = 1/32 metzel (E)	Austrian
gaillarde	–	length, distance	L	1 gaillarde $= 3.007772220 \times 10^{-3}$ m	Obsolete French unit of length employed in typography. It was a submultiple of the point (Didot). 1 gaillarde = 8 points Didot (E)	French
gal (Galileo)	Gal, gal	acceleration	LT^{-2}	1 gal $= 10^{-2}$ m.s^{-2} (E)	Obsolete unit of acceleration in the cgs system. The unit is named after the famous Italian scientist G. Galileo (1564–1642). The gal is employed extensively in geophysics and geodesy to express acceleration due to gravity.	cgs
gallon (Canadian, liq.)	gal (Canadian)	capacity, volume	L^3	1 gallon (Canadian) $= 4.546092 \times 10^{-3}$ m^3 (E)	Obsolete Canadian unit of capacity. It was equal to the imperial gallon. 1 gallon (Canadian) = 1 gallon (UK) (E)	CAN
gallon (UK) (imperial gallon)	gal (UK), imp. gal	capacity, volume	L^3	1 gallon (UK) $= 4.546092 \times 10^{-3}$ m^3	British unit used for capacity measurements for all merchandise (solids, liquids, foodstuffs, etc). Before 1976, the imperial gallon was equal to the volume occupied by 580 328.886 grains of water at 62.02°F (16.7°C). Since the *Weights and Measures Act* of 1976, the imperial gallon is equal to the volume of 10 lb of distilled water with density $\rho = 0.998859$ g.cm^{-3} measured in air of density 1.217 g.dm^{-3}.	UK

					1 gallon (UK) = 4 quarts (UK) (E) 1 gallon (UK) = 8 pints (UK) (E) 1 gallon (UK) = 32 gills (UK) (E) 1 gallon (UK) = 160 fl oz (UK) (E) 1 gallon (UK) = 1.200950454 gallon (US, liq.) 1 gallon (UK) = 1.032057144 gallons (US, dry)	
gallon (UK) per foot	gal (UK).ft^{-1}	surface, area	L^2	1 gal (UK).ft^{-1} $= 1.461500000 \times 10^{-2}$ m^2 (E)	Obsolete British unit of area used for geometric shape measurements.	UK
gallon (UK) per hour	gph (UK)	volume flow rate	L^3T^{-1}	1 gal (UK).h^{-1} $= 1.262803333 \times 10^{-6}$ $m^3.s^{-1}$	Obsolete British unit of volume flow rate employed in chemical engineering.	UK
gallon (UK) per mile (statute)	gal (UK).mi^{-1}(stat.)	surface, area	L^2	1 gal (UK).mi^{-1} (stat.) $= 2.824810606 \times 10^{-6}$ m^2	Obsolete British unit of area used for geometric shape measurements.	UK
gallon (UK) per minute	gpm (UK)	volume flow rate	L^3T^{-1}	1 gal (UK).min^{-1} $= 7.576820000 \times 10^{-5}$ $m^3.s^{-1}$	Obsolete British unit of volume flow rate employed in chemical engineering.	UK
gallon (UK) per pound	$in^3.lb^{-1}$	specific volume	L^3M^{-1}	1 (UK) gal.lb^{-1} $= 1.002241726 \times 10^{-2}$ $m^3.kg^{-1}$	Obsolete British unit of specific volume.	UK
gallon (UK) per second	gps (UK)	volume flow rate	L^3T^{-1}	1 gal (UK).s^{-1} $= 4.546092 \times 10^{-3}$ $m^3.s^{-1}$	Obsolete British unit of volume flow rate employed in chemical engineering.	UK
gallon (US, dry)	gal (US, dry)	capacity, volume	L^3	1 gallon (US, dry) $= 4.404884 \times 10^{-3}$ m^3	American unit of capacity used for dry powdered and divided solids (flour, sand, ores, etc) capacity measurements. 1 gallon (US, dry) = 4 quarts (US, dry) (E) 1 gallon (US, dry) = 8 pints (US, dry) (E) 1 gallon (US, dry) = 0.9689385959 gallon (UK) 1 gallon (US, dry) = 1.163647247 gallons (US, liq.)	US

Unit (synonym, acronym)	Symbol	Physical quantity	Dimension	Conversion factor (SI equivalent unit)	Notes, definitions, other conversion factors	System
gallon (US, liq.)	gal (US, liq.)	capacity, volume	L^3	1 gallon (US, liq.) $= 3.785411784 \times 10^{-3}$ m^3	American unit used for capacity measurements of liquids. The US gallon is equal to the volume of 8.32828 lb of water at 60°F. It is employed to express only the capacity of liquids. 1 gallon (US) = 231 in^3 (E) 1 gallon (US) = 4 quarts (US, liq.) (E) 1 gallon (US) = 8 pints (US, liq.) (E) 1 gallon (US) = 32 gills (US, liq.) (E) 1 gallon (US) = 128 fl oz (US) (E) 1 gallon (US) = 1024 fl dr (US, liq.) (E) 1 gallon (US) = 0.8326738183 gallon (UK)	US
gallon (US) per foot	gal (US, liq.).ft^{-1}	surface, area	L^2	1 gal (US, liq.).ft^{-1} $= 1.241933000 \times 10^{-2}$ m^2 (E)	Obsolete American unit of area used for geometric shape measurements.	US
gallon (US) per hour	gph (US)	volume flow rate	L^3T^{-1}	1 gal (US).h^{-1} $= 1.051503273 \times 10^{-6}$ $m^3.s^{-1}$	Obsolete American unit of volume flow rate employed in chemical engineering.	US
gallon (US) per hour-horsepower	gal (US).h^{-1}.HP^{-1}	specific fuel consumption-energy ratio	$M^{-1}LT^2$	1 gal (US).h^{-1}-HP^{-1} $= 1.410089117 \times 10^{-9}$ $m^3.J^{-1}$	American unit to expressed the specific fuel consumption (SFC) of motor oil engines. In this unit 1 HP = 550 lbf-ft.s^{-1}	US
gallon (US) per mile	gal (US).mi^{-1}	surface, area	L^2	1 gal (US).mi^{-1} $= 2.352145833 \times 10^{-6}$ m^2	Obsolete American unit of area used for geometric shape measurements.	US
gallon (US) per minute	gpm (US)	volume flow rate	L^3T^{-1}	1 gal (US).min^{-1} $= 6.309019640 \times 10^{-5}$ $m^3.s^{-1}$	Obsolete American unit of volume flow rate employed in chemical engineering.	US

gallon (US) per second	gps (US)	volume flow rate	L^3T^{-1}	1 gal (US).s $= 3.785411784 \times 10^{-3}$ $m^3.s^{-1}$	Obsolete American unit of volume flow rate employed in chemical engineering.	US
gamma	γ	magnetic induction field, magnetic flux density	$MI^{-1}T^{-2}$	$1\ \gamma = 10^{-9}$ T (E)	Obsolete unit of magnetic induction employed in geophysics and magnetohydrodynamics. $1\ \gamma = 10^{-5}$ Œ (E) $1\ \gamma = 10^{-5}$ emu cgs (E)	
gamma	γ	mass	M	$1\ \gamma = 10^{-9}$ kg (E)	Obsolete unit of mass employed in analytical chemistry. $1\ \gamma = 10^{-6}$ g (E) $1\ \gamma = 1$ µg (E)	
gar (Persian)	–	surface, area	L^2	1 gar (Persian) = 14.7456 m^2	Obsolete Persian unit of area used in ancient times. 1 gar (Persian) = 144 sq. ft (E)	Persian
gariba (Arabic) (den)	–	capacity, volume	L^3	1 gariba (Arabic) $= 216.120 \times 10^{-3}$ m^3 (of water)	Obsolete Arabic unit of capacity used in ancient times. Measured by weight. 1 gariba (Arabic) = 8 cafiz (E)	Arabic
gariba (Persian)	–	capacity, volume	L^3	1 gariba (Persian) $= 260.8 \times 10^{-3}$ m^3 (of water)	Obsolete unit of capacity of the Assyrio-Chaldean-Persian system used in ancient times. Measured by weight. 1 gariba (Persian) = 8 amphora (E)	Persian
garnetz (Russian, dry)	–	capacity, volume	L^3	1 garnetz (Russian, dry) $= 3.279842 \times 10^{-3}$ m^3	Obsolete Russian unit of capacity for dry substances used before 1917.	Russian

Unit (synonym, acronym)	Symbol	Physical quantity	Dimension	Conversion factor (SI equivalent unit)	Notes, definitions, other conversion factors	System
gauss	G, Gs, Γ	magnetic induction field, magnetic flux density	$MI^{-1}T^{-2}$	1 G = 10^{-4} T (E)	Obsolete cgs unit of magnetic induction.	cgs
gavyuti	–	length, distance	L	1 gavyuti =7312 m	Obsolete Indian unit of length used in ancient times. 1 gavyuti = 16 000 hasta (E)	Indian
Gay-Lussac (Gay-Lussac degree)	°GL	percentage of alcohol in wines and spirits	nil	1 °GL = 1% (v/v) ethanol in water	Obsolete French unit used in oenology and introduced in 1824. It serves to express percentage of ethanol in wines and spirits.	French
Gay-Lussac degree	–	specific gravity of liquids, hydrometer index, hydrometer degree	nil	°Gay-Lussac = 100 – (100/d)	Obsolete unit of specific gravity.	
geepound (slug)	slug	mass	M	1 slug = 14.59390294 kg	Obsolete unit of mass. It was equal to the mass which under an acceleration of 1 $ft.s^{-2}$ gives a force of 1 lbf. The name geepound sometimes used for the unit is derived from English: **g pound**.	UK, FPS
geira (Portuguese)	–	surface, area	L^2	1 geira (Portuguese) = 5803.281 m^2	Obsolete Portuguese unit of area. 1 geira (Portuguese) = 605 square vara (E)	Portuguese
gemmho	–	electrical conductance	$M^{-1}L^{-2}T^3I^2$	1 gemmho = 10^{-6} S (E)	The name derived from the reverse writing of the two words meg-ohm.	
gerah (Hebrew) [Hebrew obol] [sacred system]	–	mass	M	1 gerah (Hebrew) = 7.083333×10^{-4} kg	Obsolete Hebrew unit of mass used in ancient times. Sacred system. 1 gerah (Hebrew) = 1/1200 mina (E)	Hebrew

ghalva (Arabic)	–	length, distance	L	1 ghalva (Arabic) = 230.40 m	Obsolete Arabic unit of length used in ancient times. 1 ghalva (Arabic) = 720 feet (Arabic) (E)	Arabic
ghalva (Persian) [stadion]	–	length, distance	L	1 ghalva (Persian) = 230.400 m	Obsolete unit of length in the Assyrio-Chaldean-Persian system used in ancient times. 1 ghalva = 720 zereths (E)	Persian
gibbs	–	surface molar concentration	$N.L^{-2}$	1 gibbs = 10^{-14} $mol.m^{-2}$	Unit of adsorbed concentration of a molecule onto a surface. The unit is named after J.W. Gibbs (1839–1903). 1 gibbs = 10^{-10} $mol.cm^{-2}$ (E)	
gigabyte (gigaoctet)	Go, GB	quantity of information	nil	1 GB = 1 073 741 824 bytes	1 GB = 2^{30}	
gigagram	Gg	mass	M	1 Gg = 10^{6} kg (E)	Multiple of the SI base unit. 1 Gg = 10^{9} g (E)	SI
gigametre	Gm	length, distance	L	1 Gm = 10^{9} m (E)	Multiple of the SI base unit	SI
gigapascal	GPa	pressure, stress	$ML^{-1}T^{-2}$	1 GPa = 10^{9} Pa (E)	Multiple of the derived SI pressure and stress unit.	SI
gilbert	Gb	magnetomotive force	I	1 Gb = 0.79577 A	Obsolete cgs unit of magnetomotive force. The unit is named after W. Gilbert (1544–1603). 1 Gb = $[10/4\pi]$ A.tr (E) 1 Gb = $[1/4\pi]$ abampere.tr (E)	cgs
gill (UK)	gi (UK)	capacity, volume	L^{3}	1 gill (UK) = $1.420653750 \times 10^{-4}$ m^{3}	British unit used for capacity measurements for all merchandise (solids, liquids, foodstuffs, etc). *(continued overleaf)*	UK

Unit (synonym, acronym)	Symbol	Physical quantity	Dimension	Conversion factor (SI equivalent unit)	Notes, definitions, other conversion factors	System
gill (UK) *(continued)*					1 gill (UK) = 1/32 gallon (UK) (E) 1 gill (UK) = 1/8 quart (UK) (E) 1 gill (UK) = 1/4 pint (UK) (E) 1 gill (UK) = 5 fl oz (UK) (E) 1 gill (UK) = 40 fl dr (UK) (E) 1 gill (UK) = 1.200950 (US) gill	
gill (US, liq.)	gi (US, liq.)	capacity, volume	L^3	1 gill (US, liq.) $= 1.182941183 \times 10^{-4}$ m^3	American unit used for capacity measurements of liquids. 1 gill (US, liq.) = 1/32 gallon (US, liq.) (E) 1 gill (US, liq.) = 1/8 lq quart (US, liq.) (E) 1 gill (US, liq.) = 1/4 lq pint (US, liq.) (E) 1 gill (US, liq.) = 4 fl oz (US, liq.) (E) 1 gill (US, liq.) = 32 fl dr (US, liq.) (E) 1 gill (US, liq.) = 1920 minims (US, liq.) (E) 1 (US) gill = 0.832675 gill (UK)	US
giornata (Italian)	–	surface, area	L^2	1 giornata (Italian) = 3 800 m^2	Obsolete Italian unit of area.	Italian
glug	–	mass	M	1 glug = 10^{-3} kg (E)	Obsolete unit of mass. It was equal to the mass which under an acceleration of 1 $cm.s^{-2}$ gives a force of 1 dyne. It was the cgs equivalent of slug. 1 glug = 1 g (E)	cgs
go	–	capacity, volume	L^3	1 go $= 0.180906837 \times 10^{-3}$ m^3	Obsolete Japanese unit of capacity. 1 go = 1/10 shô (E)	Japanese

gô (Japanese)	–	surface, area	L^2	go (Japanese)= 0.3305785124 m^2	Obsolete Japanese unit of area. 1 gô = 1/10 bu (E)	Japanese
gomor (Hebrew)	–	capacity, volume	L^3	1 gomor (Hebrew) $= 2.142 \times 10^{-3}$ m^3 (of water)	Obsolete Hebrew unit of capacity used in ancient times. Measured by weight. 1 gomor (Hebrew) = 1/10 ephah (new)	Hebrew
gon (grade)	gr, ^{g}	plane angle	α	1 gon $= 1.570796327 \times 10^{-2}$ rad	1 gon = $\pi/200$ rad (E) 1 gon = 100 centesimal minutes (E) 1 gon $= 10^4$ centesimal seconds (E)	INT
gouy	–	electrokinetic potential	$ML^2T^{-3}I^{-1}$	(see note)	Unit of electrokinetic potential employed in electrokinetics. It was suggested in 1956. The unit is named after L.G. Gouy (1854–1926).	
grade (gon)	gr, ^{g}	plane angle	α	1 gr $= 1.570796327 \times 10^{-2}$ rad	1 grade = $\pi/200$ rad (E) 1 grade = 100 centesimal minutes (E) 1 grade $= 10^4$ centesimal seconds (E)	INT
gradus (Roman) [Roman pace]	–	length, distance	L	1 gradus (Roman) = 0.736 m	Obsolete Roman unit of length employed in ancient times. 1 gradus (Roman) = 5/2 pes (E)	Roman
grain (apothecary)	gr (apoth.), gr (ap.)	mass	M	1 gr (apoth.) $= 6.479891000 \times 10^{-5}$ kg	Obsolete British and American unit of weight employed in pharmacy. It was employed at one time as a weight unit for drugs and medicines (lotions, potions, ointments, plant extracts, etc). 5760 grains (apoth.) = 1 lb (apoth.) (E)	UK, US
grain (avoirdupois)	gr (avdp), gr (av.)	mass	M	1 gr (av.) $= 6.479891000 \times 10^{-5}$ kg (E)	Legal American and British unit of weight since the WMA of 1963. 7000 grains (av.) = 1 lb (WMA, 1963) (E)	UK, US

Unit (synonym, acronym)	Symbol	Physical quantity	Dimension	Conversion factor (SI equivalent unit)	Notes, definitions, other conversion factors	System
grain (de Paris)	gr.	mass	M	1 grain (de Paris) $= 5.311478407 \times 10^{-5}$ kg	Obsolete French unit of mass employed before the French Revolution (1789). 1 grain (de Paris) = 1/9216 livre (de Paris) (E)	French
grain (Egyptian)	–	mass	M	1 grain (Egyptian) $= 0.910 \times 10^{-3}$ kg	Obsolete Egyptian unit of mass used in ancient times. 1 grain (Egyptian) = 1/15 deben (E)	Egyptian
grain (jeweller's)	gr (jew.)	mass	M	1 gr (jeweller's) $= 5 \times 10^{-5}$ kg (E)	Obsolete British unit of mass employed in jewellery for the weighing of precious metals, precious stones and gems (diamond, ruby, sapphire). 1 gr (jeweller's) = 0.25 carat (metric) (E)	UK
grain (Russian)	–	mass	M	1 grain (Russian) $= 6.220891642 \times 10^{-5}$ kg	Obsolete Russian unit of mass used before 1917 in pharmacy. 1 grain (Russian) = 1.4 doli (E)	Russian
grain (troy)	gr (troy)	mass	M	1 gr (troy) $= 6.479891000 \times 10^{-5}$ kg	Obsolete British unit of mass employed for the weighing of precious metals, precious stones and gems (diamond, ruby, sapphire) in the United Kingdom. Now obsolete in the UK but this unit remains in common use in the USA. 5760 grains (troy) = 1 lb (troy) (E)	UK, US
gram	g	mass	M	1 g $= 10^{-3}$ kg (E)	Submultiple of the SI base unit; cgs base unit of mass.	cgs

gram-rad	–	radiation absorbed dose, specific energy, kerma, index of absorbed dose	L^2T^{-2}	1 gram-rad = 0.01 Gy (E)	Obsolete unit. 1 gram-rad = 10^2 erg.g^{-1} (E)	
gramcentimetre per second	g.cm.s^{-1}	linear momentum, momentum	MLT^{-1}	1 g.cm.s^{-1} = 10^{-5} kg.m.s^{-1} (E)	Obsolete cgs linear momentum unit.	cgs
gran (Austrian, apothecary)	–	mass	M	1 gran (Austrian, apothecary) = 9.722395833 × 10^{-5} kg	Obsolete Austrian unit of mass used in pharmacy. 1 gran (Austrian, apothecary) = 1/5760 pfund (Austrian, apothecary) (E)	Austrian
grano (Italian)	–	mass	M	1 grano (Italian) = 4.441550926 × 10^{-5} kg	Obsolete Italian unit of mass. 1 grano (Italian) = 1/6912 libbra (E)	Italian
grano (Spanish)	–	mass	M	1 grano (Spanish) = 4.992328559 × 10^{-5} kg	Obsolete Spanish unit of mass. 1 grano (Spanish) = 1/9216 libra (E)	Spanish
grao (Portuguese)	–	mass	M	1 grao (Portuguese) = 4.980468750 × 10^{-5} kg	Obsolete Portuguese unit of mass. 1 grao (Portuguese) = 1/ 9216 libra (E)	Portuguese
gray	Gy	radiation absorbed dose, specific energy, kerma, index of absorbed dose	L^2T^{-2}	**SI derived unit** 1 Gy = 1 J.kg^{-1} (E)	The gray is the absorbed dose when the energy per unit mass imparted to matter by ionizing radiation is one joule per kilogram (15th CGPM, 1976). The unit is named after L.H. Gray (1905–1965).	SI
Greenwich mean solar time	GMT	time, period, duration	T	(see note)	Obsolete time standard. The GMT is defined as the average passage of the Sun across the meridian zero every day of the year.	INT
grein (Dutch)	–	mass	M	1 grein (Dutch) = 6.760545604 × 10^{-5} kg	Obsolete Dutch unit of mass. 1 grein (Dutch) = 1/7680 pond (Dutch) (E)	Dutch

Unit (synonym, acronym)	Symbol	Physical quantity	Dimension	Conversion factor (SI equivalent unit)	Notes, definitions, other conversion factors	System
gros (de Paris) [French drachm]	–	mass	M	1 gros (de Paris) $= 3.824264453 \times 10^{-3}$ kg	Obsolete French unit of mass employed before the French Revolution (1789). 1 gros (de Paris) = 1/128 livre (de Paris) (E)	French
gur (Persian)	–	surface, area	L^2	1 gur (Persian) = 14 745.6 m^2	Obsolete Persian unit of area used in ancient times. 1 gur (Persian) = 1000 gar (E)	Persian
habbah (Arabic)	–	surface, area	L^2	1 habbah (Arabic) = 81.944444 m^2	Obsolete Arabic unit of area used in ancient times. 1 habbah (Arabic) = 1/72 feddan (E)	Arabic
halbe (Austrian)	–	capacity, volume	L^3	1 halbe (Austrian) = 7.0755×10^{-4} m^3	Obsolete Austrian unit of capacity used for liquid substances. 1 halbe (Austrian) = 1/2 mass (E)	Austrian
hand (UK)	hd	length, distance	L	1 hand (UK) = 10.16×10^{-2} m (E)	Obsolete British and American unit of length. 1 hand (UK) = 4 inches (E) 1 hand (UK) = 1/3 foot (E) 1 hand (UK) = 1/9 yard (E)	UK, US
hao (Chinese)	–	length, distance	L	1 hao (Chinese) = 2.3×10^{-5} m	Obsolete Chinese unit of length used in ancient times. 1 hao = 1/10 000 tchi (E)	Chinese
hao (Chinese)	–	surface, area	L^2	1 hao (Chinese) = 0.6144 m^2	Obsolete Chinese unit of area used in ancient times. 1 hao (Chinese) = 1/1000 meou (E)	Chinese

					ancient times. 1 hara (Indian) = 200 pala (E)	
hartree (a.u. of energy)	E_h	energy, work, heat	$ML^{2T^{-2}}$	1 a.u. of energy = 4.3597482 × 10^{-18} J	The unit is named after D.R. Hartree (1897–1958) 1 a.u. of energy = 1 hartree (E) 1 a.u. of energy = 2 rydbergs (E) 1 a.u. of energy = $\hbar^2.m\text{-}1_0 a_0{}^2$	a.u.
hasta	–	length, distance	L	1 hasta = 0.457 m	Obsolete Indian unit of length used in ancient times. Base unit of the Indian system of length.	Indian
heat flux unit (HFU)	HFU	thermal flux	MT^{-3}	1 HFU = 41.855 × 10^{-3} W.m^{-2}	Obsolete unit of thermal flux used in Earth science, especially in geophysics. 1 HFU = 1 μcal$_{15}$.cm^{-2}.s^{-1} (E)	
hectare	ha	surface, area	L^2	1 hectare = 10^4 m^2 (E)	French metric unit of area used in surveyor's measurements. It was introduced after the French Revolution. The unit is still used in agriculture.	French
hecte (Egyptian)	–	capacity, volume	L^3	1 hecte (Egyptian) = 2.125 × 10^{-3} m^3 (of water)	Obsolete Egyptian unit of capacity used in ancient times. Measured by weight. 1 hecte (Egyptian) = 1/16 khar (E)	Egyptian
hecteus (Greek, Attic) [εχτευζ]	–	capacity, volume	L^3	1 hecteus (Greek, Attic) = 8.64 × 10^{-3} m^3	Obsolete Greek unit of volume employed in ancient times. It was used for capacity measurements of dry substances. 1 hecteus (Greek, Attic) = 8 chenix (E)	Attic
hectogram	hg	mass	M	1 hg = 10^{-1} kg (E)	Submultiple of the SI base unit. 1 hg = 10^2 g (E)	SI
hectolitre	hl, hL	capacity, volume	L^3	1 hl = 0.1 m^3 (E)	Multiple of the litre.	

Unit (synonym, acronym)	Symbol	Physical quantity	Dimension	Conversion factor (SI equivalent unit)	Notes, definitions, other conversion factors	System
hectometre	hm	length, distance	L	1 hm = 10^2 m (E)	Multiple of the SI base unit	SI
hectopascal	hPa	pressure, stress	$ML^{-1}T^{-2}$	1 hPa = 10^2 Pa (E)	Multiple of the derived SI pressure and stress unit. Unit of pressure commonly employed in meteorology. It was adopted January 1st, 1986 by the World Meteorological Organization. 1 hPa = 1 mbar (E) 1 atm = 1 013.25 hPa (E)	SI
hefner unit (hefnerkerze)	HK	luminous intensity	J	1 hefner unit = 0.903 cd	Obsolete German unit of luminous intensity. It was equal to the horizontal intensity of the hefner lamp burning amyl acetate, with a flame 4 cm high. If the flame is l (mm) high, the intensity is $I = 1 + 0.027(l - 40)$. The unit is named after Hefner-Altenack (1845–1904). It was used in Germany before World War II.	German
hefnerkerze (hefner unit)	HK	luminous intensity	J	1 hefner unit = 0.903 cd	Obsolete German unit of luminous intensity. It was equal to the horizontal intensity of the hefner lamp burning amyl acetate, with a flame 4 cm high. If the flame is l mm high, the intensity is $I = 1 + 0.027(l - 40)$. The unit is named after Hefner-Altenack (1845-1904). It was used in Germany before World War II.	German
helmholtz	–	surface density of dipolar moment	ITl^{-1}	1 helmholtz = 3.336×10^{-10} C.m^{-1}	Unit suggested by Guggenheim for surface charge density of dipole measurements in the double-layer. The unit is named after	

					H.L.F. Helmholtz (1821–1894). Anecdotal. 1 helmoltz = 1 D.Å^{-2} (E)	
hemina (Roman)	–	capacity, volume	L^3	1 hemina (Roman) = 0.274 × 10^{-3} m^3	Obsolete Roman unit of volume employed in ancient times. It was used for capacity measurements of dry substances. 1 hemina (Roman) = 1/2 sextarius (E)	Roman
henry	H	electric inductance	$ML^2T^{-2}I^{-2}$	**SI derived unit** 1 H = 1 Wb.A^{-1} = 1 kg.m^2.s^{-2}.A^{-2} (E)	The henry is the inductance of a closed circuit in which an electromotive force of one volt is produced when the electric current in the circuit varies uniformly at a rate of one ampere per second. The unit is named after the American scientist J. Henry (1797–1878).	SI
henry (int. mean)	H (int. mean)	electric inductance	$ML^2T^{-2}I^{-2}$	1 H (int. mean) = 1.00049 H	Obsolete American unit of electric inductance.	US
henry (int. US)	H (US)	electric inductance	$ML^2T^{-2}I^{-2}$	1 H (int. US) = 1.000495 H	Obsolete American unit of electric inductance.	US, IEUS
henry per metre	H.m^{-1}	magnetic permeability	$MLT^{-2}I^{-1}$	**SI derived unit** 1 H.m^{-1} = 1 kg.m.s^{-2}.A^{-1} (E)		SI
heredium (Roman)	–	surface, area	L^2	1 heredium (Roman) = 5053.503744 m^2	Obsolete Roman unit of area employed in ancient times. 1 heredium (Roman) = 57 600 quadratus pes (E)	Roman
hertz	Hz	frequency	T^{-1}	**SI derived unit** 1 Hz = 1 s^{-1} (E)	The hertz is the frequency of a periodic phenomenon of which the period is one second. The unit is named after R. Hertz (1857–1894).	SI

Unit (synonym, acronym)	Symbol	Physical quantity	Dimension	Conversion factor (SI equivalent unit)	Notes, definitions, other conversion factors	System
hin (Hebrew)	–	capacity, volume	L^3	1 hin (Hebrew) $= 3.570 \times 10^{-3}$ m^3 (of water)	Obsolete Hebrew unit of capacity used in ancient times. Measured by weight. 1 hin (Hebrew) = 1/6 bath (new) (E)	Hebrew
hiro (Japanese)	–	length, distance	L	1 hiro (Japanese) = 1.515151515 m	Obsolete Japanese unit of length. 1 hiro = 5 shaku (E)	Japanese
hoé (Chinese)	–	length, distance	L	1 hoé (Chinese) $= 2.3 \times 10^{-7}$ m	Obsolete Chinese unit of length used in ancient times. 1 hoé = 1/1 000 000 tchi (E)	Chinese
hogshead (US, liq.)	hhd, hgs	capacity, volume	L^3	1 hogshead (US, liq.) $= 238.480942400 \times 10^{-3}$ m^3	Obsolete American unit of capacity. 1 hogshead (US, liq.) = 63 gallons (US) (E) 1 hogshead (US, liq.) = 2 barrels (US, liq.) (E) 1 hogshead (US, liq.) = 1 butt (UK) (E)	US
horsepower (cheval-vapeur)	cv, HP	power	ML^2T^{-3}	1 HP = 735.498750 W	Obsolete unit of power introduced by James Watt in 1782 to allow to describe the power of steam machinery. It was equal to the work effort of a horse needed to raise vertically 528 cubic feet of water to one metre in one minute (formal definition published in 1809). In 1850, the following conversion factor was adopted: 1 HP = 75 $kgfm.s^{-1}$ (E)	
horsepower (boiler)	HP	power	ML^2T^{-3}	1 HP (boiler) = 9809.500 W	Obsolete American unit of power. 1 HP (boiler) = 1000 $kgfm.s^{-1}$ (E)	US
horsepower (metric)	HP	power	ML^2T^{-3}	1 HP (metric) = 735.498750 W	Obsolete American unit of power.	US

horsepower (electric)	HP	power	[illegible]	1 HP (electric) = [illegible] W (E)	[illegible]	
horsepower (water)	HP	power	ML^2T^{-3}	1 HP (water) = 746.043 W	Obsolete American unit of power.	US
horsepower (550ft.lbf.s^{-1})	HP	power	ML^2T^{-3}	1 HP = 745.6998715 W	Obsolete American unit of power.	US
horsepower (British)	BHP, hp	power	ML^2T^{-3}	1 BHP = 745.70 W	Obsolete British unit of power. 1 BHP = 1.014 CV	UK, US
hou (Chinese)	–	capacity, volume	L^3	1 hou (Chinese) = 51.772×10^{-3} m^3	Obsolete Chinese unit of capacity used in ancient times. 1 hou (Chinese) = 50 tcheng (E)	Chinese
hour	h	time, period, duration	T	1 hour = 3600 seconds (E)	1 hour = 60 minutes (E)	
hour (sidereal)	–	time, period, duration	T	1 hour (sidereal) = 3590.170 s		@
hour of angle	–	plane angle	α	1 hour of angle = 0.2617993878 rad	Unit of plane angle employed in aeronautical navigation. 1 hour of angle = $(2\pi/24)$ rad (E) 1 hour of angle = 15° (E)	
hubble	–	length, distance	L	1 hubble = $9.46052973 \times 10^{24}$ m	Obsolete unit of distance used in astronomy. The unit is named after E.P. Hubble (1889–1953). 1 hubble = 10^9 light-years (E) 1 hubble = 9.46052973 Ym	@
hundredweight (gross or long)	cH, cwt, lg cwt	mass	M	1 cwt (UK) = 50.80234544 kg (E)	Obsolete British unit of mass. 1 cwt (UK) = 4 quarters (UK) (E) 1 cwt (UK) = 8 stones (UK) (E) 1 cwt (UK) = 112 lb (av.) (E)	UK, CAN

Unit (synonym, acronym)	Symbol	Physical quantity	Dimension	Conversion factor (SI equivalent unit)	Notes, definitions, other conversion factors	System
hundredweight (net or short)	sh. cwt	mass	M	1 sh. cwt (US) = 45.359237 kg (E)	Obsolete American unit of mass. 1 sh. cwt (US) = 100 lb (av.) (E)	US, CAN
hundredweight (UK, avoirdupois)	cwt (av.)	mass	M	1 cwt (UK) = 50.80234544 kg (E)	Obsolete British unit of mass. 1 cwt (UK) = 4 quarters (UK) (E) 1 cwt (UK) = 8 stones (UK) (E) 1 cwt (UK) = 112 lb (av.) (E)	UK, CAN
hyakume (Japanese)	–	mass	M	1 hyakume (Japanese) = 0.375 kg (E)	Obsolete Japanese unit of mass. 1 hyakume (Japanese) = 1/10 kwan (E)	Japanese
hydrotimeter degree (American)	°TH (US)	hardness of water, concentration of calcium and magnesium in water	nil	1 °TH (US) = 1 mg of $CaCO_3$. $(1000\ cm)^{-3}$ of water	American unit employed to express hardness of water. It was introduced by the US Geological Survey. 1°TH(US) = 1 ppm of $CaCO_3$	US
hydrotimeter degree (British) (Clarke degree)	°THB, °e	hardness of water, concentration of calcium and magnesium in water	nil	1 °THB = 14.3 mg of $CaCO_3$. $(1000\ cm)^{-3}$ of water	British unit employed to express hardness of water. 1°THB = 1 grain of $CaCO_3$/UK gallon of water.	UK
hydrotimeter degree (German)	°THG, °d	hardness of water, concentration of calcium and magnesium in water	nil	1 °THG = 10 mg of CaO.$(1000\ cm)^{-3}$ of water	German unit employed to express hardness of water.	German
hyl (metric slug, mug, par, TME)	Hyl, hyl	mass	M	1 Hyl = 9.80665 kg	Obsolete technical metric unit of mass employed by mechanical engineers (base unit of the metric gravitational system). It was equal to the mass which under an acceleration of 1 $m.s^{-2}$ gives a force of 1 kgf. The name TME derived from the	German

					Einheit.	
inch	in	length, distance	L	1 inch = 2.54×10^{-2} m (E)	Legal unit in the UK system (since the WMA, 1963) and US system (since the USMB, 1959). 1 inch = 2.54 cm (E)	UK, US
inch (old)	in	length, distance	L	1 in (old) = 2.539998×10^{-2} m (E)	Obsolete British unit of length before the WMA of 1963.	UK
inch (Prussian)	–	length, distance	L	1 zoll (Prussian) = $2.615416667 \times 10^{-2}$ m	Obsolete German unit of length. 1 zoll = 1/12 fuss (E)	German
inch (US, Survey)	in	length, distance	L	1 in (US, Survey) = $2.54000508 \times 10^{-2}$ m (E)	Obsolete American unit of length used in the USA for geodetic measurements. Discontinued since the USMB, 1959. 1 inch (US Survey) = (1/39.37) cm (E.)	US
inch of mercury (0°C)	inHg (0°C)	pressure, stress	$ML^{-1}T^{-2}$	1 inHg (0°C) = 3.386383658×10^3 Pa	British and American unit of pressure employed in physics to measure small pressures. It is equal to the pressure exerted by a column of mercury one inch high, measured at 0°C (32°F). 1 atm = 29.92143165 inHg (0°C)	UK, US
inch of mercury (15.56°C)	inHg (15.56°C)	pressure, stress	$ML^{-1}T^{-2}$	1 inHg (15.56°C) = 3.377182314×10^3 Pa	British and American unit of pressure employed in physics to measure small pressures. It is equal to the pressure exerted by a column of mercury one inch high, measured at 15.56°C (60°F). 1 atm = 30.00282204 inHg (15.5°C)	UK, US

Unit (synonym, acronym)	Symbol	Physical quantity	Dimension	Conversion factor (SI equivalent unit)	Notes, definitions, other conversion factors	System
inch of mercury (32°F)	inHg (32°F)	pressure, stress	$ML^{-1}T^{-2}$	1 inHg (32°F) $= 3.386383658 \times 10^3$ Pa	British and American unit of pressure employed in physics to measure small pressures. It is equal to the pressure exerted by a column of mercury one inch high, measured at 0°C (32°F). 1 atm = 29.92143165 inHg (0°C)	UK, US
inch of mercury (60°F)	inHg (60°F)	pressure, stress	$ML^{-1}T^{-2}$	1 inHg (60°F) $= 3.377182314 \times 10^3$ Pa	British and American unit of pressure employed in physics to measure small pressures. It is equal to the pressure exerted by a column of mercury one inch high, measured at 15.56°C (60°F). 1 atm = 30.00282204 inHg (60°F)	UK, US
inch of water (4°C)	inH_2O (4°C)	pressure, stress	$ML^{-1}T^{-2}$	1 inH_2O (4°C) $= 2.490819355 \times 10^2$ Pa	British and American unit of pressure employed in physics to measure small pressures. It is equal to the pressure exerted by a column of water one inch high, measured at 4°C (39.2°F). 1 atm = 406.7938519 inH_2O (39.2°F)	UK, US
inch of water (15.56°C)	inH_2O (15.56°C)	pressure, stress	$ML^{-1}T^{-2}$	1 inH_2O (15.5°C) $= 2.488453010 \times 10^2$ Pa	British and American unit of pressure employed in physics to measure small pressures. It is equal to the pressure exerted by a column of water one inch high, measured at 15.56°C (60°F). 1 atm = 407.1806844 inH_2O (15.5°C)	UK, US
inch of water (39.2°F)	inH_2O (39.2°F)	pressure, stress	$ML^{-1}T^{-2}$	1 inH_2O (39.2°F) $= 2.490819355 \times 10^2$ Pa	British and American unit of pressure employed in physics to measure small pressures. It is equal to the pressure	UK, US

					high, measured at 4°C (39.2°F). 1 atm = 406.7938519 inH_2O (4°C)	
inch of water (60°F)	inH_2O (60°F)	pressure, stress	$ML^{-1}T^{-2}$	1 inH_2O (60°F) = 2.488453010×10^2 Pa	British and American unit of pressure employed in physics to measure small pressures. It is equal to the pressure exerted by a column of water one inch high, measured at 15.56°C (60°F). 1 atm = 407.1806844 inH_2O (60°F)	UK, US
inch per second	in.s^{-1}, ips	velocity, speed	LT^{-1}	1 ips = 2.54×10^{-2} m.s^{-1} (E)	British and American unit of speed. 1 ips = 9.144×10^{-2} km.h^{-1} (E)	UK, US
inferno	–	temperature	Θ	1 inferno = 10^6 K (E)	Unit of temperature suggested in 1968. Anecdotal.	
international atomic time (temps atomique international)	TAI	time, period, duration	T	(see note)	The TAI corresponds to a time scale established by the Bureau International de l'Heure (BIH) given by atomic clocks in several locations through the world. The international abbreviation TAI is employed in all languages.	INT
international unit	UI, IU	enzymatic activity	NT^{-1}	1 IU = 1.666667×10^{-7} mol.s^{-1}	Unit employed in biochemistry to describe enzymatic activity. 1 UI is equal to the amount of enzyme which transforms 1 micromole of substrate per minute at 298 K (25°C). 1 IU = 1 μmol.min^{-1} (E)	EUR.
iron	–	length, distance	L	1 iron = 5.3×10^{-4} m	Obsolete British unit of length employed in length measurements of shoes.	UK

Unit (synonym, acronym)	Symbol	Physical quantity	Dimension	Conversion factor (SI equivalent unit)	Notes, definitions, other conversion factors	System
jacobi	–	electric resistance	$ML^2T^{-3}I^{-2}$	1 jacobi = 0.64 Ω	Obsolete unit of electric resistance defined in 1848. It is the resistance of a copper wire of length 25 feet which has a mass equal to 345 grains.	
jansky	–	electromagnetic radiation flux density	MT^{-4}	1 jansky = 10^{-26} W.m^{-2}.Hz^{-1}	Unit of adopted by International Union of Astronomy in 1973. The unit is named after K.G. Jansky (1905–1950).	@
jar	–	electric capacitance	$M^{-1}L^{-2}T^4I^2$	1 jar =1/9 × 10^{-8} F (E)	Obsolete unit of electric capacitance. One jar is equal approximately to the electric capacitance stored in the famous bottle capacitor of the Dutch city of Leyden. 1 jar = 1000 'cm' (E)	
jéroboam (jeroboam)	–	capacity, volume	L^3	1 jéroboam = 3.030728 × 10^{-3} m^3	Obsolete British unit which expressed the capacity of wine containers. Still employed in oenology, especially in France. 1 jéroboam = 2/3 gallon (UK) (E) 1 jéroboam = 4 bouteilles (E)	UK, French
jin (tchin)	–	mass	M	1 jin = 250 × 10^{-3} kg	Obsolete Chinese unit of mass.	Chinese
jô (Japanese)	–	length, distance	L	1 jô (Japanese) = 3.030303030 m	Obsolete Japanese unit of length. 1 jô = 10 shaku (E)	Japanese
joch (Austrian)	–	surface, area	L^2	1 joch (Austrian) = 5754.618224 m^2	Obsolete Austrian unit of area. 1 joch (Austrian) = 1600 sq. klafter (E)	Austrian
Jordan's elementary time	–	time, period, duration	T	1 Jordan's time = 9.3996 × 10^{-24} s	1 Jordan's time = $\alpha^3\tau_0 = r_e/c_0$	

				1 J = 1 kg. $m^2.s^{-2}$ (E)	of application of a force of one newton is displaced a distance of one metre in the direction of the force. The unit is named after J.P. Joule (1818–1889).	
joule (int.)	J (int.)	energy, work, heat	ML^2T^{-2}	1 J (int.) = 1.000165 J (E)	Obsolete international unit of energy.	INT
joule per cubic metre	$J.m^{-3}$	energy density	$ML^{-1}T^{-2}$	**SI derived unit** 1 $J.m^{-3}$ = 1 $kg.m^{-1}.s^{-2}$ (E)		SI
joule per cubic metre per hertz	$J.m^{-3}.Hz^{-1}$	spectral radiant energy density in terms of frequency	$ML^{-1}T^{-1}$	**SI derived unit** 1 $J.m^{-3}.Hz^{-1}$ = 1 $kg.m^{-1}.s^{-1}$ (E)		SI
joule per kelvin	$J.K^{-1}$	entropy	$ML^2T^{-2}\Theta^{-1}$	**SI derived unit** 1 $J.K^{-1}$ = 1 $kg.m^2.s^{-2}.K^{-1}$ (E)		SI
joule per mole	$J.mol^{-1}$	molar energy	$ML^2T^{-2}N^{-1}$	**SI derived unit** 1 $J.mol^{-1}$ = 1kg. m^2. s^{-2}. mol^{-1} (E)		SI
joule per mole per kelvin	$J.mol^{-1}K^{-1}$	molar entropy, molar heat capacity	$ML^2T^{-2}N^{-1}\Theta^{-1}$	**SI derived unit** 1 $J.mol^{-1}K^{-1}$ = 1 kg. m^2. s^{-2}. $mol^{-1}K^{-1}$ (E)		SI
joule per quadratic metre	$J.m^{-4}$	spectral radiant energy density in terms of wavelength	$ML^{-2}T^{-2}$	**SI derived unit** 1 $J.m^{-4}$ = 1 $kg.m^{-2}.s^{-2}$ (E)		SI
joule per square metre	$J.m^{-2}$	spectral radiant energy density in terms of wavenumber	MT^{-2}	**SI derived unit** 1 $J.m^{-2}$ = 1 $kg.s^{-2}$ (E)		SI
joule-second	J.s	action, angular momentum	ML^2T^{-1}	**SI derived unit** 1 J.s = 1 $kg.m^2.s^{-1}$		SI

Unit (synonym, acronym)	Symbol	Physical quantity	Dimension	Conversion factor (SI equivalent unit)	Notes, definitions, other conversion factors	System
journal (French)	–	surface, area	L^2	1 journal (French) = 4000 m^2	Obsolete French unit of area used in surveyor's measurements before the French revolution (1789).	French
jugerum (Roman)	–	surface, area	L^2	1 jugerum (Roman) = 2526.751872 m^2	Obsolete Roman unit of area employed in ancient times. 1 jugerum (Roman) = 28 800 quadratus pes (E)	Roman
jun (kwan)	–	mass	M	1 jun = 7.5 kg	Obsolete Chinese unit of mass. 1 jun = 30 jin (E)	Chinese
jungfrur (Swedish)	–	capacity, volume	L^3	1 jungfrur (Swedish) = $8.178631250 \times 10^{-5}$ m^3	Obsolete Swedish unit of capacity for liquid substances. 1 jungfrur (Swedish) = 1/32 kanna (E)	Swedish
junkfra (Swedish, dry)	–	capacity, volume	L^3	1 junkfra (Swedish, dry) = 8.178125×10^{-4} m^3	Obsolete Swedish unit of capacity for dry substances. 1 junkfra (Swedish, dry) = 1/32 kanna (E)	Swedish
kanna (Swedish)	–	capacity, volume	L^3	1 kanna (Swedish) = 2.617162×10^{-3} m^3	Obsolete Swedish unit of capacity for liquid substances.	Swedish
kanna (Swedish, dry)	–	capacity, volume	L^3	1 kanna (Swedish, dry) = 2.617×10^{-3} m^3	Obsolete Swedish unit of capacity for dry substances.	Swedish
kanne	–	capacity, volume	L^3	1 kanne = 10^{-3} m^3 (E)	Name proposed for the litre. Anecdotal interest. 1 kanne = 1 dm^3 (E)	
kapp line	–	magnetic induction flux	$ML^2T^{-2}I^{-1}$	1 kapp line = 6×10^{-5} Wb (E)	Obsolete British and American unit of magnetic flux. 1 kapp line = 6000 Mx (E)	UK, US

kappar (Swedish, dry)	–	capacity, volume	L	1 kappar (Swedish, dry) $= 4.579750 \times 10^{-3}$ m^3	Obsolete Swedish unit of capacity for dry substances. 1 kappar (Swedish, dry) = 7/4 kanna (E)	Swedish
kappland (Swedish)	–	surface, area	L^2	1 kappland (Swedish) = 154.261817 m^2	Obsolete Swedish unit of area. 1 kappland (Swedish) = 1750 square fot (E)	Swedish
karch (Austrian)	–	mass	M	1 karch (Austrian) = 224.04 kg	Obsolete Austrian unit of mass for general uses. 1 karch (Austrian) = 400 pfund (E)	Austrian
karus hiri-ichi-da (Japanese)	–	mass	M	1 karus hiri-ichi-da (Japanese) = 67.5 kg (E)	Obsolete Japanese unit of mass. 1 karus hiri-ichi-da (Japanese) = 18 kwan (E)	Japanese
katal	kat	enzymatic activity	NT^{-1}	1 kat = 1 mol.s^{-1} (E)	Unit employed in biochemistry to describe enzymatic activity. 1 IU is equal to the amount of enzyme which transforms 1 micromole of substrate per minute at $T = 298$ K (25°C).	SI
kayser	Ky	wavenumber	L^{-1}	1 Ky = 10^2 m^{-1} (E)	Obsolete wavenumber unit adopted in 1952 and used in spectroscopy. The unit is named after J.H.G. Kayser (1853–1940). 1 Ky = 1 cm^{-1} (E)	
kedet	–	mass	M	1 kedet $= 136.5 \times 10^{-3}$ kg	Obsolete Egyptian unit of mass used in ancient times. 1 kedet = 10 debens (E)	Egyptian
kelvin	K	absolute thermodynamic temperature	Θ	**SI base unit**	The kelvin, unit of thermodynamic temperature, is the fraction 1/273.16 of the thermodynamic temperature of the triple point of water [13th CGPM (1967), Resolution 4]. The unit is named after Lord Kelvin (1824–1907).	SI, MKSA

Unit (synonym, acronym)	Symbol	Physical quantity	Dimension	Conversion factor (SI equivalent unit)	Notes, definitions, other conversion factors	System
kelvin	–	energy, work, heat	ML^2T^{-2}	1 kelvin = 3.6×10^6 J (E)	Obsolete British name for the kilowatt-hour. Anecdotal. 1 kelvin = 3.6 MJ	UK
kelvin per pascal	$K.Pa^{-1}$	Joule-Thomson coefficient	$M^{-1}LT^2\Theta$	**SI derived unit**		SI
kelvin per watt	$K.W^{-1}$	thermal resistance	$M^{-1}L^{-2}T^3\Theta$	**SI derived unit** 1 $K.W^{-1}$ = 1 $kg^{-1}.m^{-2}.s^3.Q$ (E)		SI
ken (Japanese)	–	length, distance	L	1 ken (Japanese) = 1.818 m	Obsolete Japanese unit of length. 1 ken = 6 shaku (E)	Japanese
keramion (Egyptian)	–	capacity, volume	L^3	1 keramion (Egyptian) = 34×10^{-3} m^3 (of water)	Obsolete Egyptian unit of capacity used in ancient times. Measured by weight. 1 keramion (Egyptian) = 1 khar (E)	Egyptian
kerma	–	radiation absorbed dose, specific energy, kerma, index of absorbed dose	L^2T^{-2}	1 kerma = 1 $J.kg^{-1}$ (E)	The kerma is a unit employed in radiology. It corresponds to the kinetic energy transferred by neutral particles to charged particles per unit of mass of matter. The name of the unit is derived from the acronym of **k**inetic **e**nergy **r**eleased in **ma**terial.	
khar (Egyptian)	–	capacity, volume	L^3	1 khar (Egyptian) = 34×10^{-3} m^3 (of water)	Obsolete Egyptian unit of capacity used in ancient times. Measured by weight.	Egyptian
kharsha (Indian)	–	mass	M	1 kharsha (Indian) = 1.410×10^{-3} kg	Obsolete Indian unit of mass used in ancient times. 1 kharsha (Indian) = 10/3 pala (E)	Indian

					ancient times. 1 khô (Chinese) = 1/10 tcheng (E)	
khoul (Arabic) (woebe)	–	capacity, volume	L^3	1 khoul (Arabic) $= 16.32 \times 10^{-3}$ m^3 (of water)	Obsolete Arabic unit of capacity used in ancient times. Measured by weight. 1 khoul (Arabic) = 1/2 cafiz (E)	Arabic
khous (Greek gallon)	–	capacity, volume	L^3	1 khous (Greek gallon) $= 3.24 \times 10^{-3}$ m^3	Obsolete Greek unit of volume employed in ancient times. It was used for capacity measurements of liquids. 1 khous (Greek gallon) = 3 chenix (E)	Attic
kikkar (Arabic)	–	mass	M	1 kikkar (Arabic) = 42.5 kg	Obsolete Arabic unit of mass used in ancient times. System of the Prophet. 1 kikkar (Arabic) = 125 rotl (E)	Arabic
kilderkin (UK)	–	capacity, volume	L^3	1 kilderkin (UK) $= 81.82965600 \times 10^{-3}$ m^3	Obsolete British unit used for capacity measurements for all merchandise (solids, liquids, foodstuffs, etc). 1 kilderkin (UK) = 18 gallons (UK) (E) 1 kilderkin (UK) = 2 firkins (UK) (E)	UK
kilobyte (kilooctet)	KB, Ko	quantity of information	nil	1 KB = 1024 bytes (E)	Unit used in computer science. $1\ KB = 2^{10}$	
kilocalorie (4°C)	$kcal_4$	energy, work, heat	ML^2T^{-2}	$1\ kcal_4 = 4204.5$ J (E)	Obsolete multiple of a unit of energy.	
kilocalorie (15°C)	$kcal_{15}$	energy, work, heat	ML^2T^{-2}	$1\ kcal_{15} = 4185.5$ J (E)	Obsolete multiple of the calorie (15°C). It was equal to the heat needed to raise the temperature of one kilogram of air-free water from 14.5°C to 15.5°C at the constant pressure of one standard atmosphere (101 325 Pa). The use of the calorie should have ceased from December 31st, 1977.	

Unit (synonym, acronym)	Symbol	Physical quantity	Dimension	Conversion factor (SI equivalent unit)	Notes, definitions, other conversion factors	System
kilocalorie (20°C)	$kcal_{20}$	energy, work, heat	ML^2T^{-2}	1 $kcal_{20}$ = 4181.90 J (E)	Obsolete multiple of a unit of energy.	
kilocalorie (mean)	$kcal_{mean}$	energy, work, heat	ML^2T^{-2}	1 $kcal_{mean}$ = 4190.02 J (E)	Obsolete multiple of a unit of energy. The kilocalorie (mean) is equal to 1/100 of the heat needed to raise the temperature of one kilogram of air-free water from 0°C to 100°C at the constant pressure of one standard atmosphere (101 325 Pa). The use of the calorie should have ceased from December 31st, 1977.	
kilocalorie (thermochemical)	$kcal_{th}$	energy, work, heat	ML^2T^{-2}	1 $kcal_{th}$ = 4184.0 J (E)	Multiple of the obsolete unit of energy which was defined by the National Bureau of Standards (NBS) in 1953.	US
kilocalorie IT (International Steam Table)	$kcal_{IT}$	energy, work, heat	ML^2T^{-2}	1 $kcal_{IT}$ = 4186.74 J (E)	Multiple of the obsolete unit of energy which was used in steam data tables of Keenan and Keyes.	INT
kilogram	kg	mass	M	**SI base unit**	The kilogram is the unit of mass; it is equal to the mass of the International prototype of the kilogram [1st CGPM (1889), 3rd CGPM (1901)] The prototype of the standard is a cylinder of iridium-platinum alloy, 39 mm in diameter and 39 mm high. 1 kg = 10^3 g (E)	SI, MKSA
kilogram force-metre	kgfm	energy, work, heat	ML^2T^{-2}	1 kgfm = 9.80665 J (E)	Obsolete MKpS unit of energy. It was the work needed to displace horizontally without friction a weight of 1 kgf by 1 m.	MKpS

metre per second	s^{-1}					
kilogram per cubic decimetre	$kg.dm^{-3}$	density, mass density	ML^{-3}	$1\ kg.dm^{-3} = 10^3\ kg.m^{-3}$ (E)		
kilogram per cubic metre	$kg.m^{-3}$	density, mass density	ML^{-3}	**SI derived unit**	The kilogram per cubic metre is the unit of mass density of a homogenous body which has a mass of one kilogram and occupies a volume of one cubic metre.	SI
kilogram per metre	$kg.m^{-1}$	linear mass density	ML^{-1}	**SI derived unit**	The kilogram per metre is the SI derived unit of linear mass density. It is equal to the mass of one kilogram of a homogenous body of uniform section which is one metre in length.	SI
kilogram per second	$kg.s^{-1}$	mass flow rate	MT^{-1}	**SI derived unit**	The kilogram per second is the SI derived unit of mass flow rate. It corresponds to a uniform flow stream of a homogenous fluid of one kilogram in one second.	SI
kilogram per square metre	$kg.m^{-2}$	surface density, surface mass density	ML^{-2}	**SI derived unit**	The kilogram per square metre is the surface density of a homogenous body having an area of one square metre and which has a mass of one kilogram.	SI
kilogram per square metre	$kg.m^{-2}$	surface mass density	ML^{-2}	**SI derived unit**	The kilogram per square metre is the unit of surface mass density of a homogenous body having a uniform thickness and which has a mass of one kilogram per square metre of area.	SI
kilogram-force (kilogram-weight)	kgf, kgp, kg'	force, weight	MLT^{-2}	1 kgf = 9.80665 N (E)	Obsolete MKpS unit of force.	MKpS

Unit (synonym, acronym)	Symbol	Physical quantity	Dimension	Conversion factor (SI equivalent unit)	Notes, definitions, other conversion factors	System
kilogram-force per square centimetre	$kgf.cm^{-2}$	pressure, stress	$ML^{-1}T^{-2}$	1 $kgf.cm^{-2} = 9.80665 \times 10^4$ Pa (E)	Obsolete pressure and stress unit. Multiple of the MKpS unit. 1 $kgf.cm^{-2} = 0.980665$ bar (E) 1 $kgf.cm^{-2} = 1$ at (E) 1 atm = 1.033227452 $kgf.cm^{-2}$	MKpS
kilogram-force per square decimetre	$kgf.dm^{-2}$	pressure, stress	$ML^{-1}T^{-2}$	1 $kgf.dm^{-2} = 9.80665 \times 10^2$ Pa (E)	Obsolete pressure and stress unit. Multiple of the MKpS unit. 1 atm = 103.3227452 $kgf.dm^{-2}$	MKpS
kilogram-force per square metre	$kgf.m^{-2}$	pressure, stress	$ML^{-1}T^{-2}$	1 $kgf.m^{-2} = 9.806\ 65$ Pa (E)	Obsolete pressure and stress unit. 1 atm = 10 332.274520 $kgf.m^{-2}$	MKpS
kilogram-force per square millimetre	$kgf.mm^{-2}$	pressure, stress	$ML^{-1}T^{-2}$	1 $kgf.mm^{-2} = 9.80665 \times 10^6$ Pa (E)	Obsolete pressure and stress unit. Submultiple of the MKpS unit. 1 $kgf.mm^{-2} = 0.980665$ GPa (E) 1 atm = $1.033227453 \times 10^{-2}$ $kgf.mm^{-2}$	MKpS
kilogram-force-second per square metre	$kgf.s.m^{-2}$	dynamic viscosity, absolute viscosity	$ML^{-1}T^{-1}$	1 $kgf.s.m^{-2} = 9.80665$ Pa.s (E)	Obsolete MKpS unit of dynamic viscosity or absolute viscosity.	MKpS
kilogram-square metre	$kg.m^2$	moment of inertia	ML^2	**SI derived unit**		SI
kilometre	km	length, distance	L	1 km = 10^3 m (E)	Multiple of the SI base unit	SI
kilotonne equivalent TNT	kt (TNT)	energy, work, heat	ML^2T^{-2}	1 kt (TNT) = 4.18×10^{18} J	Obsolete unit of energy commonly used in seismology and in military applications. It serves to express the ratio of an explosion or seismic intensity in terms of the energy released by the explosion of one kilotonne of trinitrotoluene.	

kilowatt-hour	kWh	energy, work, heat	ML^2T^{-2}	1 kWh = 3.6×10^{6} J (E)	Unit of power, for business use. It was adopted in 1882 by the Board of Trade Orders. It was equal to the energy produced by an electric current of one thousand amperes flowing under a potential difference of one volt during one hour. Sometimes, the Board of Trade kelvin is used as synonym. 1 kWh = 3.6 MJ (E)	
kin (Japanese)	–	mass	M	1 kin (Japanese) = 6×10^{-1} kg (E)	Obsolete Japanese unit of mass. 1 kin (Japanese) = 0.16 kwan (E)	Japanese
king (Chinese)	–	surface, area	L^2	1 king (Chinese) = 6144 m^2	Obsolete Chinese unit of area used in ancient times. 1 king (Chinese) = 10 meou (E)	Chinese
kintal (cental, centner, hundredweight)	cH, cwt	mass	M	1 kintal = 45.359237 kg (E)	Obsolete British and American unit of mass used in business transactions, especially in agriculture. It served to measure the weight of grains. 1 cental = 1 cwt (E) 1 cental = 100 lb (av.) (E) 1 cental = 1 sh. cwt (US) (E)	UK, US
kip (kilopound)	kip	mass	M	1 kip = 453.59237 kg (E)	American unit of mass used by mechanical and civil engineers to express the weight of a construction. The name of the unit derived from the English acronym Kilo Imperial Pound. 1 kip = 1000 lb (av.) (E) 1 kip = 16 000 oz (av.) (E)	US

Unit (synonym, acronym)	Symbol	Physical quantity	Dimension	Conversion factor (SI equivalent unit)	Notes, definitions, other conversion factors	System
kip per square inch (kilopound-force per square inch)	$kip.in^{-2}$, ksi, KSI	pressure, stress	$ML^{-1}T^{-2}$	1 ksi = 6.89475729×10^{6} Pa	Obsolete British and American pressure and stress unit. 1 ksi = 1000 psi (E) 1 atm = $1.46959488 \times 10^{-2}$ ksi	UK, US
kip-force (kilopound-force)	kipf	force, weight	MLT^{-2}	1 kipf = 4448.2216152605 N	British and American unit of force employed in civil engineering. The name of the unit is derived from the acronym Kilo Imperial Pound-force. 1 kipf = 1000 lbf (E) 1 kip = 16 000 ozf (E)	UK, US
kish (Chinese)	–	surface, area	L^2	1 kish (Chinese) = 153.60 m^2	Obsolete Chinese unit of area used in ancient times. 1 kish (Chinese) = 1/4 meou (E)	Chinese
kist (Arabic)	–	capacity, volume	L^3	1 kist (Arabic) =1.36×10^{-3} m^3 (of water)	Obsolete Arabic unit of capacity used in ancient times. Measured by weight. 1 kist (Arabic) = 1/24 cafiz (E)	Arabic
kiyak-kin (Japanese)	–	mass	M	1 kiyak-kin (Japanese) = 60 kg (E)	Obsolete Japanese unit of mass. 1 kiyak-kin (Japanese) = 16 kwan (E)	Japanese
klafter (Austrian) [Austrian line]	–	length, distance	L	1 klafter (Austrian) = 1.896480 m	Obsolete Austrian unit of length. 1 klafter (Austrian) = 6 fuss (E)	Austrian
knot (noeud, naut. mile per hour)	kn, knot	velocity, speed	LT^{-1}	1 knot = $5.144444444 \times 10^{-1}$ $m.s^{-1}$	International unit of velocity employed in navigation. The unit dates from the late sixteenth century, when the speed of a ship was found by dropping a float tied to a knotted line (knotted log) over the side of the vessel. The knots were originally seven	INT

					fathoms apart. The number of knots passing in 30 seconds gave the speed of the ship in nautical miles per hour. 1 knot = 1 nautical mile per hour (E) 1 knot = 1852 $m.h^{-1}$ (E) 1 knot = 6080 $ft.h^{-1}$ (E)	
koku	–	capacity, volume	L^3	1 koku = 180.906837 × 10^{-3} m^3	Obsolete Japanese unit of capacity. 1 koku = 100 shô (E)	Japanese
kolläst (Swedish, dry)	–	capacity, volume	L^3	1 kolläst (Swedish, dry) = 1.978452 m^3	Obsolete Swedish unit of capacity for dry substances. 1 kolläst (Swedish, dry) = 756 kanna (E)	Swedish
koltunna (Swedish, dry)	–	capacity, volume	L^3	1 koltunna (Swedish, dry) = 164.871 × 10^{-3} m^3	Obsolete Swedish unit of capacity for dry substances. 1 koltunna (Swedish, dry) = 63 kanna (E)	Swedish
komma-ichi-da (Japanese)	–	mass	M	1 komma-ichi-da (Japanese) = 150 kg (E)	Obsolete Japanese unit of mass. 1 komma-ichi-da (Japanese) = 40 kwan (E)	Japanese
kona (Indian)	–	mass	M	1 kona (Indian) = 7.050 × 10^{-3} kg	Obsolete Indian unit of mass used in ancient times. 1 kona (Indian) = 20/3 pala (E)	Indian
kop (Dutch, dry)	–	capacity, volume	L^3	1 kop (Dutch, dry) = 0.851875 × 10^{-3} m^3	Obsolete Dutch unit of capacity used for dry substances. 1 kop (Dutch, dry) = 1/32 schepel (E)	Dutch
korn (Swedish)	–	mass	M	1 korn = 4.250797024 × 10^{-4} kg	Obsolete Swedish unit of weight. 1 korn = 1/1000 skålpund (E)	Swedish
krouchka (Russian)	–	capacity, volume	L^3	1 krouchka (Russian) = 1.229941 × 10^{-3} m^3	Obsolete Russian unit of capacity for liquid substances used before 1917. 1 krouchka (Russian) = 10 tcharka (E)	Russian

Unit (synonym, acronym)	Symbol	Physical quantity	Dimension	Conversion factor (SI equivalent unit)	Notes, definitions, other conversion factors	System
krushky (Russian, dry)	–	capacity, volume	L^3	1 krushky (Russian, dry) = $1.31193680 \times 10^{-3}$ m^3	Obsolete Russian unit of capacity for dry substances used before 1917. 1 krushky (Russian, dry) = 2/5 garnetz (E)	Russian
kung (Chinese)	–	surface, area	L^2	1 kung (Chinese) = 2.560 m^2	Obsolete Chinese unit of area used in ancient times. 1 kung (Chinese) = 1/240 meou (E)	Chinese
kunitz	–	enzymatic activity of ribonuclease	L^{-3}	(see note)	The unit was proposed in 1946 and employed in biochemistry. It is used to expressed the enzymatic activity of ribonuclease. One kunitz is the amount of ribonuclease required to cause a decrease of 100% per minute in the UV light (300 nm) absorbed at 25°C by a 0.05% solution of yeast nucleic acid in a 0.05 M solution of acetate buffer (pH 5). The unit is named after the American biochemist M. Kunitz.	
kwan	–	mass	M	1 kwan (Japanese) = 3.75 kg (E)	Obsolete Japanese unit of mass used in business transactions for pearls. 1 kwan = 1000 mommes (E)	Japanese
kwan (jun)	–	mass	M	1 kwan = 7.5 kg	Obsolete Chinese unit of mass. 1 kwan = 30 jin (E)	Chinese
kyne	$cm.s^{-1}$	velocity, speed	LT^{-1}	1 kyne = 10^{-2} $m.s^{-1}$ (E)	Name proposed in 1888 by the British Association for the cgs unit of velocity. Anecdotal.	

kyo (Chinese)	–	length, distance	L	1 kyo (Chinese) $= 7.666666667 \times 10^{-7}$ m	Obsolete Chinese unit of length used in ancient times. 1 kyo = 300 tchi (E)	Chinese
lambda	λ	capacity, volume	L^3	1 lambda $= 10^{-9}$ m^3 (E)	Obsolete unit of capacity employed in analytical chemistry. 1 lambda = 1 μl (E)	
lambert	L	luminous luminance	JL^{-2}	1 L $= 3.183 \times 10^3$ $cd.m^{-2}$	Obsolete American unit of luminous luminance. It was equal to the luminance of a surface equal to the emission of one lumen per square centimetre. The unit is named after J.H. Lambert (1728–1777). 1 L $= [1/\pi] \times 10^4$ $cd.m^{-2}$ (E) 1 L $= 10^4$ asb. (E)	US
lana (Russian)	–	mass	M	1 lana $= 3.412643160 \times 10^{-2}$ kg	Obsolete Russian unit of mass used before 1917 for general use. 1 loth = 1/12 funt (E)	Russian
langley	–	surface power density	MT^{-3}	1 langley $= 6.975833333 \times 10^2$ $W.m^{-2}$	Obsolete unit of surface power density. It was equal to half the solar radiation surface density on Earth. The unit is named after S.P. Langley (1834–1906). 1 Langley = 1 $cal_{15°}.cm^{-2}.min^{-1}$ (E)	
last (Dutch, dry)	–	capacity, volume	L^3	1 last (Dutch, dry) $= 2944 \times 10^{-3}$ m^3	Obsolete Dutch unit of capacity used for dry substances. 1 last (Dutch, dry) = 108 schepel (E)	Dutch
last (UK)	–	capacity, volume	L^3	1 last (UK) = 2.909498880 m^3	Obsolete British unit used for capacity measurements for all merchandise (solids, liquids, foodstuffs, etc). 1 last (UK) = 64 gal (UK) (E)	UK

Unit (synonym, acronym)	Symbol	Physical quantity	Dimension	Conversion factor (SI equivalent unit)	Notes, definitions, other conversion factors	System
league (Canadian)	–	length, distance	L	1 league (Canadian) = 4827 m	Obsolete Canadian unit of length.	CAN
league (international nautical)	leag. (int. naut.)	length, distance	L	1 league (int. naut.) = 5556 m	Obsolete international unit of length employed in navigation. 1 league (int. naut.) = 3 miles (int. naut.) (E)	INT, UK, US
league (statute, land)	leag. (stat.)	length, distance	L	1 league (statute) = 4828.032 m	Obsolete British and American unit of length. 1 league (statute) = 3 miles (statute) (E) 1 league (statute) = 960 rods (E)	UK, US
league (UK, nautical)	leag. (UK, naut.)	length, distance	L	1 league (UK, naut.) = 5559.552 m	Obsolete British unit of length used in navigation. 1 league (UK naut.) = 3 miles (UK naut.) (E) 1 league (UK naut.) = 3040 fathoms (E) 1 league (UK naut.) = 6080 yards (E)	UK
league (US, nautical)	leag. (US, naut.)	length, distance	L	1 league (US, naut.) = 5559.552 m	Obsolete American unit of length employed in navigation. 1 league (US, naut.) = 3 miles (US, naut.) (E) 1 league (US, naut.) = 3040 fathoms (E) 1 league (US, naut.) = 6080 yards (E)	US
legoa (Portuguese)	–	length, distance	L	1 legoa (Portuguese) = 6192 m	Obsolete Portuguese unit of length. 1 legoa (Portuguese) = 24 estadio (E)	Portuguese
legua (Spanish)	–	length, distance	L	1 legua (Spanish) = 4179.525 m	Obsolete Spanish unit of length. 1 legua (Spanish) = 5000 vara (E)	Spanish

lentor (Stokes)	St	kinematic viscosity	L^2T^{-1}	1 St = 10^{-4} $m^2.s^{-1}$ (E)	Obsolete cgs unit of kinematic viscosity. 1 St = 1 $mm^2.s^{-1}$ (E)	cgs
lenz	–	magnetic field strength	IL^{-1}	1 lenz = 1 $A.m^{-1}$ (E)	Obsolete unit.	
lenz (resistance)	–	electric resistance	$ML^2T^{-3}I^{-2}$	1 lenz = $8 \times 10^4 \Omega$	Obsolete unit of electrical resistance defined in 1838. It is the resistance of a no. 11 copper wire one foot in length.	
leo	–	acceleration	LT^{-2}	1 leo = 10 $m.s^{-2}$ (E)	Obsolete metric unit of acceleration. 1 leo = 1 $dam.s^{-2}$ (E)	
letech (Egyptian)	–	capacity, volume	L^3	1 letech (Egyptian) = 143.4375×10^{-3} m^3 (of water)	Obsolete Egyptian unit of capacity used in ancient times. Measured by weight. 1 letech (Egyptian) = 135/32 khar (E)	Egyptian
li (Chinese)	–	length, distance	L	1 li (Chinese) = 414 m	Obsolete Chinese unit of length. 1 li = 1800 tchi (E)	Chinese
liang	–	mass	M	1 liang = 15.625×10^{-3} kg	Obsolete Chinese unit of mass. 1 liang = 1/16 jin (E)	Chinese
libbra (Italian)	–	mass	M	1 libbra (Italian) = 0.307 kg	Obsolete Italian unit of mass.	Italian
libra (Portuguese)	–	mass	M	1 libra (Portuguese) = 0.459 kg	Obsolete Portuguese unit of mass.	Portuguese
libra (Roman pound)	–	mass	M	1 libra (Roman pound) = 327×10^{-3} kg	Obsolete Roman unit of mass used in ancient times. 1 libra (Roman pound) = 12 unciae (Roman) (E)	Roman
libra (Spanish)	–	mass	M	1 libra (Spanish) = 0.460093 kg	Obsolete Spanish unit of mass.	Spanish
liespund (Swedish)	–	mass	M	1 liespund = 8.501594048 kg	Obsolete Swedish unit of weight. 1 liespund = 20 skålpund (E)	Swedish

Unit (synonym, acronym)	Symbol	Physical quantity	Dimension	Conversion factor (SI equivalent unit)	Notes, definitions, other conversion factors	System
lieue (de Poste)	–	length, distance	L	1 lieue (de Poste) = 3898 m	Obsolete French unit of length used before the French Revolution (1789). 1 lieue (de Poste) = 2000 toises (de Paris) (E)	French
lieue (metric) (French metric league)	–	length, distance	L	1 lieue (metric) = 4000 m (E)	Obsolete French metric unit of length used in France from 1812 to 1840. 1 lieue (metric) = 2000 toises (metric) (E)	French
lieue marine [French nautical league]	–	length, distance	L	1 lieue marine = 2850.2 m	Obsolete French unit of length used in navigation before the French Revolution (1789). 1 lieue marine = 2850.4 toises (de Perou) (E)	French
light-year (année-lumière)	ly (AL, al)	length, distance	L	1 ly = $9.46052973 \times 10^{15}$ m	Unit of distance employed in astronomy and astrophysics. The light-year corresponds to the distance travelled in one tropical year by electromagnetic radiation in vacuum.	@
ligne (de Paris) (French line)	ligne	length, distance	L	1 ligne (de Paris) = $2.255829282 \times 10^{-3}$ m	Obsolete French unit of length. It was employed before the French Revolution (1789). It is still used in botany to describe the dimensions of plants. 1 ligne (de Paris) = 1/864 toises (de Paris) (E) 1 ligne (de Paris) = 12 points (de Paris) (E) 1 ligne (de Paris) = 1/12 pouce (de Paris) (E)	French

ligne (metric) (French metric line)		length, distance	L	1 ligne (metric) = [illegible] m	Obsolete French unit of length used in France from 1812 to 1840. 1 ligne (metric) = 1/12 pouce (metric) (E)	French
line	–	magnetic induction flux	$ML^2T^{-2}I^{-1}$	1 line = 10^{-8} Wb (E)	Obsolete British and American unit of magnetic flux. 1 line = 1 maxwell (E)	US
line (Prussian)	–	length, distance	L	1 linie (Prussian) = $2.179562500 \times 10^{-3}$ m	Obsolete German unit of length. 1 linie = 1/144 fuss (E)	German
line (UK) (UK button)	line (UK)	length, distance	L	1 line (UK) = 2.1166667×10^{-3} m	Obsolete British unit of length. 1 line (UK) = 1/12 inch (E) 1 line (UK) = 1/144 foot (E) 1 line (UK) = 1/432 yard (E) 1 line (UK) $\approx$ 2.117 mm	UK
line (US) (US button)	line (US)	length, distance	L	1 line (US) = 6.35×10^{-4} m (E)	Obsolete American unit of length. It was used in botany for plant measurements. 1 line (US) = 1/40 inch (E) 1 line (US) = 1/480 foot (E) 1 line (US) = 1/1440 yard (E) 1 line (US) = 0.635 mm (E)	US
linea (Spanish)	–	length, distance	L	1 linea (Spanish) = $1.451223959 \times 10^{-3}$ m	Obsolete Spanish unit of length. 1 linea (Spanish) = 1/576 vara (E)	Spanish
linen (lea)	–	specific length	$M^{-1}L$	1 linen = 604.772571 $m.kg^{-1}$	Obsolete American and British unit employed in the textile industry. 1 linen = 300 $yd.lb^{-1}$ (E)	UK, US
linha (Portuguese)	–	length, distance	L	1 linha (Portuguese) = 2.281250×10^{-3} m	Obsolete Portuguese unit of length. 1 linha (Portuguese) = 1/144 pe (E)	Portuguese
linie (Austrian) [Austrian line]	–	length, distance	L	1 linie (Austrian) = 2.195000×10^{-3} m	Obsolete Austrian unit of length. 1 linie (Austrian) = 1/144 fuss (E)	Austrian

Unit (synonym, acronym)	Symbol	Physical quantity	Dimension	Conversion factor (SI equivalent unit)	Notes, definitions, other conversion factors	System
linie (German) [Prussian line]	–	length, distance	L	1 linie (Prussian) $= 2.179562500 \times 10^{-3}$ m	Obsolete German unit of length. 1 linie = 1/144 fuss (E)	German
linie (Swedish) [Swedish line]	–	length, distance	L	1 linie (Swedish) $= 2.061805556 \times 10^{-3}$ m	Obsolete Swedish unit of length.	Swedish
link (engineer's)	–	length, distance	L	1 link (engineer's) = 0.3048 (E)	Obsolete American surveyor's unit of length. 1 link (engineer's) = 1/100 chain (engineer's) (E)	US
link (Gunter's)	–	length, distance	L	1 link (Gunter's) = 0.201168 m (E)	Obsolete British surveyor's unit of length. 1 link (Gunter's) = 1/100 chain (Gunter's) (E)	UK
link (Ramsden's)	–	length, distance	L	1 link (Ramsden's) = 0.3048 m (E)	Obsolete American surveyor's unit of length. 1 link (Ramsden's) = 1/100 chain (Ramsden's) (E)	US
link (surveyor's)	–	length, distance	L	1 link (surveyor's) = 0.201168 m (E)	Obsolete British surveyor's unit of length. 1 link (surveyor's) = 1/100 chain (surveyor's) (E)	UK
litre (1964)	L, l	capacity, volume	L^3	1 l (1964) $= 10^{-3}$ m^3 (E)	Unit approved by the 12th CGPM (1964). 1 l = 1 dm^3 (E)	INT
litre (Obsolete)	L, l	capacity, volume	L^3	1 l (old) $= 1.0000028001 \times 10^{-3}$ m^3	The obsolete litre was equal to the volume occupied by one cubic decimetre of pure water measured at the temperature of maximum density (4°C).	French

litre per kilogram	$l.kg^{-1}$	specific volume	L^3M^{-1}	$1\ l.kg^{-1} = 10^{-3}\ m^3.kg^{-1}$		
litre per minute	$l.min^{-1}$	volume flow rate	L^3T^{-1}	$1\ l.min^{-1} = 1.666667 \times 10^{-5}\ m^3.s^{-1}$	$1\ l.min^{-1} = (10^{-3}/60)\ m^3.s^{-1}$ (E)	
litre per second	$l.s^{-1}$	volume flow rate	L^3T^{-1}	$1\ l.s^{-1} = 10^{-3}\ m^3.s^{-1}$ (E)		
litre-atmosphere	l.atm	energy, work, heat	ML^2T^{-2}	1 l.atm = 101.325 J (E)	Obsolete unit of energy usually used in chemistry.	
litron (de Paris, dry)	–	capacity, volume	L^3	1 litron (de Paris, dry) $= 1.164237500 \times 10^{-3}\ m^3$	Obsolete French unit of volume employed before the French Revolution. It was used for capacity measurements of dry substances. 1 litron = 1/16 boisseau (E)	French
livre (de Paris) [French pound]	–	mass	M	1 livre (de Paris) $= 489.505850 \times 10^{-3}$ kg	Obsolete French unit of mass employed before the French Revolution (1789). 1 livre (de Paris) = 16 onces (de Paris).	French
livre de Charlemagne	–	mass	M	1 livre de Charlemagne = 0.367128 kg	Obsolete French unit of mass employed before the French Revolution (1789).	French
load (UK)	–	mass	M	1 load (UK) = 587.8556701 kg	Obsolete British unit of mass used by farmers. 1 load (UK) = 1296 pounds (UK, straw) (E)	UK
load (UK, wool)	–	mass	M	1 load (UK, wool) = 952.543977 kg	Obsolete British unit of mass used in the weighing of wool. 1 load (UK, wool) = 2100 lb (av.) (E) 1 load (UK, wool) = 6 sacks (UK, weight) (E) 1 load (UK, wool) = 12 weys (UK, weight) (E)	UK

Unit (synonym, acronym)	Symbol	Physical quantity	Dimension	Conversion factor (SI equivalent unit)	Notes, definitions, other conversion factors	System
lof (Russian, dry)	–	capacity, volume	L^3	1 lof (Russian, dry) $= 64.722216 \times 10^{-3}$ m^3	Obsolete Russian unit of capacity for dry substances used before 1917. 1 lof (Russian, dry) = 592/30 garnetz (E)	Russian
log (Hebrew)	–	capacity, volume	L^3	1 log (Hebrew) $= 2.975 \times 10^{-4}$ m^3 (of water)	Obsolete Hebrew unit of capacity used in ancient times. Measured by weight. 1 log (Hebrew) = 1/72 ephah (new)	Hebrew
long ton (UK)	lg ton (UK)	mass	M	1 long ton (UK) = 1016.046909 kg	Obsolete British unit of mass. 1 long ton (UK) = 2240 lb (E)	UK
lorentz	–	wavelength resolution for spectral lines per unit of magnetic induction	$M^{-1}L^{-1}T^2I$	1 lorentz = 46.7 $m^{-1}.T^{-1}$	Obsolete unit of spectral resolution employed in atomic spectroscopy (Zeeman effect). It was equal to the wavenumber per unit of magnetic field induction. The unit is named after the Dutch physicist H.A. Lorentz (1853–1928). It is equal to the ratio of the Bohr magneton divided by *hc*. 1 lorentz $= \mu_B.h^{-1}c = e/4\pi mc^2$	
lot (de Paris)	–	mass	M	1 lot (de Paris) $= 15.297057810 \times 10^{-3}$ kg	Obsolete French unit of mass employed under the Ancien Régime before the French Revolution (1789). 1 lot (de Paris) = 1/32 livre (de Paris) (E)	French
loth (Austrian)	–	mass	M	1 loth (Austrian) $= 1.750031250 \times 10^{-2}$ kg	Obsolete Austrian unit of mass for general uses. 1 loth (Austrian) = 1/32 pfund (E)	Austrian
loth (Prussian)	–	mass	M	1 loth (Prussian) $= 1.461596875 \times 10^{-2}$ kg	Obsolete German unit of mass. 1 loth (Prussian) = 1/32 pfund (E)	German

[Russian lot]					1917 for general use. 1 loth = 1/32 funt (E)	
lumberg (lumerg)	–	radiant intensity	$ML^2T^{-3}\Omega^{-1}$	1 lumberg = 10^{-7} W.sr^{-1} (E)	Obsolete cgs unit of radiant intensity. 1 lumberg = 1 erg.s^{-1}.sr^{-1} (E)	cgs
lumen	lm	luminous flux	$J\Omega$	**SI derived unit** 1 lm = 1 cd.sr (E)	The lumen is the luminous flux emitted in a solid angle of one steradian by a point source having a uniform intensity of one candela.	SI
lusec	–	volume flow rate	L^3T^{-1}	1 lusec = 10^{-3} m^3.s^{-1} (E)	Obsolete British unit of flow rate used in vacuum technology. It is equal to the flow rate of pumping under a pressure of 0.1133 322 Pa. The name of the unit derived from the acronym of litre per **second**. 1 lusec = 1 dm^3.s^{-1} (E)	UK
lux	lx	illuminance	$JL^{-2}\Omega^{-1}$	**SI derived unit** 1 lx = 1 lm.m^{-2} = 1 cd.m^{-2}.sr^{-1} (E)	The lux is the illuminance produced by a luminous flux of one lumen uniformly distributed over a surface of one square metre.	SI
luxon (troland)	–	luminous luminance	$J.L^{-2}$	1 luxon = 10^4 cd.m^{-2} (E)	Obsolete unit of luminous luminance employed in ophtalmology. It was equal to the retinal luminous luminance received by eyes by one surface having a luminous luminance of one candela per square metre. The optical aperture of eyes is about one square millimetre.	
luxon (photon)	hν	energy of electromagnetic radiation	$ML^{-2}T^{-2}$	1 hν = $6.626075540 \times 10^{-34}$ ν(Hz)	Obsolete unit of quantum of luminous energy transported by an electromagnetic radiation of frequency ν (Hz).	

Unit (synonym, acronym)	Symbol	Physical quantity	Dimension	Conversion factor (SI equivalent unit)	Notes, definitions, other conversion factors	System
lyi (Chinese)	–	surface, area	L^2	1 lyi (Chinese) = 6.144 m^2	Obsolete Chinese unit of area used in ancient times. 1 lyi (Chinese) = 1/100 meou (E)	Chinese
lyne (Dutch) [Dutch line]	–	length, distance	L	1 lyne (Dutch) = 2.144×10^{-3} m	Obsolete unit of length used in Amsterdam (Netherlands).	Dutch
ma (Chinese)	–	capacity, volume	L^3	1 ma (Chinese) = 3.276800 m^3	Obsolete Chinese unit of capacity used in ancient times. 1 ma (Chinese) = cubic tchi (E)	Chinese
Mache unit	–	radioactivity	T^{-1}	1 Mache unit = 13.2 Bq	Obsolete unit of radioactivity employed in radiochemistry. It was defined in 1930 by The International Radium Standards Committee. One Mache unit is equal to the amount of radon $^{222}_{86}Rn$ (emanation) needed to produce a saturated current equal to 10^{-3} esu cgs. The unit is named after the Austrian scientist H. Mache (1876-1954). 1 Mache unit = 3.6×10^{-10} Ci	
magnum	–	capacity, volume	L^3	1 magnum = $1.515364000 \times 10^{-3}$ m^3	Obsolete British unit which expressed the capacity of a wine container. Still employed in oenology, especially in France. 1 magnum = 1/3 gallons (UK) (E) 1 magnum = 2 bouteilles (E)	UK, French
makuk (Arabic)	–	capacity, volume	L^3	1 makuk (Arabic) = 4.08×10^{-3} m^3 (of water)	Obsolete Arabic unit of capacity used in ancient times. Measured by weight. 1 makuk (Arabic) = 1/8 cafiz (E)	Arabic

makuk (Persian)	–	capacity, volume	L^3	1 makuk (Persian) $= 4.075 \times 10^{-3}$ m^3 (of water)	Obsolete unit of capacity of the Assyrio-Chaldean-Persian system used in ancient times. Measured by weight. 1 makuk (Persian) = 1/8 amphora (E)	Persian
man (Arabic)	–	mass	M	1 man (Arabic) = 0.680 kg	Obsolete Arabic unit of mass used in ancient times. System of the Prophet. 1 man (Arabic) = 2 rotl (E)	Arabic
man (Egyptian)	–	capacity, volume	L^3	1 man (Egyptian) $= 8.5 \times 10^{-4}$ m^3 (of water)	Obsolete Egyptian unit of capacity used in ancient times. Measured by weight. 1 man (Egyptian) = 1/40 khar (E)	Egyptian
mansion (Persian) [stathmos]	–	length, distance	L	1 mansion (Persian) $= 2.56 \times 10^4$ m	Obsolete unit of length in the Assyrio-Chaldean-Persian system used in ancient times. 1 mansion = 80 000 zereths (E)	Persian
marc (de Paris)	–	mass	M	1 marc (de Paris) $= 244.752925 \times 10^{-3}$ kg	Obsolete French unit of mass employed under the Ancien Régime (1789). The name of the unit is derived from Francic *marca*. 1 marc (de Paris) = 8 onces (de Paris) (E) 1 marc (de Paris) = 1/2 livre (de Paris) (E)	French
marco (Spanish)	–	mass	M	1 marco (Spanish) $= 230.046500 \times 10^{-3}$ kg	Obsolete Spanish unit of mass. 1 marco (Spanish) = 1/2 libra (E)	Spanish
marhala (Arabic)	–	length, distance	L	1 marhala (Arabic) $= 4.608 \times 10^4$ m	Obsolete Arabic unit of length used in ancient times. 1 marhala (Arabic) = 144 000 feet (Arabic) (E)	Arabic

Unit (synonym, acronym)	Symbol	Physical quantity	Dimension	Conversion factor (SI equivalent unit)	Notes, definitions, other conversion factors	System
maris (Greek, Attic)	–	capacity, volume	L^3	1 maris (Greek, Attic) $= 2.16 \times 10^{-3}$ m^3	Obsolete Greek unit of volume employed in ancient times. It was used for capacity measurements of dry substances. 1 maris (Greek, Attic) = 2 chenix (E)	Attic
mark (Austrian)	–	mass	M	1 mark (Austrian) = 0.280005 kg	Obsolete Austrian unit of mass for general use. 1 mark (Austrian) = 1/2 pfund (E)	Austrian
mark (Dutch)	–	mass	M	1 mark (Dutch) $= 2.460838600 \times 10^{-1}$ kg	Obsolete Dutch unit of mass. 1 mark (Dutch) = 1/2 pond (Dutch) (E)	Dutch
marok (Hungarian)	–	length, distance	L	1 marok (Hungarian) = 0.105360 m	Obsolete Hungarian unit of length.	Hungarian
masha (Indian)	–	mass	M	1 masha (Indian) $= 8.812500 \times 10^{-4}$ kg	Obsolete Indian unit of mass used in ancient times. 1 masha (Indian) = 160/3 pala (E)	Indian
mass (Austrian)	–	capacity, volume	L^3	1 mass (Austrian) $= 1.4151 \times 10^{-3}$ m^3	Obsolete Austrian unit of capacity used for liquid substances.	Austrian
mathusalem (methuselah)	–	capacity, volume	L^3	1 mathusalem $= 6.819138000 \times 10^{-3}$ m^3	Obsolete British unit which expressed the capacity of wine container. Still employed in oenology, especially in France. 1 mathusalem = 3/2 gallons (UK) (E) 1 mathusalem = 9 bouteilles (E)	UK, French
maxwell	Mx	magnetic induction flux	$ML^2T^{-2}I^{-1}$	1 Mx $= 10^{-8}$ Wb (E)	Obsolete cgs unit of magnetic flux. The unit is named after J.C. Maxwell (1831–1879).	cgs

mayer	–	specific heat capacity	$L^2T^{-2}\Theta^{-1}$	1 Mayer = 10^3 J.kg^{-1}.K^{-1} (E)	Obsolete unit of heat capacity used in heat transfer engineering. The unit is named after J.R. Mayer (1814–1878). 1 mayer = 1 J.g^{-1}.°C^{-1} (E)	
mean solar time	–	time, period, duration	T	(see note)	Time scale related to solar time. It corresponds to the universal time scale deduced from mean solar time by corrections of its secular inequalities and periodicity and which has its reference at twelve o'clock.	
mease (UK) (cran)	cran	capacity, volume	L^3	1 mease (UK) = 170.47845×10^{-3} m^3	Obsolete British unit of capacity employed in the fishing industry. It describes an amount of herring and it is equal to the number of herring which can be packed into a standard box with a volume of 37.5 gallons (UK). 1 mease ≈ 750 herrings	UK
medimne (Greek, Attic)	–	capacity, volume	L^3	1 medimne (Greek, Attic) = 51.84×10^{-3} m^3	Obsolete Greek unit of volume employed in ancient times. It was used for capacity measurements of dry substances. 1 medimne (Greek, Attic) = 48 chenix (E)	Attic
medio (Spanish, dry)	–	capacity, volume	L^3	1 medio (Spanish, dry) = $2.312541670 \times 10^{-3}$ m^3	Obsolete Spanish unit of capacity used for dry substances. 1 medio (Spanish, dry) = 1/24 fanega (E)	Spanish
megabyte (mégaoctet)	Mo, MB	quantity of information	nil	1 MB = 1 048 576 byte	Unit used in computer science. 1 MB = 2^{20}	
megagram	Mg	mass	M	1 Mg = 10^3 kg (E)	Multiple of the SI base unit. 1 Mg = 10^6 g (E)	SI
megametre	Mm	length, distance	L	1 Mm = 10^6 m (E)	Multiple of the SI base unit.	SI

Unit (synonym, acronym)	Symbol	Physical quantity	Dimension	Conversion factor (SI equivalent unit)	Notes, definitions, other conversion factors	System
megapascal	MPa	pressure, stress	$ML^{-1}T^{-2}$	1 MPa = 10^6 Pa (E)	Multiple of the SI derived unit. 1 atm = 0.101325 MPa (E)	SI
megatonne equivalent TNT	Mt (TNT)	energy, work, heat	ML^2T^{-2}	1 Mt (TNT) = 4.18×10^{21} J	Obsolete unit of energy commonly used in seismology and in military applications. It serves to express the ratio of an explosion or seismic intensity with the energy release by the explosion of one megatonne of trinitrotoluene.	
mehah (Hebrew) [Talmudic system]	–	mass	M	1 mehah (Hebrew) = 5.903333×10^{-4} kg	Obsolete Hebrew unit of mass used in ancient times. Rabbinacal or Talmudic system. 1 mehah (Hebrew) = 1/600 mina (E)	Hebrew
meile (Austrian) [Austrian mile]	–	length, distance	L	1 meile (Austrian) = 7585.92 m	Obsolete Austrian unit of length. 1 meile = 24 000 fuss (E)	Austrian
meile (German) [Prussian mile]	–	length, distance	L	1 meile (Prussian) = 7532.4 m	Obsolete German unit of length. 1 meile = 24 000 fuss (E)	German
meile (Hungarian)	–	length, distance	L	1 meile (Hungarian) = 8353.6 m	Obsolete Hungarian unit of length.	Hungarian
meio (Portuguese)	–	capacity, volume	L^3	1 meio (Portuguese) = 0.6875×10^{-3} m^3	Obsolete Portuguese unit of capacity used for liquid substances. 1 meio (Portuguese) = 1/24 almude (E)	Portuguese
meio (Portuguese)	–	mass	M	1 meio (Portuguese) = 0.2295 kg	Obsolete Portuguese unit of mass. 1 meio (Portuguese) = 1/2 libra (E)	Portuguese
meio (Portuguese, dry)	–	capacity, volume	L^3	1 meio (Portuguese, dry) = 6.75×10^{-3} m^3	Obsolete Portuguese unit of capacity used for dry substances. 1 meio (Portuguese, dry) = 1/8 fanga (E)	Portuguese

		pitch			threshold is taken to be 1000 mels. The pitch of any sound judged to be double that pitch is taken to be 2000 mels, etc. The name mel is derived from the first three letters of the word **mel**ody.	
meou (Chinese)	–	surface, area	L^2	1 meou (Chinese) = 614.4 m^2	Obsolete Chinese unit of area used in ancient times.	Chinese
mertföld (Hungarian) [Hungarian mile]	–	length, distance	L	1 mertföld (Hungarian) = 8353.6 m	Obsolete Hungarian unit of length.	Hungarian
mesh	mesh	aperture of sieves opening of testing sieves, mesh	nil	(see note)	The mesh is the unit which describes the number of apertures of a sieve per linear inch. It depends on the geometry of the open area, on wire diameter, etc. It exists in several series (ISO, Tyler, AFNOR, ASTM, BS, DIN, IMM, etc). In general, in the same series the number of meshes follows a geometric progression with a defined ratio which depends on the scale. 10 mesh (US) = 2 mm (approx.) 100 mesh (US) = 150 μm (approx.) 400 mesh (US) = 38 μm (approx.)	UK, US
metre	m	length, distance	L	**SI base unit**	The metre is the length of the path travelled by light in vacuum during a time interval of 1/299792458 of a second (17th CGPM (1983), Resolution 1). The unit is named after the Greek word *metron* meaning measure.	SI, MKSA, MTS

Unit (synonym, acronym)	Symbol	Physical quantity	Dimension	Conversion factor (SI equivalent unit)	Notes, definitions, other conversion factors	System
metre of water (4°C)	mH_2O, mCE (4°C)	pressure, stress	$ML^{-1}T^{-2}$	1 mH_2O (4°C) = 9806.375413 Pa	Unit of pressure employed in physics to measure small pressures. It is equal to the pressure exerted by a column of water one metre high, measured at 4°C (39.2°F). 1 atm = 10.33256383 mH_2O (4°C)	UK, US
metre of water (15.56°C)	mH_2O, mCE (15.56°C)	pressure, stress	$ML^{-1}T^{-2}$	1 mH_2O (15.56°C) = 9797.059096 Pa	Unit of pressure employed in physics to measure small pressures. It is equal to the pressure exerted by a column of water one metre high, measured at 15.56°C (60°F). 1 atm = 10.334238938 mH_2O (60°F)	UK, US
metre of water (39.2°F)	mH_2O, mCE (39.2°F)	pressure, stress	$ML^{-1}T^{-2}$	1 mH_2O (39.2°F) = 9806.375413 Pa	Unit of pressure employed in physics to measure small pressures. It is equal to the pressure exerted by a column of water one metre high, measured at 4°C (39.2°F). 1 atm = 10.33256383 mH_2O (39.2°F)	UK, US
metre of water (60°F)	mH_2O, mCE (60°F)	pressure, stress	$ML^{-1}T^{-2}$	1 mH_2O (60°F) = 9797.059096 Pa	Unit of pressure employed in physics to measure small pressures. It is equal to the pressure exerted by a column of water one metre high, measured at 15.56°C (60°F). 1 atm = 10.334238938 mH_2O (60°F)	UK, US
metre per second	$m.s^{-1}$	velocity, speed	LT^{-1}	**SI derived unit**	The metre per second is the SI derived unit of linear velocity. It is the speed of a body which moves according to a uniform displacement and which covers a distance of one metre in one second.	SI

second					derived unit of acceleration of an object which undergoes a variation of speed of one metre per second in one second.	
metrete (Greek, Attic)	–	capacity, volume	L^3	1 metrete (Greek, Attic) $= 38.38 \times 10^{-3}\ m^3$	Obsolete Greek unit of volume employed in ancient times. It was used for capacity measurements of liquids. 1 metrete (Greek, Attic) = 36 chenix (E)	Attic
metrete of Heron (Egyptian)	–	capacity, volume	L^3	1 metrete of Heron (Egyptian) $= 42.5 \times 10^{-3}\ m^3$ (of water)	Obsolete Egyptian unit of capacity used in ancient times. Measured by weight. 1 metrete of Heron (Egyptian) = 5/4 khar (E)	Egyptian
metric	–	specific length	$M^{-1}L$	1 metric $= 10^{-3}$ kg.m^{-1} (E)	Obsolete metric unit employed in the textile industry. 1 metric = 1 kg.(1000 m)$^{-1}$ (E) 1 metric = 1 g.m^{-1} (E)	
metric slug (hyl, mug, par, TME, techma)	mug	mass	M	1 mug = 9.80665 kg (E)	Obsolete technical metric unit of mass employed by mechanical engineers (base unit of the metric gravitational system). It was equal to the mass which under an acceleration of 1 m.s^{-2} gives a force of 1 kgf. The name TME is derived from the German acronym Technische Mass Einheit. The name mug is derived from the abbreviation of **metric slug**.	German
metze (Austrian)	–	surface, area	L^2	1 metze (Austrian) = 1918.206074 m^2	Obsolete Austrian unit of area. 1 metze (Austrian) = 1/3 joch (E)	Austrian
metze (Austrian, dry)	–	capacity, volume	L^3	1 metze (Austrian, dry) $= 61.489 \times 10^{-3}\ m^3$	Obsolete Austrian unit of capacity used for dry substances.	Austrian

Unit (synonym, acronym)	Symbol	Physical quantity	Dimension	Conversion factor (SI equivalent unit)	Notes, definitions, other conversion factors	System
metze (Prussian, dry)	–	capacity, volume	L^3	1 metze (Prussian, dry) $= 3.435890 \times 10^{-3}$ m^3	Obsolete German unit of capacity used for dry substances.	German
mho	mho	electric conductance	$M^{-1}L^{-2}T^3I^2$	1 mho = 1 S (E)	Obsolete unit of conductance. The name derived from the reverse writing of the word ohm.	INT
mic	–	electric inductance	$ML^2T^{-2}I^{-2}$	1 mic $= 10^{-6}$ H (E)	Obsolete unit of electric inductance employed by the Royal Navy at the beginning of the century.	UK
microbar (barye, barrie, dyn.cm^{-2})	μbar	pressure, stress	$ML^{-1}T^{-2}$	1 μbar $= 10^{-1}$ Pa (E)	Obsolete cgs derived unit of pressure and stress. 1 μbar = 1 dyn.cm^{-2} (E) 1 μbar = 1 barye (E) 1 atm = 1 013 250 μbar	cgs
microgamma (picogram)	γγγ	mass	M	1 γγγ $= 10^{-15}$ kg (E)	Obsolete unit of mass employed in analytical chemistry. 1 γγγ $= 10^{-12}$ g (E) 1 γγγ = 1 pg (E)	
microgram (gamma)	μg	mass	M	1 μg $= 10^{-9}$ kg (E)	Submultiple of the SI base unit. 1 μg $= 10^{-6}$ g (E)	SI
microinch	μin	length, distance	L	1 μin $= 2.54 \times 10^{-8}$ m (E)	American and British submultiple of the inch.	UK, US
microlitre	μl	capacity, volume	L^3	1 μl $= 10^{-9}$ m^3 (E)	1 μl = 1 mm^3 (E)	
micrometre	μμ	length, distance	L	1 μm $= 10^{-6}$ m (E)	Submultiple of the SI base unit.	SI

(bicron)					atomic spectroscopy. 1 μμ = 1 pm (E)	
micron	μ	length, distance	L	1 μ = 10^{-6} m (E)	Obsolete unit of wavelength unit used in atomic spectroscopy. 1 μ = 1 μm (E)	
micron of mercury (millitorr)	μHg (0°C)	pressure, stress	$ML^{-1}T^{-2}$	1 μHg (0°C) = 0.1333223684 Pa	Obsolete unit of pressure employed in physics to measure small pressures. It is equal to the pressure exerted by a column of mercury one micrometre high, measured at 0°C (32°F). 1 μHg (0°C) = 10^{-3} mmHg (0°C) (E) 1 μHg (0°C) = 10^{-3} torr (E) 1 atm = 760 000 μHg (0°C) (E) 1 μHg (0°C) = (101325/760000) Pa (E)	
miglio (Italian)	–	length, distance	L	1 miglio (Italian) = 2226.336667 m	Obsolete Italian unit of length. 1 miglio (Italian) = 13 000/3 piedi liprando (E)	Italian
mignonne	–	length, distance	L	1 mignonne = 2.631800695 × 10^{-3} m	Obsolete French unit of length employed in typography. It was a submultiple of the point (Didot). 1 mignonne = 7 points (Didot) (E)	French
mil (Swedish)	–	length, distance	L	1 mil (Swedish) = 5344.20 m	Obsolete Swedish unit of length.	Swedish
mil (thou)	mil	length, distance	L	1 mil = 2.54 × 10^{-5} m (E)	American and British submultiple of the inch. 1 mil = 1 thou (E) 1 mil = 10^{-1} calibre (E) 1 mil = 10^{-3} inch (E) 1 mil = 25.4 μm (E)	UK, US

Unit (synonym, acronym)	Symbol	Physical quantity	Dimension	Conversion factor (SI equivalent unit)	Notes, definitions, other conversion factors	System
mil per year	mpy	rate of corrosion	LT^{-1}	1 mpy = 8.043269×10^{-13} $m.s^{-1}$	American and British unit used by corrosion engineers to express the rate of corrosion of metals and alloys. 1 mpy = 25.4 μm/year (E)	UK, US
mile (geographical)	mile (geogr.)	length, distance	L	1 mile (geographical) = 7421.591 m	Obsolete British and American unit of length. It was equal to the length which is subtended by an arc of 4 minutes at the Equator.	UK, US
mile (international nautical)	mi (int. naut.)	length, distance	L	1 mile (int. naut.) = 1852 m (E)	Legal international unit of length temporarily maintained with the SI. It is still used in navigation (mercantile marine, aviation) . It is equal to the length of an arc of one minute measured at N45° latitude. The int. nautical mile has been taken equal to the nautical mile since 1970. 1 mile (int. naut.) = 6076.11 feet	INT
mile (international)	mi (int.)	length, distance	L	1 mile (int.) = 1609.347 m	Obsolete International unit of length used in navigation (mercantile marine, aviation) 1 mile (int.) = 1760 yards (E)	INT
mile (naut.) per hour (knot, noeud)	knot	velocity, speed	LT^{-1}	1 knot = $5.144444444 \times 10^{-1}$ $m.s^{-1}$	International unit of velocity employed in navigation. The unit dates from the late sixteenth century, when the speed of a ship was found by dropping a float tied to a knotted line (knotted log) over the side of the vessel. The knots were originally seven fathoms apart. The number of knots passing in 30 seconds gave the speed of the	INT

					1 knot = 1852 $m.h^{-1}$ (E) 1 knot = 6080 $ft.h^{-1}$ (E)	
mile (Prussian) [meile]	–	length, distance	L	1 meile (Prussian) = 7532.4 m	Obsolete German unit of length. 1 meile = 24 000 fuss (E)	German
mile (statute, land)	mi (stat.)	length, distance	L	1 mile (statute) = 1609.344 m	Obsolete British unit of length employed for land distance measurements. 1 statute mile = 1.603344 km 1 statute mile = 5280 feet (E) 1 statute mile = 320 rods (E)	UK
mile (telegraph, nautical)	mi (teleg., naut.)	length, distance	L	1 mile (teleg., naut.) = 1855.3176 m	Obsolete British and American unit of length employed in navigation. It is equal to the length of a minute of arc measured at Equator. 1 mile (teleg. naut.) = 6087 feet (E)	INT, U.K; US
mile (UK, nautical)	mi (UK, naut.)	length, distance	L	1 mile (UK, naut.) = 1853.184 m (E)	Obsolete British unit of length employed in navigation. It is equal to the length of a minute of arc measured at a latitude of N48°. The British Admiralty used the round figure of 6080 feet. 1 mile (UK Naut.) = 6080 feet (E)	UK
mile (US, nautical)	mi (US, naut.)	length, distance	L	1 mile (US, naut.) = 1853.184 m (E)	Obsolete American unit of length employed in navigation. It is equal to the length of a minute of arc measured at a latitude of N48°. 1 mile (US naut.) = 6080 feet (E)	US
mile (US, survey)	mi (US, survey)	length, distance	L	1 mile (US, survey) = 1609.3472187 m	Obsolete American unit of length used in geodetic and surveyor's measurements.	US

Unit (synonym, acronym)	Symbol	Physical quantity	Dimension	Conversion factor (SI equivalent unit)	Notes, definitions, other conversion factors	System
mile per hour (stat.)	mph, $mi.h^{-1}$	velocity, speed	LT^{-1}	$1\ mi.h^{-1} = 4.470400000 \times 10^{-1}\ m.s^{-1}$	American and British unit of linear velocity employed to express the terrestrial speed of vehicles. $1\ mi.h^{-1} = 1.609344\ km.h^{-1}$	UK, US
milha (Portuguese)	–	length, distance	L	1 milha (Portuguese) = 2064 m	Obsolete Portuguese unit of length. 1 milha (Portuguese) = 8 estadio (E)	Portuguese
milla (Spanish)	–	length, distance	L	1 milla (Spanish) = 1393.175 m	Obsolete Spanish unit of length. 1 milla (Spanish) = 5000/3 vara (E)	Spanish
mille (Arabic)	–	length, distance	L	1 mille (Arabic) = 1.920×10^3 m	Obsolete Arabic unit of length used in ancient times. 1 mille (Arabic) = 6000 feet (Arabic) (E)	Arabic
mille (Egyptian) [Egyptian mile]	–	length, distance	L	1 mille (Egyptian) = 1.745×10^3 m	Obsolete Egyptian unit of length used in ancient times. 1 mile = 10 000/3 Royal cubit (E)	Egyptian
mille (Greek, Attic) [Greek mile]	–	length, distance	L	1 mille (Greek, Attic) = 1388.520 m	Obsolete Greek unit of length employed in ancient times. 1 mille (Attic) = 4500 pous (E)	Attic
mille (Persian) [Babylonian mile]	–	length, distance	L	1 mille (Persian) = 1728 m	Obsolete unit of length in the Assyrio-Chaldean-Persian system used in ancient times. 1 mille = 5400 zereths (E)	Persian
mille marin [French nautical mile]	–	length, distance	L	1 mille marin = 950.13 m	Obsolete French unit of length used in navigation before the French Revolution (1789).	French

millennium	–	time, period, duration	T	1 millennium = 3.15576 × 10[illegible] s	1 millennium = 1000 years (E)	
millia (Roman) [Roman mile]	–	length, distance	L	1 millia (Roman) = 1472 m	Obsolete Roman unit of length employed in ancient times. 1 milia (Roman) = 5000 pes (E)	Roman
milliampere per square centimetre	$mA.cm^{-2}$	electric current density	IL^{-2}	1 $mA.cm^{-2}$ = 10 $A.m^{-2}$ (E)	Unit of electric current density usually employed in electrochemistry in laboratory experiments.	
millibar	mbar	pressure, stress	$ML^{-1}T^{-2}$	1 mbar = 10^2 Pa (E)	At one time it was a unit of pressure commonly employed in meteorology. It was replaced by the hPa on January 1st 1986 by the World Meteorological Organization. 1 mbar = 1 hPa (E) 1 atm = 1013.25 mbar (E)	
millidarcy	–	hydrodynamic permeability	L^2	1 millidarcy = 9.869233×10^{-16} m^2	Obsolete permeability unit employed in hydrology and civil engineering.	
millième (French)	‰ (FRA)	plane angle	α	1‰(FRA) = 9.9999966×10^{-4} rad	Unit of plane angle employed in the French artillery. One millième is equal to a plane angle under which it is possible to observe a difference of height of 1 m at 1000 m. 1‰(FRA) = 0.05729576041° 1‰(FRA) = 0.063662 grade	French
millième (NATO)	‰ (NATO)	plane angle	α	1‰(NATO) = $9.81747704 \times 10^{-4}$ rad	Unit of plane angle employed in the artillery. It is equal to the 6400th part of the circle. 1‰(NATO) = 0.05625° 1‰(NATO) = 0.0625 grade	INT

Unit (synonym, acronym)	Symbol	Physical quantity	Dimension	Conversion factor (SI equivalent unit)	Notes, definitions, other conversion factors	System
millième (US before 1945)	‰(US)	plane angle	α	1‰(US) = $1.570796327 \times 10^{-3}$ rad	Obsolete unit of plane angle employed in the American artillery before World War II. It is equal to the 4000th part of the circle. 1‰(US) = 0.09° 1‰(US) = 0.1 grade	US
millième (USSR)	‰ (USSR)	plane angle	α	1‰(USSR) = $9.973310011 \times 10^{-4}$ rad	Obsolete unit of plane angle employed in the Soviet artillery. It is equal to the 6300th part of the circle. 1‰(USSR) = 0.05714285714° 1‰(USSR) = 0.06349206349 grade	Russian
milligamma (nanogram)	$\gamma\gamma$	mass	M	1 $\gamma\gamma = 10^{-12}$ kg (E)	Obsolete unit of mass employed in analytical chemistry. 1 $\gamma\gamma = 10^{-9}$ g (E) 1 $\gamma\gamma$ = 1 ng (E)	
milligram	mg	mass	M	1 mg = 10^{-6} kg (E)	Submultiple of the SI base unit. 1 mg = 10^{-3} g (E)	SI
milligram per square decimetre per day	mg. dm^{-2}. d^{-1}	rate of corrosion	LT^{-1}	1 mg.dm^{-2}.d^{-1} = 8.680555×10^{-9} m.s^{-1} (for a material with a specific gravity 8)	Unit used by corrosion engineers to express the rate of corrosion of metals and alloys. 1 mg.dm^{-2}.d^{-1} = 4.565625×10^{-3} mm per year for a material of a density of 8000 kg.m^{-3}	
millik	millik	reactivity of nuclear power reactor	nil	1 millik = 10^{-5} (E)	Obsolete Canadian unit of nuclear reactivity employed in nuclear engineering.	CAN

millimetre	mm	length, distance	L	1 mm = 10^{-3} m (E)	Submultiple of the SI base unit.	SI
millimetre of mercury (0°C)	mmHg, torr, Torr (0°C)	pressure, stress	$ML^{-1}T^{-2}$	1 mmHg (0°C) = 1333223684 Pa	Obsolete unit of pressure employed in physics to measure small pressures. It is equal to the pressure exerted by a column of mercury one millimetre high, measured at 0°C (32°F). 760 mmHg (0°C) = 1 atm (E) 1 mmHg = 1 torr (E) 1 mmHg (0°C) = (101 325/760) Pa (E)	
millimetre of water (4°C)	mmH_2O, mmCE (4°C)	pressure, stress	$ML^{-1}T^{-2}$	1 mmH_2O (4°C) = 9806375414 Pa	Unit of pressure employed in physics to measure small pressures. It is equal to the pressure exerted by a column of water one millimetre high, measured at 4°C (39.2°F). 1 atm = 10 332.56383 mmH_2O (4°C)	
millimetre of water (15.56°C)	mmH_2O, mmCE (15.56°C)	pressure, stress	$ML^{-1}T^{-2}$	1 mmH_2O (15.56°C) = 9.797059096 Pa	Obsolete unit of pressure employed in physics to measure small pressures. It is equal to the pressure exerted by a column of water one millimetre high, measured at 15.56°C (60°F). 1 atm = 10 342.38938 mmH_2O (60°F)	
millimetre of water (39.2°F)	mmH_2O, mmCE (39.2°F)	pressure, stress	$ML^{-1}T^{-2}$	1 mmH_2O (39.2°F) = 9.806375414 Pa	Unit of pressure employed in physics to measure small pressures. It is equal to the pressure exerted by a column of water one millimetre high, measured at 4°C (39.2°F). 1 atm = 10 332.56383 mmH_2O (4°C)	

Unit (synonym, acronym)	Symbol	Physical quantity	Dimension	Conversion factor (SI equivalent unit)	Notes, definitions, other conversion factors	System
millimetre of water (60°F)	mmH_2O, mmCE (60°F)	pressure, stress	$ML^{-1}T^{-2}$	1 mmH_2O (60°F) = 9.797059096 Pa	Obsolete unit of pressure employed in physics to measure small pressures. It is equal to the pressure exerted by a column of water one millimetre high, measured at 15.56°C (60°F). 1 atm = 10 342.38938 mmH_2O (60°F)	
millimicron (nanometre)	mμ, nm	length, distance	L	1 mμ = 10^{-9} m (E)	Obsolete unit of wavelength employed in spectroscopy. 1 mμ = 1 nm (E)	
MIM	MIM	number of instructions computed per unit of time	T^{-1}	1 MIM = 6×10^6 instructions.s^{-1}	Unit used in computer science. The unit is named after the English acronym: Million Instructions per Minute.	
mina (Hebrew) [Sacred system]	–	mass	M	1 mina (Hebrew) = 0.850 kg	Obsolete Hebrew unit of mass used in ancient times. Sacred system.	Hebrew
mina (Hebrew) [Talmudic system]	–	mass	M	1 mina (Hebrew) = 0.3542 kg	Obsolete Hebrew unit of mass used in ancient times. Rabbinacal or Talmudic system.	Hebrew
mine (de Paris, dry)	–	capacity, volume	L^3	1 mine (de Paris, dry) = 11.176680×10^{-3} m^3	Obsolete French unit of volume employed before the French Revolution. It was used for capacity measurements of dry substances. 1 mine = 6 boisseaux (E)	French
mine (Greek, Attic)	–	mass	M	1 mine (Greek, Attic) = 0.432 kg	Obsolete Greek unit of mass used in ancient times. 1 mine (Greek, Attic) = 1/60 Greek talents (E)	Attic

miner's inch	–	volume flow rate	L^3T^{-1}	1 miner's inch = 1.018227048 $m^3.s^{-1}$ (h = 5 inches)	Obsolete American and British unit of flow rate of water. It was equal to the volume of water which flows through an aperture of one square inch section under a difference of height ranging from 4 to 6.5 inches. An Act of the California legislature, May 23, 1901 makes the standard miner's inch 1.5 cubic feet per minute through any aperture or orifice.	UK, US
mingelen (Dutch)	–	capacity, volume	L^3	1 mingelen (Dutch) = 1.200×10^{-3} m^3	Obsolete Dutch unit of capacity used for liquid substances.	Dutch
minim (UK)	min (UK), ♍	capacity, volume	L^3	1 minim (UK) = $5.919398438 \times 10^{-8}$ m^3	British unit used for capacity measurements for all merchandise (solids, liquids, foodstuffs, etc). 1 minim (UK) = 1/76800 gal (UK) (E)	UK
minim (US)	min (US), ♍	capacity, volume	L^3	1 minim (US) = $6.161151992 \times 10^{-8}$ m^3	American unit used for capacity measurements of liquids. 1 minim (US) = 1/61400 gal (US) (E)	US
minot (de Paris, dry)	–	capacity, volume	L^3	1 minot (de Paris, dry) = 5.588340×10^{-3} m^3	Obsolete French unit of volume employed before the French Revolution. It was used for capacity measurements of dry substances. 1 minot = 3 boisseaux (E)	French
minute	min, mn	time, period, duration	T	1 minute = 60 seconds (E)	Hexadecimal multiple of the second.	INT
minute of angle	′	plane angle	α	$1' = 2.908882 \times 10^{-4}$ rad	Hexadecimal submultiple of the plane angle degree. $1' = \pi/(60 \times 180)$ rad (E)	INT

Unit (synonym, acronym)	Symbol	Physical quantity	Dimension	Conversion factor (SI equivalent unit)	Notes, definitions, other conversion factors	System
minute of angle (new)	c	plane angle	α	$1\ c = 1.570796 \times 10^{-4}$ rad	Obsolete French unit of plane angle introduced after the French Revolution (1789). 1 c = 1/100 gon (grade) (E)	French
mio (Portuguese, dry)	–	capacity, volume	L^3	1 mio (Portuguese, dry) $= 810 \times 10^{-3}\ m^3$	Obsolete Portuguese unit of capacity used for dry substances. 1 mio (Portuguese, dry) = 15 fanga (E)	Portuguese
mired	–	temperature of colour	Θ^{-1}	1 mired $= 10^6\ K^{-1}$ (E)	Unit of temperature of colour employed in photography. TC (mired) = 1 000 000/T (K). The name of the unit derived from the English acronym: **Mi**cro **Re**ciprocal **D**egrees.	
MKpS unit of mass	–	mass	M	1 MKps = 9.806 65 kg (E)	Obsolete MKpS unit of mass.	MKps
mô (Japanese)	–	length, distance	L	1 mô (Japanese) $= 3.030303030 \times 10^{-5}$ m	Obsolete Japanese unit of length. 1 mô = 1/10 000 shaku (E)	Japanese
modius (Arabic)	–	capacity, volume	L^3	1 modius (Arabic) $= 40.8 \times 10^{-3}\ m^3$ (of water)	Obsolete Arabic unit of capacity used in ancient times. Measured by weight. 1 modius (Arabic) = 5/4 cafiz (E)	Arabic
modius (Roman muid)	–	capacity, volume	L^3	1 modius (Roman muid) $= 8.788 \times 10^{-3}\ m^3$	Obsolete Roman unit of volume employed in ancient times. It was used for capacity measurements of dry substances. 1 modius (Roman muid) = 16 sextarius (E)	Roman
moio (Spanish)	–	capacity, volume	L^3	1 moio (Spanish) $= 250.290592 \times 10^{-3}\ m^3$	Obsolete Spanish unit of capacity used for liquid substances. 1 moio (Spanish) = 16 arroba (water) (E)	Spanish

mole	mol	amount of substance	N	**SI base unit**	The mole is the amount of substance of a system which contains as many elementary entities as there are atoms in 0.012 kg of carbon 12. When the mole is used, the elementary entities must be specified and may be atoms, molecules, ions, electrons, other particles, or specified groups of such particles [14th CGPM (1971), Resolution 3]. In this definition, it is understood that the carbon 12 atoms are unbound, at rest and in their ground state.	SI
mole per centicubic metre	$mol.cm^{-3}$ mM	molarity, molar concentration	$N.L^{-3}$	$1\ mol.cm^{-3} = 10^6\ mol.m^{-3}$ (E)		
mole per cubic decimetre	$mol.dm^{-3}$, M	molarity, molar concentration	$N.L^{-3}$	$1\ mol.dm^{-3} = 10^3\ mol.m^{-3}$ (E)		
mole per cubic metre	$mol.m^{-3}$	molarity, molar concentration	$N.L^{-3}$	**SI derived unit**		SI
mole per litre	$mol.l^{-1}$, M	molarity, molar concentration	$N.L^{-3}$	$1\ mol.l^{-1} = 10^3\ mol.m^{-3}$ (E)	Unit employed in chemistry.	
momme	Me	mass	M	$1\ momme = 3.75 \times 10^{-3}$ kg (E)	Obsolete Japanese unit of mass used in the pearl trade. 1 momme = 3.75 g (E)	Japanese
mon	–	flatness of rolled steel plate	L	$1\ mon = 2.54 \times 10^{-5}$ m (E)	Obsolete British unit employed in metallurgy for describing the flatness of rolled steel plates. A surface has a flatness of one mon if no part of it is more than one mil above or below a straight line drawn between any two points one metre apart on the surface.	UK

Unit (synonym, acronym)	Symbol	Physical quantity	Dimension	Conversion factor (SI equivalent unit)	Notes, definitions, other conversion factors	System
month (30 days)	–	time, period, duration	T	1 month (30 d) = 2.592×10^6 s (E)	1 month = 30 days (E)	INT
month (lunar)	–	time, period, duration	T	(see note)	Unit of time employed in navigation for tidal computations. The lunar month is equal to the time interval between two successive full Moons. 1 month (lunar) = 28 days (E)	
month (solar mean)	–	time, period, duration	T	1 month (solar mean) = 2.628×10^6 s (E)		
morgen (Dutch)	–	surface, area	L^2	1 morgen (Dutch) = 8244.346 m^2	Obsolete Dutch unit of area.	Dutch
morgen (Prussian)	–	surface, area	L^2	1 morgen (Prussian) = 2532.24 m^2	Obsolete German unit of area. 1 morgen (Prussian) = 180 square Ruthe (E)	German
Moszkowski (Weißkopf unit)	–	probability of transition	nil	(see note)	Obsolete unit employed in nuclear physics to express the nuclear quantum state's transition probability.	
mud (Dutch, dry)	–	capacity, volume	L^3	1 mud (Dutch, dry) = 109.040×10^{-3} m^3	Obsolete Dutch unit of capacity used for dry substances. 1 mud (Dutch, dry) = 4 schepel (E)	Dutch
mudd (Arabic)	–	capacity, volume	L^3	1 mudd (Arabic) = 0.68×10^{-3} m^3 (of water)	Obsolete Arabic unit of capacity used in ancient times. Measured by weight. 1 mudd (Arabic) = 1/48 cafiz (E)	Arabic
mug (hyl, metric slug, par, TME)	mug	mass	M	1 mug = 9.80665 kg (E)	Obsolete technical metric unit of mass employed by mechanical engineers (base unit of the metric gravitational system). It was equal to the mass which under an	German

					acceleration of 1 m.s^{-2} gives a force of 1 kgf. The name TME derives from the German acronym **T**echnische **M**asse **E**inheit. The name mug derives from the abbreviation of **m**etric sl**ug**.	
muid (de Paris)	–	capacity, volume	L^3	1 muid (de Paris) $= 274.239 \times 10^{-3}$ m^3	Obsolete French unit of capacity employed before the French Revolution. It served to express the capacity of liquids and grains. It varied according to the location and merchandise. 1 muid (de Paris) = 288 pintes (de Paris)	French
muid (de Paris, dry)	–	capacity, volume	L^3	1 muid (de Paris, dry) $= 268.240320 \times 10^{-3}$ m^3	Obsolete French unit of capacity employed before the French Revolution. It served to express the capacity of liquids and grains. It varies according to the location and merchandise. 1 muid (de Paris, dry) = 144 boisseaux (de Paris) (E)	French
musec (cumec)	cumec	volume flow rate	L^3T^{-1}	1 cumec = 1 m^3.s^{-1} (E)	Obsolete British and American unit of flow rate. It was used in vacuum pump technology. The name of the unit derived from the acronym of **cu**bic **me**tre per second.	UK, US
musti (Indian)	–	capacity, volume	L^3	1 musti (Indian) $= 5.156250 \times 10^{-5}$ m^3 (of water)	Obsolete Indian unit of capacity used in ancient times. Measured by weight. 1 musti (Indian) = 1/256 drona (E)	Indian
muth (Austrian, dry)	–	capacity, volume	L^3	1 muth (Austrian, dry) $= 1.844670$ m^3	Obsolete Austrian unit of capacity used for dry substances. 1 muth (Austrian, dry) = 30 metzel (E)	Austrian

Unit (synonym, acronym)	Symbol	Physical quantity	Dimension	Conversion factor (SI equivalent unit)	Notes, definitions, other conversion factors	System
muthmassel (Austrian, dry)	–	capacity, volume	L^3	1 muthmassel (Austrian, dry) $= 3.8430625 \times 10^{-3}$ m^3	Obsolete Austrian unit of capacity used for dry substances. 1 muthmassel (Austrian, dry) = 1/16 metzel (E)	Austrian
mutsje (Dutch)	–	capacity, volume	L^3	1 mutsje (Dutch) $= 0.150 \times 10^{-3}$ m^3	Obsolete Dutch unit of capacity used for liquid substances. 1 mutsje (Dutch) = 1/8 mingelen (E)	Dutch
myriagram	–	mass	M	1 myriagram = 10 kg (E)	Obsolete metric designation for gramultiple. 1 myriagram $= 10^4$ g (E)	
myriametre	–	length, distance	L	1 mym $= 10^4$ m (E)	Obsolete name for a multiple of the metre. 1 mym = 10 km (E)	
n unit	–	dose of neutrons	L^2T^{-2}	(see note)	Obsolete unit of neutron dose employed in nuclear physics. One n unit is equal to the dose of fast neutrons which produces the same amount of ionization in a ionizing chamber as one röntgen of X-rays.	
nabuchodonosor (nabuchadnezzar)	–	capacity, volume	L^3	1 nabuchodonosor $= 15.153640 \times 10^{-3}$ m^3	Obsolete British unit used to express the capacity of a wine container. Still employed in oenology, especially in France. 1 nabuchodonosor = 10/3 gallons (UK) (E) 1 nabuchodonosor = 20 bouteilles (E)	UK, French
nail (UK)	–	length, distance	L	1 nail (UK) $= 2.286 \times 10^{-1}$ m (E)	Obsolete British unit of length. 1 nail (UK) = 9 inches (E)	UK

					1 ng = 10^{-9} g (E)	
nanometre	nm	length, distance	L	1 nm = 10^{-9} m (E)	Submultiple of the SI base unit.	SI
nanon (nanometre)	–	length, distance	L	1 nanon = 10^{-9} m (E)	Obsolete unit of wavelength employed in spectroscopy. 1 nanon = 1 nm (E)	
nasch (Arabic)	–	mass	M	1 nasch (Arabic) = 56.666667 × 10^{-3} kg	Obsolete Arabic unit of mass used in ancient times. System of the Prophet. 1 nasch (Arabic) = 1/6 rotl (E)	Arabic
nautical mile (international)	mi (int., naut.)	length, distance	L	1 mile (int., naut.) = 1 852 m (E)	Legal international unit of length temporarily maintained with the SI. It is still used in navigation (mercantile marine, aviation). It is equal to the length of an arc of one minute measured at a latitude of N45°. The int. nautical mile has been taken equal to the nautical mile since 1970. 1 mile (int., naut.) = 6076.11 feet	INT
nautical mile (UK)	mi (UK, naut.)	length, distance	L	1 mile (UK, naut.) = 1853.184 m (E)	Obsolete British unit of length employed in navigation. It is equal to the length of a minute of arc measured at a latitude of N48°. The British Admiralty used the round figure of 6080 feet. 1 mile (UK, naut.) = 6080 feet (E)	UK
nautical mile (US)	mi (US, naut.)	length, distance	L	1 mile (US, naut.) = 1853.184 m (E)	Obsolete American unit of length employed in navigation. It is equal to the length of a minute of arc measured at a latitude of N48°. 1 mile (US, naut.) = 6080 feet (E)	US

Unit (synonym, acronym)	Symbol	Physical quantity	Dimension	Conversion factor (SI equivalent unit)	Notes, definitions, other conversion factors	System
neper	Np	logarithm (Napierian) of a ratio of two sound powers	nil	1 Np = 8.685890 dB (see note)	The neper expresses the ratio of two sound powers as a natural logarithm difference according to the equation $N(N_P) = \ln P_1/P_2$. The unit is named after John Napier (1550–1617).	
nevat (Arabic)	–	mass	M	1 nevat (Arabic) = 14.166667×10^{-3} kg	Obsolete Arabic unit of mass used in ancient times. System of the Prophet. 1 nevat (Arabic) = 1/24 rotl (E)	Arabic
newton	N	force, weight	MLT^{-2}	**SI derived unit** 1 N = 1 $kg.m.s^{-2}$ (E)	The newton is that force which, when applied to a body having a mass of one kilogram, gives it an acceleration of one metre per second squared (9th CGPM, 1948). The newton is approximately equal to the force applied by the Earth's gravity field on an apple. The unit is named after Isaac Newton (1642–1727).	SI
newton per metre	$N.m^{-1}$	surface tension, interfacial tension	MT^{-2}	**SI derived unit** 1 $N.m^{-1}$ = 1 $J.m^{-2}$ = 1$kg.s^{-2}$		SI
newton per square metre	$N.m^{-2}$	pressure, stress	$ML^{-1}T^{-2}$	**SI derived unit** 1 $N.m^{-2}$ = 1 Pa (E)	1 atm = 101 325 $N.m^{-2}$ (E)	SI
newton second	N.s	linear momentum, momentum	MLT^{-1}	**SI derived unit** 1 $N.s^{-1}$ = 1$kg.m.s^{-1}$		SI
newton-metre	N.m	moment of a force, torque	ML^2T^{-2}	**SI derived unit**		SI
nile	nile	reactivity of nuclear power reactor	nil	1 nile = 10^{-5} (E)	Obsolete British unit of nuclear reactivity employed in nuclear engineering.	UK

nisoku-ichi-nin (Japanese)	–	mass	M	1 nisoku-ichi-nin (Japanese) = 26.250 kg (E)	Obsolete Japanese unit of mass. 1 nisoku-ichi-nin (Japanese) = 7 kwan (E)	Japanese
nit	nt	luminous luminance	JL^{-2}	1 nit = 1 $cd.m^{-2}$ (E)	Obsolete MKSA unit of luminous luminance. The unit was adopted by the *Comité International de L'Éclairage* (CIE) in 1948. 1 nit = 10^{-4} stilb (E)	MKSA
niyo (Japanese)	–	mass	M	1 niyo (Japanese) = 1.875×10^{-2} kg (E)	Obsolete Japanese unit of mass. 1 niyo (Japanese) = 0.004 kwan (E)	Japanese
noeud (knot, naut. mile per hour)	n. mi. h^{-1}, knot, nmph	velocity, speed	LT^{-1}	1 noeud = $5.144444444 \times 10^{-1}$ $m.s^{-1}$	International unit of velocity employed in navigation. The unit dates from the late sixteenth century, when the speed of a ship was found by dropping a float tied to a knotted line (knotted log) over the side of the vessel. The knots were originally seven fathoms apart. The number of knots passing in 30 seconds gave the speed of the ship in nautical miles per hour. 1 knot = 1 nautical mile per hour (E) 1 knot = 1852 $m.h^{-1}$ (E) 1 knot = 6080 $ft.h^{-1}$ (E)	INT
noggin (UK)	nogg (UK)	capacity, volume	L^3	1 noggin (UK) = $1.420653750 \times 10^{-4}$ m^3	Obsolete British unit of capacity. 1 noggin (UK) = 1/4 pint (UK)	U.K
normal acceleration, (standard free-fall acceleration, normal gravity, standard gravity)	g_n	acceleration	LT^{-2}	1 g_n = 9.80665 $m.s^{-2}$ (E)	Physical fundamental constant adopted by the *Bureau International des Poids et Mesures* (BIPM) in 1892 and approved definitively by the 5th CGPM (1913). The acceleration due to gravity varies according to the location, but for *(continued overleaf)*	INT

Unit (synonym, acronym)	Symbol	Physical quantity	Dimension	Conversion factor (SI equivalent unit)	Notes, definitions, other conversion factors	System
normal acceleration *(continued)*					convenience in computation, the value was standardized in 1913. The value has been adopted as a standard or accepted value by the CIPM. 1 g_n = 32.1740 ft.s^{-2} (E) in the FPS system.	
nox	nox	illuminance	$JL^{-2}\Omega^{-1}$	1 nox = 10^{-3} lx (E)	Obsolete unit of illuminance introduced in Germany during World War II for small illumination measurements. 1 nox = 10^{-3} apostilb (E)	German
noy	–	noisiness	nil	(see note)	American unit of perceived noise. One noy is defined as the perceived noisiness of the frequency band 910–1090 Hz of random noise at a sound pressure level of 40 dB above 2×10^{-4} barye. The unit derived from the first syllable of the word **noise**.	US
nyläst (Swedish)	–	mass	M	1 nyläst (Swedish) = 5100.956429 kg	Obsolete Swedish unit of weight. 1 nyläst (Swedish) = 12 000 skålpund (E)	Swedish
obol (Greek, Attic)	–	mass	M	1 obol (Greek, Attic) = 0.72×10^{-3} kg	Obsolete Greek unit of mass used in ancient times. 1 obol (Greek, Attic) = 1/6 drachma (Greek, Attic)	Attic
ochava (Spanish)	–	mass	M	1 ochava (Spanish) = $3.583278816 \times 10^{-4}$ kg	Obsolete Spanish unit of mass. 1 ochava (Spanish) = 1/128 libra (E)	Spanish
ochavillo (Spanish, dry)	–	capacity, volume	L^3	1 ochavillo (Spanish, dry) = $72.2669271 \times 10^{-6}$ m^3	Obsolete Spanish unit of capacity used for dry substances. 1 ochavillo (Spanish, dry) = 1/768 fanega (E)	Spanish

ocque (Arabic)	–	mass	M	1 ocque (Arabic) = 1.360 kg	Obsolete Arabic unit of mass used in ancient times. System of the Prophet. 1 ocque (Arabic) = 4 rotl (E)	Arabic
octave	–	logarithmic musical interval	nil	1 octave = $\log_2 f_1/f_2$ (E) (see note)	The octave is the interval between two musical sounds having as a basic frequency ratio the 1st root of two. The number of octaves between frequencies f_1 and f_2 is: $I = \log_2(f_1/f_2)$ 1 octave = 301 savart (E) 1 octave = 1200 cent (E)	
octet (byte)	o, B	quantity of information	nil	1 byte = 8 bits (E)	Unit employed in computer science. It is equal to a sequence of adjacent binary digits operated upon as a unit in a computer and usually shorter than a word. Multiple of 8 bits. 1 KB = 1024 B ($K = 2^{10}$) 1 MB = 1 048 576 B ($M = 2^{20}$)	
oersted	Œ, œ	magnetic field strength	IL^{-1}	1 Œ = 79.57747 $A.m^{-1}$	Obsolete cgs of magnetic field induction in the emu subsystem. One oersted is the magnetic field strength that would exert a force of one dyne on a unit magnetic dipole in vacuo. The unit is named after H.C. Oersted (1777–1851). 1 Œ = $1000/4\pi$ $A.m^{-1}$ (E)	cgs
ohm	Ω	electric resistance	$ML^2T^{-3}I^{-2}$	**SI derived unit** 1 Ω = 1 $kg.m^2.s^{-3}.A^{-2}$ (E)	The ohm is the electric resistance between two points of a conductor when a constant difference of potential of one volt, applied between these two points, produces in this *(continued overleaf)*	SI

Unit (synonym, acronym)	Symbol	Physical quantity	Dimension	Conversion factor (SI equivalent unit)	Notes, definitions, other conversion factors	System
ohm *(continued)*					conductor a current of one ampere, this conductor not being the source of electromotive force. The unit is named after G.S. Ohm (1787–1854).	
ohm (acoustic, cgs)	–	acoustic impedance	$ML^{-4}T^{-1}$	1 cgs acoustic ohm $= 10^{-5}$ Pa.s.m^{-3} (E)	Obsolete cgs unit of acoustic impedance. An acoustic impedance (including acoustic resistance and reactance) has a magnitude of 1 cgs acoustic ohm when a sound pressure of 1 barye produces a volume velocity of 1 cm^3.s^{-1}. 1 cgs acoustic ohm = 1 dyn.s.cm^{-5} (E)	cgs
ohm (acoustic, SI)	–	acoustic impedance	$ML^{-4}T^{-1}$	1 SI acoustic ohm = 1 Pa.s.m^{-3} (E)	SI unit of acoustic impedance. An acoustic impedance (including acoustic resistance and reactance) has a magnitude of 1 SI acoustic ohm when a sound pressure of 1 pascal produces a volume velocity of 1 m^3.s^{-1}. 1 SI acoustic ohm = 1 N.s.m^{-5} (E)	SI
ohm (int. mean)	–	electric resistance	$ML^2T^{-3}I^{-2}$	1 ohm (int. mean) = 1.00049 Ω	Obsolete IEUS unit of electric resistance defined in 1908. It is the resistance measured at 0°C of a mercury column of 106.300 cm length which has a mass equal to 14.4521 grams.	US, IEUS
ohm (int. US)	–	electric resistance	$ML^2T^{-3}I^{-2}$	1 ohm (int. US) = 1.000495 Ω	Obsolete IEUS unit of electric resistance.	US, IEUS

ohm (legal)	ohm legal	electric resistance	$ML^2T^{-3}I^{-2}$	1 legal ohm = 0.9972 Ω	Obsolete unit of electric resistance defined in 1883. It was the resistance measured at 0°C of a mercury column of 106 cm length which has a cross section of one square millimetre.	US
ohm (mechanical, cgs)	–	mechanical impedance	$ML^{-4}T^{-1}$	1 ohm (mechanical, cgs) $= 10^{-5}$ kg.m^{-4}.s^{-1} (E)	Obsolete cgs unit for mechanical impedance used in acoustical engineering. 1 ohm (mechanical, cgs) = 1 g.cm^{-4}.s^{-1} (E)	cgs
ohm (mechanical, SI)	–	mechanical impedance	$ML^{-4}T^{-1}$	**SI derived unit** 1 ohm (mechanical, SI) = 1 kg.m^{-4}.s^{-1} (E)		SI
ohm (Prussian)	–	capacity, volume	L^3	1 ohm (Prussian) $= 137.403600 \times 10^{-3}$ m^3	Obsolete German unit of capacity used for liquid substances. 1 ohm (Prussian) = 120 quart (E)	German
ohm (Swedish)	–	capacity, volume	L^3	1 ohm (Swedish) $= 157.029720 \times 10^{-3}$ m^3	Obsolete Swedish unit of capacity for liquid substances. 1 ohm (Swedish) = 60 kanna (E)	Swedish
ohm circular mil per foot	Ω.cmi .ft^{-1}	electric resistivity	$MLT^{-3}I^{-2}$	1 Ω.cmi.ft^{-1} $= 1.662426113 \times 10^{-9}$ Ω.m	Obsolete British and American unit of electric resistivity. 1 Ω.cmi.ft^{-1} = 16.624261130 μΩ.cm	UK, US
ohm metre	Ω.m	electric resistivity	$MLT^{-3}I^{-2}$	**SI derived unit** 1 Ω.m= 1 kg.m.s^{-3}.A^{-2} (E)		SI
onca (Portuguese)	–	mass	M	1 onca (Portuguese) $= 2.868750 \times 10^{-2}$ kg	Obsolete Portuguese unit of mass. 1 onca (Portuguese) = 1/16 libra (E)	Portuguese
once (de Paris) [French ounce]	–	mass	M	1 once (de Paris) $= 30.594115630 \times 10^{-3}$ kg	Obsolete French unit of mass employed before the French Revolution (1789). 1 once (de Paris) = 1/16 livre (de Paris) (E) 1 once (de Paris) = 8 gros (de Paris) (E)	French

Unit (synonym, acronym)	Symbol	Physical quantity	Dimension	Conversion factor (SI equivalent unit)	Notes, definitions, other conversion factors	System
once (Russian)	–	mass	M	1 once = $2.559482370 \times 10^{-2}$ kg	Obsolete Russian unit of mass used before 1917 for general use. 1 once = 1/16 funt (E)	Russian
once (Russian)	–	mass	M	1 once (Russian) = $2.986027988 \times 10^{-2}$ kg	Obsolete Russian unit of mass used before 1917 in pharmacy. 1 once (Russian) = 672 doli (E)	Russian
oncia (Italian)	–	length, distance	L	1 oncia (Italian) = 1.284425×10^{-1} m	Obsolete Italian unit of length. 1 oncia (Italian) = 1/12 piedi liprando (E)	Italian
oncia (Italian)	–	mass	M	1 oncia (Italian) = 2.558333×10^{-2} kg	Obsolete Italian unit of mass. 1 oncia (Italian) = 1/12 libbra (E)	Italian
onza (Spanish)	–	mass	M	1 onza (Spanish) = $28.755812500 \times 10^{-3}$ kg	Obsolete Spanish unit of mass. 1 onza (Spanish) = 1/16 libra (E)	Spanish
orguia (Greek, Attic) [Greek fathom]	–	length, distance	L	1 orguia (Greek, Attic) = 1.851360 m	Obsolete Greek unit of length employed in ancient times. 1 orguia (Attic) = 6 pous (E)	Attic
orgye (Arabic) [Arabic fathom]	–	length, distance	L	1 orgye (Arabic) = 1920 m	Obsolete Arabic unit of length used in ancient times. 1 orgye (Arabic) = 6 feet (Arabic) (E)	Arabic
orgye (Egyptian) [Egyptian fathom]	–	length, distance	L	1 orgye (Egyptian) = 2094 m	Obsolete Egyptian unit of length used in ancient times. 1 orgye = 4 Royal cubit (E)	Egyptian
ort (Swedish)	–	mass	M	1 ort = $4.250797024 \times 10^{-3}$ kg	Obsolete Swedish unit of weight. 1 ort = 1/100 skålpund (E)	Swedish

ort (Swedish, dry)		capacity, volume	L^3	1 ort (Swedish, dry) $= 8.178125 \times 10^{-5}$ m^3	Obsolete Swedish unit of capacity for dry substances. 1 ort (Swedish, dry) = 1/32 kanna (E)	Swedish
osmini (Russian, dry)	–	capacity, volume	L^3	1 osmini (Russian, dry) $= 104.954944 \times 10^{-3}$ m^3	Obsolete Russian unit of capacity for dry substances used before 1917. 1 osmini (Russian, dry) = 32 garnetz (E)	Russian
ottavo (Italian)	–	mass	M	1 ottavo (Italian) $= 3.197916667 \times 10^{-3}$ kg	Obsolete Italian unit of mass. 1 ottavo (Italian) = 1/96 libbra (E)	Italian
oukia (Arabic)	–	mass	M	1 oukia (Arabic) $= 1.133333 \times 10^{-1}$ kg	Obsolete Arabic unit of mass used in ancient times. System of the Prophet. 1 oukia (Arabic) = 1/3 rotl (E)	Arabic
ounce (apothecary)	oz (apoth.), oz (ap.)	mass	M	1 oz (apoth.) $= 3.110347680 \times 10^{-2}$ kg	Obsolete British and American unit of weight employed in pharmacy. It was used at one time as a unit of weight for drugs and other medical preparations (lotions, potions, ointments, plants extracts, etc). 1 oz (apoth.) = 1/12 lb (apoth.) (E) 1 oz (apoth.) = 8 drachms (apoth.) (E) 1 oz (apoth.) = 24 scruples (apoth.) (E) 1 oz (apoth.) = 480 grains (apoth.) (E)	UK, US
ounce (av.)-force per square inch	ozf.in^{-2}, osi, OSI	pressure, stress	$ML^{-1}T^{-2}$	1 ounce(av.)-force per square inch = 430.9223308 Pa	Obsolete British and American unit of pressure and stress. 1 atm = 235.1351803 osi 1 osi = 1/16 psi (E)	UK, US
ounce (avoirdupois)	oz avdp, oz (av.)	mass	M	1 oz (av.) $= 2.834952313 \times 10^{-2}$ kg (E)	Legal American and British unit of weight since the WMA of 1963. 1 oz (av.) = 1/16 lb (av.) (E) 1 oz (av.) = 16 drachms (av.) (E) 1 oz (av.) = 437.5 grains (av.) (E)	UK, US

Unit (synonym, acronym)	Symbol	Physical quantity	Dimension	Conversion factor (SI equivalent unit)	Notes, definitions, other conversion factors	System
ounce (troy)	oz (troy)	mass	M	1 oz (troy) = $3.110347680 \times 10^{-2}$ kg	Obsolete British unit of mass employed for the weighing of precious metals, precious stones and gems (diamond, ruby, sapphire) in the United Kingdom. Now obsolete in the UK, but this unit remains in common use in the USA. 1 oz (troy) = 1/12 lb (troy) (E) 1 oz (troy) = 20 pennyweights (troy) (E) 1 oz (troy) = 480 grains (troy) (E)	UK, US
ounce (UK, liquid)	fl oz (UK)	capacity, volume	L^3	1 fluid ounce (UK) = $28.41307500 \times 10^{-6}$ m^3	British unit used for capacity measurements for all merchandise (solids, liquids, foodstuffs, etc). 1 fl oz (UK) = 1/160 gallon (UK) 1 fl oz (UK) = 1/40 quart (UK) (E) 1 fl oz (UK) = 1/20 pint (UK) (E) 1 fl oz (UK) = 1/5 gill (UK) (E) 1 fl oz (UK) = 8 fl dr (UK) (E) 1 fl oz (UK) = 480 minims (UK) (E) 1 fl oz (UK) = 0.960759 fl oz (US)	UK
ounce (US, liquid)	fl oz (US)	capacity, volume	L^3	1 fluid ounce (US) = $29.57352958 \times 10^{-6}$ m^3	American unit used for capacity measurements of liquids. 1 fl oz (US) = 1/128 gall (US, liq.) (E) 1 fl oz (US) = 1/32 quart (US, liq.) (E) 1 fl oz (US) = 1/16 pint (US, liq.) (E) 1 fl oz (US) = 1/4 gill (US, liq.) (E) 1 fl oz (US) = 8 fl dr (US) (E) 1 fl oz (US) = 480 minims (US, liq.) (E) 1 fl oz (US) = 1.040840 fl oz (UK)	US

					1 ozf = 1/16 000 kip (E) 1 ozf = 1/16 lbf (E)	CAN
outava (Portuguese, dry)	–	capacity, volume	L^3	1 outava (Portuguese, dry) $= 1.6875 \times 10^{-3}$ m^3	Obsolete Portuguese unit of capacity used for dry substances. 1 outava (Portuguese, dry) = 1/32 fanga (E)	Portuguese
outavada (Portuguese)	–	mass	M	1 outavada (Portuguese) $= 3.585937500 \times 10^{-3}$ kg	Obsolete Portuguese unit of mass. 1 outavada (Portuguese) = 1/128 libra (E)	Portuguese
outen (Egyptian)	–	capacity, volume	L^3	1 outen (Egyptian) $= 2.125 \times 10^{-4}$ m^3 (of water)	Obsolete Egyptian unit of capacity used in ancient times. Measured by weight. 1 outen (Egyptian) = 1/160 khar (E)	Egyptian
oxhoft (Prussian)	–	capacity, volume	L^3	1 oxhoft (Prussian) $= 206.105400 \times 10^{-3}$ m^3	Obsolete German unit of capacity used for liquid substances. 1 oxhoft (Prussian) = 180 quart (E)	German
oxhooft (Dutch)	–	capacity, volume	L^3	1 oxhooft (Dutch) $= 230.400 \times 10^{-3}$ m^3	Obsolete Dutch unit of capacity used for liquid substances. 1 oxhooft (Dutch) = 192 mingelen (E)	Dutch
oxhufud (Swedish)	–	capacity, volume	L^3	1 oxhufud (Swedish) $= 235.544580 \times 10^{-3}$ m^3	Obsolete Swedish unit of capacity for liquid substances. 1 oxhufud (Swedish) = 90 kanna (E)	Swedish
oxybaphon (Greek, Attic)	–	capacity, volume	L^3	1 oxybaphon (Greek, Attic) $= 6.75 \times 10^{-5}$ m^3	Obsolete Greek unit of volume employed in ancient times. It was used for capacity measurements of dry substances. 1 oxybaphon (Greek, Attic) = 1/16 chenix (E)	Attic

Unit (synonym, acronym)	Symbol	Physical quantity	Dimension	Conversion factor (SI equivalent unit)	Notes, definitions, other conversion factors	System
pace (Persian)	–	length, distance	L	1 pace (Persian) = 1.875 m	Obsolete unit of length in the Assyrio-Chaldean-Persian system used in ancient times. 1 pace = 6 zereths (E)	Persian
pace (UK)	–	length, distance	L	1 pace (UK) = 7.62×10^{-1} m (E)	Obsolete British unit of length. 1 pace (UK) = 2.5 feet (E) 1 pace (UK) = 30 inches (E)	UK
pala (Indian)	–	capacity, volume	L^3	1 pala (Indian) = 5.156250×10^{-5} m^3 (of water)	Obsolete Indian unit of capacity used in ancient times. Measured by weight. 1 pala (Indian) = 1/256 drona (E)	Indian
pala (Indian)	–	mass	M	1 pala (Indian) = 4.70×10^{-3} kg	Obsolete Indian unit of mass used in ancient times.	Indian
palestra (Greek, Attic) [Greek palm]	–	length, distance	L	1 palestra (Greek, Attic) = 7.714×10^{-2} m	Obsolete Greek unit of length employed in ancient times. 1 palestra (Attic) = 1/4 pous (E)	Attic
paletz (Russian)	–	length, distance	L	1 paletz (Russian) = 1.27×10^{-2} m	Obsolete Russian unit of length used before 1917. 1 duïme (Russian) = 1/24 foute (E)	Russian
palm (Egyptian, short)	–	length, distance	L	1 palm (Egyptian, short) = $7.485714286 \times 10^{-2}$ m	Obsolete Egyptian unit of length used in ancient times.	Egyptian
palm (Hebrew)	–	length, distance	L	1 palm (Hebrew) = 9.25×10^{-2} m	Obsolete Hebrew unit of length used in ancient times. 1 palm = 1/6 cubit (E)	Hebrew

palm (Persian)	–	length, distance	L	1 palm (Persian) = 8×10^{-2} m	Obsolete unit of length in the Assyrio-Chaldean-Persian system used in ancient times. 1 palm = 1/4 zereth (E)	Persian
palm (UK)	–	length, distance	L	1 palm (UK) = 7.62×10^{-2} m (E)	Obsolete British unit of length. 1 palm (UK) = 3 inches (E) 1 palm (UK) = 10 paces (UK) (E) 1 palm (UK) = 36 lines (UK) (E)	UK
palma (Spanish)	–	length, distance	L	1 palma (Spanish) = $2.089762500 \times 10^{-1}$ m	Obsolete Spanish unit of length. 1 palma (Spanish) = 1/4 vara (E)	Spanish
palmipes (Roman)	–	length, distance	L	1 palmipes (Roman) = 0.368 m	Obsolete Roman unit of length employed in ancient times. 1 palmipes (Roman) = 5/4 pes (E)	Roman
palmo (Portuguese)	–	length, distance	L	1 palmo (Portuguese) = 2.190×10^{-1} m	Obsolete Portuguese unit of length. 1 palmo (Portuguese) = 2/3 pe (E)	Portuguese
palmus (Roman) [Roman palm]	–	length, distance	L	1 palmus (Roman) = 7.36×10^{-2} m	Obsolete Roman unit of length employed in ancient times. 1 palmus (Roman) = 1/4 pes (E)	Roman
panilla (Spanish)	–	capacity, volume	L^3	1 panilla (Spanish) = $0.15643162 \times 10^{-3}$ m^3	Obsolete Spanish unit of capacity used for liquid substances. 1 panilla (Spanish) = 1/100 arroba (water) (E)	Spanish
par (hyl, mug, metric slug, TME, techma)	par	mass	M	1 par = 9.80665 kg (E)	Obsolete technical metric unit of mass employed by mechanical engineers (base unit of the metric gravitational system). It was equal to the mass which under an acceleration of $1 m.s^{-2}$ gives a force of 1 kgf. The name TME derives from the *(continued overleaf)*	German

Unit (synonym, acronym)	Symbol	Physical quantity	Dimension	Conversion factor (SI equivalent unit)	Notes, definitions, other conversion factors	System
par *(continued)*					German acronym Technische Mass Einheit. The name of the unit *par* derives from the French word *paresseux* = sluggish.	
parasang (Arabic)	–	length, distance	L	1 parasang (Arabic) = 5.760×10^3 m	Obsolete Arabic unit of length used in ancient times. 1 parasang (Arabic) = 18 000 feet (Arabic) (E)	Arabic
parasang (Persian)	–	length, distance	L	1 parasang (Persian) = 6400 m	Obsolete unit of length in the Assyrio-Chaldean-Persian system used in ancient times. 1 parasang = 20 000 zereths (E)	Persian
parasange (Egyptian)	–	length, distance	L	1 parasange (Egyptian) = 6.980×10^3 m	Obsolete Egyptian unit of length used in ancient times. 1 parasange = 40 000/3 Royal cubit (E)	Egyptian
parsec	pc	length, distance	L	1 parsec = $3.085677567 \times 10^{16}$ m	Astronomical unit of length employed to describe stellar distance. The parsec is the distance from which one AU would subtend an angle of 1 second of arc. The parsec is derived from the English acronym **par**allax **sec**ond. 1 pc = 2.062648×10^5 AU 1 pc = 3.261633 l.y.	
part per billion	ppb	fraction, relative values, yields, efficiencies, abundance	nil	1 ppb = 10^{-9} (E)		INT

part per hundred (percent)	pph, %, pct	fraction, relative values, yields, efficiencies, abundance	nil	1 pph = 1% = 10[illegible] (E)		INT
part per hundred million	pphm	fraction, relative values, yields, efficiencies, abundance	nil	1 pphm = 10^{-8} (E)		INT
part per hundred thousand	ppht	fraction, relative values, yields, efficiencies, abundance	nil	1 ppht = 10^{-5} (E)		INT
part per million	ppm	fraction, relative values, yields, efficiencies, abundance	nil	1 ppm = 10^{-6} (E)		INT
part per quadrillion	ppq	fraction, relative values, yields, efficiencies, abundance	nil	1 ppt = 10^{-15} (E)		INT
part per tera	ppt	fraction, relative values, yields, efficiencies, abundance	nil	1 ppt = 10^{-12} (E)		INT
part per thousand	ppt, ‰	fraction, relative values, yields, efficiencies, abundance	nil	1 ppt = 1‰ = 10^{-3} (E)		INT
pas (de Paris) (French pace)	–	length, distance	L	1 pas (de Paris) = 0.624 m	Obsolete French unit of length used before the French Revolution (1789).	French

Unit (synonym, acronym)	Symbol	Physical quantity	Dimension	Conversion factor (SI equivalent unit)	Notes, definitions, other conversion factors	System
pascal	Pa	pressure, stress	$ML^{-1}T^{-2}$	**SI derived unit** 1 Pa = 1 $N.m^{-2}$ (E)	The pascal is the uniform pressure which applied on a plane area of one square metre, exerts perpendicularly at this surface a total force of one newton (14th CGPM, 1971). The unit is named after the French scientist B. Pascal (1623–1662). 1 atm = 101 325 Pa (E)	SI
pascal second	Pa.s	dynamic viscosity, absolute viscosity	$ML^{-1}T^{-1}$	**SI derived unit** 1 Pa.s = 1 $kg.m^{-1}.s^{-1}$ (E)	The pascal second is the unit of dynamic viscosity or absolute viscosity. It describes the laminar flow of an homogeneous fluid which undergoes a shear stress normal to it of one pascal when it undergoes a velocity gradient of one reciprocal second.	SI
passo (Spanish)	–	length, distance	L	1 passo (Spanish) = 1.393175 m	Obsolete Spanish unit of length. 1 passo (Spanish) = 5/3 vara (E)	Spanish
passus (Roman) [Roman double pace]	–	length, distance	L	1 passus (Roman) = 1.472 m	Obsolete Roman unit of length employed in ancient times. 1 passus (Roman) = 5 pes (E)	Roman
pastille dose (B unit)	–	exposure	ITM^{-1}	1 pastille dose = 1.290×10^{-1} $C.kg^{-1}$	Obsolete unit of exposure of ionizing radiations. It was equal to the radiation dose required to change the variation colour of a basic barium platinocyanide pastille from apple green (tint A) to reddish brown (tint B). 1 B unit = 500 röntgens (E)	

payok (Russian, dry)	–	capacity, volume	L^3	1 payok (Russian, dry) $= 52.477472 \times 10^{-3}$ m^3	Obsolete Russian unit of capacity for dry substances used before 1917. 1 payok (Russian, dry) = 16 garnetz (E)	Russian
pe (Portuguese)	–	length, distance	L	1 pe (Portuguese) = 0.3285 m	Obsolete Portuguese unit of length.	Portuguese
peck (UK)	pk (UK)	capacity, volume	L^3	1 peck (UK) $= 9.092184000 \times 10^{-3}$ m^3	Obsolete British unit used for capacity measurements for all merchandise (solids, liquids, foodstuffs, etc). 1 peck (UK) = 2 gallons (UK) (E) 1 peck (UK) = 4 pottle (UK) (E) 1 peck (UK) = 8 quarts (UK) (E) 1 peck (UK) = 16 pints (UK) (E) 1 peck (UK) = 32 gills (UK) (E) 1 peck (UK) = 1.032060 peck (US)	UK
peck (US, dry)	pk (US, dry)	capacity, volume	L^3	1 peck (US, dry) $= 8.809768000 \times 10^{-3}$ m^3	Obsolete American unit of capacity used to measure the volume of powdered or divided solid materials (flour, sand, cement, ores, etc). 1 peck (US, dry) = 2 gallons (US, dry) (E) 1 peck (US, dry) = 16 pints (US, dry) (E) 1 peck (US, dry) = 8 quarts (US, dry) (E) 1 peck (US, dry) = 0.968940 peck (UK)	US
pekeis (Egyptian)	–	surface, area	L^2	1 pekeis (Egyptian) = 27.405 m^2	Obsolete Egyptian unit of area used in ancient times.	Egyptian
pennyweight (troy)	dwt (troy)	mass	M	1 dwt (troy) $= 1.555173840 \times 10^{-3}$ kg	Obsolete British and American unit of mass. 1 dwt (troy) = 1/240 lb (troy) (E) 1 dwt (troy) = 24 grains (troy) (E)	UK, US

Unit (synonym, acronym)	Symbol	Physical quantity	Dimension	Conversion factor (SI equivalent unit)	Notes, definitions, other conversion factors	System
percent	%	plane angle	α	$1\% = 9.999666687 \times 10^{-3}$ rad	Unit of plane angle employed in civil engineering. One percent describes a plane angle which corresponds to a difference of height of one metre for 100 metres of length. $1\% = 0.5729386977^\circ$ 1% = 0.636598553 gon	INT
perch (rod, pole)	rd	length, distance	L	1 perch = 5.029200 m (E)	Obsolete American and British unit of length used in surveyor's measurements. 1 perch = 1 pole (E) 1 perch = 1 rod (E) 1 perch = 5.5 yards (E) 1 perch = 16.5 feet (E) 1 perch = 198 inches (E)	UK, US
perche (de Paris)	–	length, distance	L	1 perche (de Paris) = 5.847109500 m	Obsolete French unit of length used before the French Revolution (1789). 1 perche (Paris) = 18 pieds (de Paris) (E) 1 perche (Paris) = 3 toises (de Pérou) (E)	French
perche (de Paris)	–	length, distance	L	1 perche (de Paris) = 6.496788334 m	Obsolete French unit of length used before the French Revolution (1789). 1 perche (de Paris) = 20 pieds (de Paris) (E) 1 perche (de Paris) = 20/3 toises (de Pérou) (E)	French

				= 34.18868950 m^2	surveyor's measurements before the French revolution in 1789. 1 perche (de Paris) = 1/100 arpent (de Paris) (E)	
perche (Eaux et Forêts)	–	length, distance	L	1 perche (Eaux et Forêts) = 7.146 m	Obsolete French unit of length used before the French Revolution (1789). 1 perche (Eaux et Forêts) = 22 pieds (de Paris) (E)	French
perche (Eaux et Forêts)	–	surface, area	L^2	1 perche (Eaux et Forêts) = 51.07199295 m^2	Obsolete French metric unit of area used in surveyor's measurements. 1 perche (Eaux et Forêts) = 1/100 arpent (Eaux et Forêts) (E)	French
perche (masonry) [of stone]	–	capacity, volume	L^3	1 perche (mansonry) = 700.8419532 × 10^{-3} m^3	Obsolete American and British unit of employed in civil engineering. It was equal to a quantity 1.5 feet thick, 1 foot high and 16.5 feet long. 1 perche (mansonry) = 24.75 ft^3	UK, US
perche (Québec)	–	length, distance	L	1 perche (Québec) = 5.8471 m	Obsolete French unit of length used before the French Revolution (1789). Still used in the Québec (Canada).	CAN
perm	–	permeability	L^2	1 perm = 10^{-4} m^2	1 perm = 1 cm^2 (E)	
perm (0°C)	perm (0°C)	permeability	L^{-1}T	1 perm (0°C) = 5.72135 × 10^{-11} kg.Pa^{-1}.m^{-2}.s^{-1}	American unit employed in gas separation by the membrane process. It serves to express the product of the Henry's constant by the gas diffusion coefficient in the mass flow rate basis. 1 kg.Pa^{-1}.m^{-2}.s^{-1} = 1 m^{-1}.s (E)	US

Unit (synonym, acronym)	Symbol	Physical quantity	Dimension	Conversion factor (SI equivalent unit)	Notes, definitions, other conversion factors	System
perm (23°C)	perm (23°C)	permeability	$L^{-1}T$	1 perm (23°C) $= 5.74525 \times 10^{-11}$ kg.Pa^{-1}.m^{-2}.s^{-1}	American unit employed in gas separation by the membrane process. It serves to express the product of the Henry's constant by the gas diffusion coefficient in the mass flow rate basis. 1 kg.Pa^{-1}.m^{-2}.s^{-1} = 1 m^{-1}.s (E)	US
perm-inch (0°C)	perm-inch (0°C)	permeability coefficient (mass flow rate)	T	1 perm-inch (0°C) $= 1.45322 \times 10^{-12}$ kg.Pa^{-1}.m^{-1}.s^{-1}	American unit employed in gas separation by membrane process. It serves to express the product of the Henry's constant by the gas diffusion coefficient divided by membrane thickness in the mass flow rate basis. 1 kg.Pa^{-1}.m^{-1}.s^{-1} = 1 s (E)	UK, US
perm-inch (23°C)	perm-inch (23°C)	permeability coefficient (mass flow rate)	T	1 perm-inch (23°C) $= 1.45929 \times 10^{-12}$ kg.Pa^{-1}.m^{-1}.s^{-1}	American unit employed in Gas separation by membrane process. It serves to express the product of the Henry's constant by the gas diffusion coefficient divided by membrane thickness in the mass flow rate basis. 1 kg.Pa^{-1}.m^{-1}.s^{-1} = 1 s (E)	UK, US
permicron	–	wavenumber	L^{-1}	1 permicron = 10^{6} m^{-1} (E)	Obsolete wavenumber unit which has been proposed in spectroscopy. 1 permicron = 1 μm^{-1} (E)	
pes (Roman) [Roman foot]	–	length, distance	L	1 pes (Roman) = 0.2944 m	Obsolete Roman unit of length employed in ancient times. Base unit of length.	Roman
petagram	Pg	mass	M	1 Pg = 10^{12} kg (E)	Multiple of the SI base unit. 1 Pg = 10^{15}g (E)	SI

petametre	Pm	length, distance	L	[illegible]	[illegible]	
petit romain	–	length, distance	L	1 petit romain $= 3.383743750 \times 10^{-3}$ m	Obsolete French unit of length employed in typography. It was a submultiple of the point Didot. 1 petit romain = 9 points Didot (E)	French
petit texte	–	length, distance	L	1 petit texte $= 2.819786459 \times 10^{-3}$ m	Obsolete French unit of length employed in typography. It was a submultiple of the point Didot. 1 petit texte = 7.5 points Didot (E)	French
pfennig (Austrian) (demat)	–	mass	M	1 pfennig (Austrian) $= 1.093769531 \times 10^{-3}$ kg	Obsolete Austrian unit of mass for general use. 1 pfennig (Austrian) = 1/512 pfund (E)	Austrian
pfiff (Austrian)	–	capacity, volume	L^3	1 pfiff (Austrian) $= 1.768875 \times 10^{-4}$ m^3	Obsolete Austrian unit of capacity used for liquid substances. 1 pfiff (Austrian) = 1/8 mass (E)	Austrian
pfund (Austrian)	–	mass	M	1 pfund (Austrian) = 0.560010 kg	Obsolete Austrian unit of mass for general use.	Austrian
pfund (Austrian, apothecary)	–	mass	M	1 pfund (Austrian, apothecary) = 0.42001 kg	Obsolete Austrian unit of mass used in pharmacy. 1 pfund (Austrian, apothecary) = 3/4 pfund (Austrian)	Austrian
pfund (Prussian)	–	mass	M	1 pfund $= 467.711 \times 10^{-3}$ kg (E)	Obsolete German unit of mass.	German
philosophia	–	length, distance	L	1 philosophia $= 4.135686806 \times 10^{-3}$ m	Obsolete French unit of length employed in typography. It was a submultiple of the point Didot. 1 philosophia = 11 points Didot (E)	French

Unit (synonym, acronym)	Symbol	Physical quantity	Dimension	Conversion factor (SI equivalent unit)	Notes, definitions, other conversion factors	System
phon	phon, P	loudness level	nil	(see note)	The phon is the level of sound one decibel above a reference sound level of 20×10^{-6} Pa of a pure tone of frequency 1000 Hz.	US, UK
phot	ph	illuminance	JL^{-2}	1 ph = 10^4 lux (E)	Obsolete cgs illuminance unit.	cgs
photon (luxon)	hν	energy of electromagnetic radiation	$ML^{-2}T^{-2}$	1 hν = $6.626075540 \times 10^{-34}$ ν (Hz)	Obsolete unit of quantum of luminous energy transported by an electromagnetic radiation of frequency ν (Hz).	
pica	–	length, distance	L	1 pica = 4.217518×10^{-3} m	American unit of length employed in typography. The pica is still used today in the printing industry. 1 pica = 12 points US (E) 1 pica = 1/6 inch (approx.)	US
picogram	pg	mass	M	1 pg = 10^{-15} kg (E)	Submultiple of the SI base unit. 1 pg = 10^{-12} g (E)	SI
picometre	pm	length, distance	L	1 pm = 10^{-12} m (E)	Submultiple of the SI base unit.	SI
pie (Spanish)	–	length, distance	L	1 pie (Spanish) = 2.786350×10^{-1} m	Obsolete Spanish unit of length. 1 pie (Spanish) = 1/3 vara (E)	Spanish
pied (de Paris)	pied	length, distance	L	1 pied (de Paris) = 0.3248394167 m	Obsolete French unit of length used before the French Revolution (1789). 1 pied = 12 pieds (de Paris) (E) 1 pied = 1/6 toise (de Pérou) (E)	French
pied (metric) (French metric foot)	–	length, distance	L	1 pied (metric) = 0.33 m (E)	Obsolete French unit of length used from 1812 to 1840.	French

					1 pied (metric) = 12 pouces (metric) (E) 1 pied (metric) = 144 lignes (metric) (E)	
piede liprando	–	length, distance	L	1 piede liprando = 0.51377 m	Obsolete Italian unit of length. Base unit of length.	Italian
pièze (sthène.m^{-2})	pz	pressure, stress	$ML^{-1}T^{-2}$	1 pz = 10^3 Pa (E)	Obsolete MTS pressure and stress derived unit with a special name. 1pz = 1 sn.m^{-2} (E) 1 hpz = 1 bar (E) 1 atm = 1.01325 hpz (E)	MTS
pigon (Egyptian)	–	length, distance	L	1 pigon (Egyptian) = 4.362500 × 10^{-1} m	Obsolete Egyptian unit of length used in ancient times. 1 pigon = 5/6 Royal cubit (E)	Egyptian
ping (Chinese)	–	capacity, volume	L^3	1 ping (Chinese) = 517.720 × 10^{-3} m^3	Obsolete Chinese unit of capacity used in ancient times. 1 ping (Chinese) = 500 tcheng (E)	Chinese
pint (Dutch)	–	capacity, volume	L^3	1 pint (Dutch) = 0.600 × 10^{-3} m^3	Obsolete Dutch unit of capacity used for liquid substances. 1 pint (Dutch) = 1/2 mingelen (E)	Dutch
pint (UK)	pt (UK)	capacity, volume	L^3	1 pint (UK) = 0.568261500 × 10^{-3} m^3	Obsolete British unit used for capacity measurements for all merchandise (solids, liquids, foodstuffs, etc). 1 pint (UK) = 1/8 gallon (UK) (E) 1 pint (UK) = 1/4 pottle(UK) (E) 1 pint (UK) = 1/2 quart (UK) (E) 1 pint (UK) = 4 gills (UK) (E) 1 pint (UK) = 20 fl oz (UK) (E) 1 pint (UK) = 160 fl dr (UK) (E) 1 pint (UK) = 1.032060 pint (US)	UK

Unit (synonym, acronym)	Symbol	Physical quantity	Dimension	Conversion factor (SI equivalent unit)	Notes, definitions, other conversion factors	System
pint (US, dry)	pt (US, dry)	capacity, volume	L^3	1 pint (US, dry) $= 0.550610500 \times 10^{-3}\ m^3$	American unit of capacity used to measure the volume of powdered or divided solid materials (flour, sand, clinker, ore, etc). 1 pint (US, dry) = 1/8 gallon (US, dry)	US
pint (US, liquid)	pt (US, liq.)	capacity, volume	L^3	1 pint (US, liq.) $= 0.4731764730 \times 10^{-3}\ m^3$	American unit used for capacity measurements of liquids. 1 pint (US, liq.) = 1/8 gallon (US, liq.) (E) 1 pint (US, liq.) = 1/2 quart (US, liq.) (E) 1 pint (US, liq.) = 4 gills (US, liq.) (E) 1 pint (US, liq.) = 16 fl oz (US, liq.) (E) 1 pint (US, liq.) = 128 fl dr (US, liq.) (E) 1 (US) pint = 0.938940 pint (UK)	US
pinte (de Paris)	–	capacity, volume	L^3	1 pinte (de Paris) $= 0.931389 \times 10^{-3}\ m^3$	Obsolete French unit of capacity employed before the French Revolution. It served to express the capacity of liquids and grains. It varied according to the location and merchandise. 1 pinte (de Paris) = 1/288 muid (de Paris) (E)	French
pipa (Portuguese)	–	capacity, volume	L^3	1 pipa (Portuguese) $= 429 \times 10^{-3}\ m^3$	Obsolete Portuguese unit of capacity used for liquid substances. 1 pipa (Portuguese) = 26 almude (E)	Portuguese

pipa (Spanish)		capacity, volume	L^3	1 pipa (Spanish) $= 422.365374 \times 10^{-3}$ m^3	Obsolete Spanish unit of capacity used for liquid substances. 1 pipa (Spanish) = 27 arroba (water) (E)	Spanish
pipe (Russian)	–	capacity, volume	L^3	1 pipe (Russian) $= 442.778760 \times 10^{-3}$ m^3	Obsolete Russian unit of capacity for liquid substances used before 1917. 1 pipe (Russian) = 36 vedro (E)	Russian
pipe (Swedish)	–	capacity, volume	L^3	1 pipe (Swedish) $= 471.089160 \times 10^{-3}$ m^3	Obsolete Swedish unit of capacity for liquid substances. 1 pipe (Swedish) = 180 kanna (E)	Swedish
pixel	–	screen resolution	nil	(see note)	Unit of screen resolution employed in computer science.	
planck	–	action, angular momentum	ML^2T^{-1}	1 planck = 1 J.s (E)	Unit of angular momentum proposed for the SI. The unit is named after the German physicist M. Planck (1858–1947). Anecdotal.	
plethron (Greek, Attic)	–	length, distance	L	1 plethron (Greek, Attic) = 30.856 m	Obsolete Greek unit of length employed in ancient times. 1 plethron (Attic) = 100 pous (E)	Attic
point (de Paris)	–	length, distance	L	1 point (de Paris) $= 1.879857639 \times 10^{-4}$ m	Obsolete French unit of length used before the French Revolution (1789). 1 point (de Paris) = 1/12 ligne (de Paris) (E) 1 point (de Paris) = 1/144 pouce (de Paris) (E) 1 point (de Paris) = 1/10 368 toise (de Pérou) (E)	French

Unit (synonym, acronym)	Symbol	Physical quantity	Dimension	Conversion factor (SI equivalent unit)	Notes, definitions, other conversion factors	System
point (Didot)	–	length, distance	L	1 point (Didot) $= 3.759715278 \times 10^{-4}$ m	French unit of length used in typography. The point Didot is still used in the French printing industry and it has been adopted in several countries. The unit is named after F.-A. Didot (1730–1804). 1 point Didot = 1/6 ligne (de Paris) (E) 1 point Didot = 1/72 pouce (de Paris) (E)	French
point (jeweller's)	–	mass	M	1 point (jeweller's) $= 2 \times 10^{-6}$ kg (E)	Obsolete British and American unit of mass employed in jewellery for the weighing of precious metals, precious stones and gems (e.g. diamond, ruby, sapphire). 1 point (jeweller's) = 0.01 carat (metric) (E) 1 point (jeweller's) = 2 mg (E)	UK, US
point (US printer's)	–	length, distance	L	1 point (US) $= 3.514598 \times 10^{-4}$ m	American unit of length used in typography. The point is still used today in the printing industry. 1 point (US printer's) $\approx$ 1/72 inch (approx.)	US
point (I.N.) (Imprimerie Nationale)	–	length, distance	L	1 point (I.N.) $= 3.9877 \times 10^{-4}$ m	French unit of length used in typography. The point (I.N.) is still used in the French printing industry.	French
poise	P, Po	dynamic viscosity, absolute viscosity	$ML^{-1}T^{-1}$	1 P $= 10^{-1}$ Pa.s (E)	Obsolete unit of dynamic viscosity or absolute viscosity in the cgs system. In the UK the written symbol was Po.	cgs

poiseuille		dynamic viscosity, absolute viscosity	$ML^{-1}T^{-1}$	SI derived unit 1 poiseuille = 1 $m^2.s^{-1}$ (E)	Unit proposed for the SI unit of dynamic viscosity. The unit is named after the French physician and physicist J.-L. Poiseuille (1799–1869).	SI
pole (rod, perch)	rd	length, distance	L	1 pole = 5.0292 m (E)	Obsolete American and British unit of length used in surveyor's measurements. 1 pole = 1 perch (E) 1 pole = 1 rod (E) 1 pole = 5.5 yards (E) 1 pole = 16.5 feet (E)	UK, US
pollegada (Portuguese)	–	length, distance	L	1 pollegada (Portuguese) = 2.737500×10^{-2} m	Obsolete Portuguese unit of length. 1 pollegada (Portuguese) = 1/12 pe (E)	Portuguese
pollen count	–	pollen surface density	L^{-2}	1 pollen count = 10^{-4} $pollen.m^{-2}$	Obsolete unit introduced in 1930 by botanists for counting pollen under a microscope. It was equal to the number of grains of pollen per unit of area 1 pollen count = 1 $pollen.cm^{-2}$ (E)	
polougarnetz (Russian, dry)	–	capacity, volume	L^3	1 polougarnetz (Russian, dry) = 1.639921×10^{-3} m^3	Obsolete Russian unit of capacity for dry substances used before 1917. 1 polougarnetz (Russian, dry) = 1/2 garnetz (E)	Russian
poncelet	–	power	ML^2T^{-3}	1 poncelet = 980.665 W	Obsolete French unit of power. It was equal to the work done during one second by a force which accelerates a mass of 100 kg on a distance of one metre. The unit is named after J.V. Poncelet (1788–1867). 1 poncelet = 100 $kgfm.s^{-1}$ (E)	French
pond (Amsterdam)	–	mass	M	1 pond (Amsterdam) = 0.49409032 kg	Obsolete Dutch unit of mass.	Dutch

Unit (synonym, acronym)	Symbol	Physical quantity	Dimension	Conversion factor (SI equivalent unit)	Notes, definitions, other conversion factors	System
pond (Dutch) [Dutch pound]	–	mass	M	1 pond (Dutch) = 0.49216772 kg	Obsolete Dutch unit of mass.	Dutch
pond (Dutch, apothecary)	–	mass	M	1 pond (Dutch, apothecary) = 0.369126 kg	Obsolete Dutch unit of mass. 1 pond (Dutch, apothecary) = 3/4 pond (Dutch)	Dutch
pond (gram-force)	p	force, weight	MLT^{-2}	1 pond = 9.80665×10^{-3} N (E)	Obsolete submultiple of the MKpS system.	MKpS
pondiuscule (Hebrew) [Talmudic system]	–	mass	M	1 pondiuscule (Hebrew) = 2.951667×10^{-4} kg	Obsolete Hebrew unit of mass used in ancient times. Rabbinical or Talmudic system. 1 pondiuscule (Hebrew) = 1/1200 mina (E)	Hebrew
pood (Russian)	–	mass	M	1 pood = 16.38068717 kg	Obsolete Russian unit of mass used before 1917 for general use. 1 poods = 40 funts (E)	Russian
posson (de Paris)	–	capacity, volume	L^3	1 posson (de Paris) = $1.164236250 \times 10^{-4}$ m^3	Obsolete French unit of capacity employed before the French Revolution. It served to express the capacity of liquids and grains. It varied according to the location and merchandise. 1 posson (de Paris) = 1/8 pinte (de Paris) (E)	French
pottle (UK)	–	capacity, volume	L^3	1 pottle (UK) = 2.273046×10^{-3} m^3	Obsolete British unit used for capacity measurements for all merchandise (solids, liquids, foodstuffs, etc). 1 pottle (UK) = 1/2 gallon (UK) (E) 1 pottle (UK) = 2 quarts (UK) (E)	UK

					1 pottle (UK) = 4 pints (UK) (E) 1 pottle (UK) = 10 gills (UK) (E)	
pou (Chinese)	–	length, distance	L	1 pou (Chinese) = 4140 m	Obsolete Chinese unit of length used in ancient times. 1 pôu = 18 000 tchi (E)	Chinese
pôu (Chinese)	–	length, distance	L	1 pôu (Chinese) = 1.150 m	Obsolete Chinese unit of length used in ancient times. 1 pôu = 5 tchi (E)	Chinese
pouce (de Paris)	–	length, distance	L	1 pouce (de Paris) $= 2.706995139 \times 10^{-2}$ m	Obsolete French unit of length used before the French Revolution (1789). 1 pouce (de Paris) = 12 lignes (de Paris) (E) 1 pouce (de Paris) = 144 points (de Paris) (E) 1 pouce (de Paris) = 1/72 toise (de Pérou) (E)	French
pouce (metric) (French metric inch)	–	length, distance	L	1 pouce (metric) $= 2.75 \times 10^{-2}$ m (E)	Obsolete French unit of length used in France from 1812 until 1840. 1 pouce (metric) = 1/3 pied (metric) (E) 1 pouce (metric) = 12 lignes (metric) (E)	French
poumar	–	linear mass density	ML^{-1}	1 poumar $= 4.960546479 \times 10^{-7}$ kg.m^{-1}	Obsolete British unit of linear mass density employed in the textile industry. The name of the unit is derived from the English acronym: **pou**nd per **m**illion y**ar**d. 1 poumar = 1 lb.10^6 yd^{-1} (E)	UK
pound (apothecary)	lb (ap.), lb (apoth.)	mass	M	1 lb (apoth.) $= 373.241721600 \times 10^{-3}$ kg (E)	Obsolete British and American unit of weight employed in pharmacy. It was employed at one time as a unit of weight for drugs and medicinal preparations *(continued overleaf)*	UK, US

Unit (synonym, acronym)	Symbol	Physical quantity	Dimension	Conversion factor (SI equivalent unit)	Notes, definitions, other conversion factors	System
pound (apothecary) *(continued)*					(lotions, potions, ointments, plant extracts, etc). 1 lb (apoth.) = 5760 grains (apoth.) (E) 1 lb (apoth.) = 240 pennyweights (apoth.) (E) 1 lb (apoth.) = 12 ounces (apoth.) (E)	
pound (Austrian) (zollpfund)	–	mass	M	1 zollpfund = 560.00448×10^{-3} kg	Obsolete Austrian unit of weight. 1 kg = 1.7857 zollpund (E)	Austrian
pound (avoirdupois, pound-mass)	lb (avdp.), lb (av.)	mass	M	1 lb (av.) = $453.5924277 \times 10^{-3}$ kg (E)	Obsolete British and American unit of mass. It was used as a base unit of mass for the imperial system before the WMA of 1963. 1 lb (av.) = 7 000 grains (av.) (E) 1 lb (av.) = 256 drams (av.) (E) 1 lb (av.) = 16 ounces (av.) (E)	UK, US, FPS
pound (imperial standard)	lb (IS)	mass	M	1 lb (IS) = $453.592338 \times 10^{-3}$ kg	Obsolete British unit of mass. It was used as a base unit of mass for the imperial system before the WMA of 1963.	UK
pound (Prussian)	–	mass	M	1 pfund = 467.711×10^{-3} kg (E)	Obsolete German unit of mass.	German
pound (Russian, apothecary)	–	mass	M	1 pound (Russian, apothecary) = 0.5332192836 kg	Obsolete Russian unit of mass used before 1917 in pharmacy.	Russian
pound (Russian)	–	mass	M	1 funt = $409.5171792 \times 10^{-3}$ kg	Obsolete Russian unit of weight employed before 1917 for general use. 1 kg = 2.4419 funts (E)	Russian

pound (Swedish) (skålpund)	–	mass	M	1 skålpund = $425.0797024 \times 10^{-3}$ kg	Obsolete Swedish unit of weight. 1 kg = 2.3525 skålpunds (E)	Swedish
pound (Swiss) (zollpfund)	–	mass	M	1 zollpfund = 500×10^{-3} kg	Obsolete Swiss unit of weight. 1 kg = 2 zollpfunds (E)	Swiss
pound (troy)	lb (tr.), lb (troy)	mass	M	1 lb (troy) = $373.241721600 \times 10^{-3}$ kg (E)	Obsolete British unit of mass employed for the weighing of precious metals, precious stones and gems (diamond, ruby, sapphire) in the United Kingdom. Now obsolete in the UK, this unit remains in common use in the USA. 1 lb (troy) = 5 760 grains (troy) (E) 1 lb (troy) = 240 pennyweights (troy) (E) 1 lb (troy) = 12 ounces (troy) (E)	UK, US
pound (UK, new hay)	lb (UK, new hay)	mass	M	1 lb (UK, new hay) = $272.1554029 \times 10^{-3}$ kg	Obsolete British unit of mass used by farmers. 1 lb (UK, new hay) = 3/5 lb (UK, straw) (E)	UK
pound (UK, obsolete hay)	lb (UK, obsolete hay)	mass	M	1 lb (UK, obsolete hay) = $291.5950745 \times 10^{-3}$ kg	Obsolete British unit of mass used by farmers. 1 lb (UK, obsolete hay) = 9/14 lb (UK, straw) (E)	UK
pound (UK, straw)	lb (UK, straw)	mass	M	1 lb (UK, straw) = $453.592338 \times 10^{-3}$ kg	Obsolete British unit of mass used by farmers. 1 lb (UK, straw) = 1 lb (av., obsolete) (E)	UK
pound (US)	lb (US)	mass	M	1 lb (US) = $453.59224277 \times 10^{-3}$ kg	Obsolete American unit of mass. 1 lb (US) = 5760 grains (av., obsolete)	US
pound (WMA, 1963) [USMB, 1959]	lb (1963)	mass	M	1 lb (WMA, 1963) = 453.59237×10^{-3} kg (E)	Legal unit of mass in the imperial British System and in the American. It was *(continued overleaf)*	UK, US, FPS

Unit (synonym, acronym)	Symbol	Physical quantity	Dimension	Conversion factor (SI equivalent unit)	Notes, definitions, other conversion factors	System
pound (WMA, 1963) [USMB, 1959] *(continued)*					adopted in 1963 by the Weights and Measures Act (WMA) in England and in the US in 1959 by the US Metric Board (USMB, 1959). It has been linked to the international kilogram standard (Sèvres).	
pound per cubic foot	$lb.ft^{-3}$	density, mass density	ML^{-3}	1 $lb.ft^{-3}$ = 565.6866954 $kg.m^{-3}$	American and British unit of density.	U.K, US, FPS
pound-centigrade-unit (centigrade-heat unit)	pcu	energy, work, heat	ML^2T^{-2}	1 pcu = 1899.18 J	The pcu is an obsolete British and American heat unit. It was equal to the heat needed to raise the temperature of one pound of air-free water by 1°C at a constant pressure of one standard atmosphere (101 325 Pa). The named was derived from the acronym **p**ound **c**entigrade **u**nit. 1 pcu = 1.8 Btu (E)	UK, US
pound-force	lbf av., lbf	force, weight	MLT^{-2}	1 pound-force = 4.4482216152605 N (E)	British and American legal unit of force and weight. It is equal to the force which is produced by the standard acceleration ($g = 9.80665\ m.s^{-2}$ or $32.174\ ft.s^{-2}$) of a body weighing one pound (1963). 1 lbf = 1 lb × 32.174 $ft.s^{-2}$ (E) 1 lbf = 10^{-3} kip (E) 1 lbf = 16 ozf (E)	UK, US, FPS
pound-force per square foot	$lbf.ft^{-2}$ (lbf/sq.ft, psf, PSF)	pressure, stress	$ML^{-1}T^{-2}$	1 $lbf.ft^{-2}$ = 47.88025898 Pa	FPS pressure and stress unit. 1 $lbf.ft^{-2}$ = 1 psf (E) 1 atm = 2116.216623 psf	UK, US, FPS

pound-force per square inch	$lbf.in^{-2}$ psi, PSI	pressure, stress	$ML^{-1}T^{-2}$	1 psi = $6.894757292 \times 10^{3}$ Pa	FPS pressure and stress unit. 1 $lbf.ft^{-2}$ = 1 psi (E) 1 atm = 14.69594877 psi	UK, US, FPS
pound-mole	lbmol	amount of substance	N	1 lbmol = 0.45359237 mol (E)	Obsolete British and American unit used to express an amount of substance. It was defined after the pound (WMA, 1963)	UK, US
poundal	pdl	force, weight	MLT^{-2}	1 pdl = 0.1382549544 N (E)	Obsolete FPS unit of force and weight. It is equal to the force which is produced by an acceleration of 1 $ft.s^{-2}$ of a body weighing one pound (1963). 1 pdl = 1 lb × 1 $ft.s^{-2}$ (E) 1 lbf = 1 lb × 32.174 $ft.s^{-2}$ (E)	FPS
poundal per square foot	$pdl.ft^{-2}$	pressure, stress	$ML^{-1}T^{-2}$	1 $pdl.ft^{-2}$ = 1.488163944 Pa	Obsolete British and American unit of pressure and stress.	UK, US
pourcentmille	pcm	reactivity of nuclear power reactor	nil	1 pcm = 10^{-5} (E)	Obsolete French unit of nuclear reactivity employed in nuclear engineering.	French
pous (Greek, Attic) [Greek foot]	–	length, distance	L	1 pous (Greek, Attic) = 0.30856 m	Obsolete Greek unit of length employed in ancient times.	Attic
pragilbert	–	magnetomotive force	I	1 pragilbert = 11 459.08 A	1 pragilbert = 4π ampere-turns	
praoersted	–	magnetic field strength	IL^{-1}	1 praoersted = 11 459.08 $A.m^{-1}$	1 praoersted = 4π ampere-turns per metre (E)	
prastha (Indian)	–	capacity, volume	L^3	1 prastha (Indian) = 8.250×10^{-4} m^3 (of water)	Obsolete Indian unit of capacity used in ancient times. Measured by weight. 1 prastha (Indian) = 1/16 drona (E)	Indian
preece	–	electric resistance	$ML^2T^{-3}I^{-2}$	1 preece = 10^6 Ω (E)	Obsolete British unit of electric resistance employed in the last century for insulating *(continued overleaf)*	UK

Unit (synonym, acronym)	Symbol	Physical quantity	Dimension	Conversion factor (SI equivalent unit)	Notes, definitions, other conversion factors	System
preece *(continued)*					material measurements. The unit is named after W. Preece (1834–1913). 1 preece = 1 MΩ (E)	
probmetze (Austrian, dry)	–	capacity, volume	L^3	1 probmetze (Austrian, dry) $= 6.0047851 \times 10^{-5}$ m^3	Obsolete Austrian unit of capacity used for dry substances. 1 probmetze (Austrian, dry) = 1/1024 metzel (E)	Austrian
prony	–	power	ML^2T^{-3}	1 prony = 98.0665 W (E)	1 prony = 10 kgfm.s^{-1} (E)	
proof (UK)	–	percentage of alcohol in wines and spirits	nil	1 UK proof = 0.5727% (v/v) ethanol 1 UK proof = 0.4950% (m/m) ethanol	British unit used in oenology. It serves to express the percentage of ethanol in wines and spirits. 1 UK proof = 0.5727% °GL	UK
proof (US)	–	percentage of alcohol in wines and spirits	nil	1 US proof = 0.5% (v/v) ethanol 1 US proof = 0.4249% (m/m) ethanol	American unit used in oenology. It serves to express the percentage of ethanol in wines and spirits. 1 US proof = 0.5% °GL	US
prout	–	energy, work, heat	ML^2T^{-2}	1 prout $= 2.9638 \times 10^{-14}$ J	Obsolete unit of energy employed in nuclear physics for measurements of binding energy of nuclides. One prout is equal to the binding energy of the deuteron. The unit is named after W. Prout (1786–1850). 1 prout ≈ 0.185 MeV	
puff	puff	electric capacitance	$M^{-1}L^{-2}T^4I^2$	1 puff $= 10^{-12}$ F (E)	Obsolete unit of electric capacitance employed in electronics. The name is derived from the pronunciation of the	

					symbol "pF" very quickly. 1 puff = 1 pF (E)	
pulgada (Spanish)	–	length, distance	L	1 pulgada (Spanish) $= 2.321958333 \times 10^{-2}$ m	Obsolete Spanish unit of length. 1 pulgada (Spanish) = 1/36 vara (E)	Spanish
puncheon (UK)	–	capacity, volume	L^3	1 puncheon (UK) $= 318.226440 \times 10^{-3}$ m^3	Obsolete British unit used for capacity measurements for all merchandise (solids, liquids, foodstuffs, etc). 1 puncheon (UK) = 70 gallons (UK) (E)	UK
punkt (German) [Prussian point]	–	length, distance	L	1 punkt (Prussian) $= 1.816302083 \times 10^{-4}$ m	Obsolete German unit of length. 1 punkt = 1/1728 fuss (E)	German
punkt(Austrian) [Austrian point]	–	length, distance	L	1 punkt (Austrian) $= 1.829166667 \times 10^{-4}$ m	Obsolete Austrian unit of length. 1 punkt (Austrian) = 1/1728 fuss (E)	Austrian
punto (Italian)	–	length, distance	L	1 punto (Italian) $= 3.567847222 \times 10^{-3}$ m	Obsolete Italian unit of length. 1 punto (Italian) = 1/144 piedi liprando (E)	Italian
punto (Spanish)	–	length, distance	L	1 punto (Spanish) $= 1.209353299 \times 10^{-4}$ m	Obsolete Spanish unit of length. 1 punto (Spanish) = 1/6912 vara (E)	Spanish
pyron	–	irradiance, radiant flux received, energy flux	MT^{-3}	1 pyron $= 6.975833333 \times 10^2$ $W.m^{-2}$	Obsolete unit of irradiance employed for pyrometry measurements. The name derived from Greek *pyros* = fire. 1 pyron = 1 langley (E) 1 pyron = 1 $cal_{15°}.cm^{-2}.min^{-1}$ (E)	
Q unit	–	energy, work, heat	ML^2T^{-2}	1 Q unit $= 1.055 \times 10^{21}$ J	Obsolete American unit of heat and energy employed in energy resource and reserve assessment. 1 Q unit = 10^{18} Btu (E) 1 Q unit = 10^3 quad (E)	US

Unit (synonym, acronym)	Symbol	Physical quantity	Dimension	Conversion factor (SI equivalent unit)	Notes, definitions, other conversion factors	System
qamha (Arabic)	–	surface, area	L^2	1 qamha (Arabic) = 61.458333 m^2	Obsolete Arabic unit of area used in ancient times. 1 qamha (Arabic) = 1/96 feddan (E)	Arabic
qasab (Arabic)	–	length, distance	L	1 qasab (Arabic) = 3.840 m	Obsolete Arabic unit of length used in ancient times. 1 qasab (Arabic) = 12 feet (Arabic) (E)	Arabic
qasab (Persian)	–	length, distance	L	1 qasab (Persian) = 3.840 m	Obsolete unit of length in the Assyrio-Chaldean-Persian system used in ancient times. 1 qasab = 12 zereths (E)	Persian
qirat (Arabic)	–	surface, area	L^2	1 qirat (Arabic) = 245.833333 m^2	Obsolete Arabic unit of area used in ancient times. 1 qirat (Arabic) = 1/24 feddan (E)	Arabic
quad [quadrillion (Btu)]	quad	energy, work, heat	ML^2T^{-2}	1 quad = $1.05505585262 \times 10^{18}$ J	Obsolete American unit of heat and energy employed in energy resource and reserve assessment. One quad is approximately equal to the energy realeased during the combustion of 180 million petroleum barrels. 1 quad = 10^{15} Btu (IT) (E)	US
quade (de Paris)	–	capacity, volume	L^3	1 quade (de Paris) = $1.862778000 \times 10^{-3}$ m^3	Obsolete French unit of capacity employed before the French Revolution. It served to express the capacity of liquids and grains. It varied according to the location and merchandise. 1 quade (de Paris) = 2 pintes (de Paris)	French

quadrans (Roman)	–	mass	M	1 quadrans (Roman) $= 81.75 \times 10^{-3}$ kg	Obsolete Roman unit of mass used in ancient times. 1 quadrans (Roman) = 3 unciae (Roman) (E)	Roman
quadrant	–	length, distance	L	1 quadrant $= 10^{7}$ m (E)	Obsolete unit of length which is equal to a quadrant of the Earth measured at the Equator. 1 quadrant = 10 000 km (E)	INT
quadrant	–	plane angle	α	1 quadrant = 1.570796327 rad (E)	1 quadrant $= \pi/2$ rad (E)	
quadratic coulomb quadratic metre per cubic joule	$C^4.m^4.J^{-3}$	2nd hyper-polarizability	$M^{-2}T^8I^4$	**SI derived unit** 1 $C^4.m^4.J^{-3} = 1\ kg^{-2}.s^8.A^4$ (E)		SI
quadratus pes (Roman)	–	surface, area	L^2	1 quadratus pes (Roman) $= 8.773444 \times 10^{-2}\ m^2$	Obsolete Roman unit of area employed in ancient times.	Roman
quadrillion (Btu) [quad]	quad	energy, work, heat	ML^2T^{-2}	1 quad $= 1.05505585262 \times 10^{18}$ J	Obsolete American unit of heat and energy employed in energy resource and reserve assessment. One quad is approximately equal to the energy realeased during the combustion of 180 million petroleum barrels. 1 quad $= 10^{15}$ Btu (IT) (E)	US
quanthar (Arabic)	–	mass	M	1 quanthar (Arabic) = 34 kg	Obsolete Arabic unit of mass used in ancient times. System of the Prophet. 1 quanthar (Arabic) = 100 rotl (E)	Arabic
quart (de Paris, dry)	–	capacity, volume	L^3	1 quart (de Paris, dry) $= 0.465695 \times 10^{-3}\ m^3$	Obsolete French unit of capacity employed before the French Revolution. It served to express the capacity of dry substances. 1 quart (de Paris) = 1/4 boisseau (de Paris) (E)	French

Unit (synonym, acronym)	Symbol	Physical quantity	Dimension	Conversion factor (SI equivalent unit)	Notes, definitions, other conversion factors	System
quart (Prussian)	–	capacity, volume	L^3	1 quart (Prussian) $= 1.145030 \times 10^{-3}$ m^3	Obsolete German unit of capacity used for liquid substances. 1 quart (Prussian) = 64 square zoll (E)	German
quart (Prussian, dry)	–	capacity, volume	L^3	1 quart (Prussian, dry) $= 1.145296667 \times 10^{-3}$ m^3	Obsolete German unit of capacity used for dry substances. 1 quart (Prussian, dry) = 1/3 metzel (E)	German
quart (UK)	qt (UK)	capacity, volume	L^3	1 quart (UK) $= 1.136523000 \times 10^{-3}$ m^3	Obsolete British unit used for capacity measurements for all merchandise (solids, liquids, foodstuffs, etc). 1 quart (UK) = 1/4 gallon (UK) (E) 1 quart (UK) = 2 pints (UK) (E) 1 quart (UK) = 8 gills (UK) (E) 1 quart (UK) = 40 fl oz (UK) (E) 1 quart (UK) = 320 fl dr (UK) (E) 1 quart (UK) = 1.032 060 quart (US)	UK
quart (US, dry)	qt (US, dry)	capacity, volume	L^3	1 quart (US, dry) $= 1.101220938 \times 10^{-3}$ m^3	Obsolete American unit of capacity used to measure the volume of powdered or divided solid materials (flour, sand, cement, ores, etc). 1 quart (US, dry) = 2 pints (US, dry) (E)	US
quart (US, liquid)	qt (US, liq.)	capacity, volume	L^3	1 quart (US, liq.) $= 0.9463529460 \times 10^{-3}$ m^3	American unit used for capacity measurements of liquids. 1 quart (US, liq.) = 1/4 gallon (US, liq.) (E) 1 quart (US, liq.) = 2 pints (US, liq.) (E) 1 quart (US, liq.) = 8 gills (US, liq.) (E) 1 quart (US, liq.)	US

					1 quart (US, liq.) = 256 fl dr (US, liq.) (E) 1 quart (US, liq.) = 0.968940 quart (UK)	
quarteau (de Paris)	–	capacity, volume	L^3	1 quarteau (de Paris) $= 67.060008 \times 10^{-3}$ m^3	Obsolete French unit of capacity employed before the French Revolution. It served to express the capacity of liquids and grains. It varies according to the location and merchandise. 1 quarteau (de Paris) = 72 pintes (de Paris) (E)	French
quarter (Swedish)	–	capacity, volume	L^3	1 quarter (Swedish) $= 3.2714525 \times 10^{-4}$ m^3	Obsolete Swedish unit of capacity for liquid substances. 1 quarter (Swedish) = 1/8 kanna (E)	Swedish
quarter (Swedish, dry)	–	capacity, volume	L^3	1 quarter (Swedish, dry) $= 3.271250 \times 10^{-4}$ m^3	Obsolete Swedish unit of capacity for dry substances. 1 quarter (Swedish, dry) = 1/8 kanna (E)	Swedish
quarter (UK capacity)	–	capacity, volume	L^3	1 quarter (UK cap) $= 290.949888 \times 10^{-3}$ m^3	British unit used for capacity measurements of liquids. 1 quarter (UK, capacity) = 64 gallons (UK)	UK
quarter (UK, mass)	–	mass	M	1 quarter (UK, mass) = 12.700586360 kg (E)	Obsolete British unit of mass. 1 quarter = 28 lb (av.) (E) 1 quarter = 2 stones (av.) (E)	UK
quarter (US, long)	–	mass	M	1 quarter (US, long) = 254.0117272 kg (E)	Obsolete American unit of mass. 1 quarter (US, long) = 560 lb (av.) (E) 1 quarter (US, long) = 5 lg cwt (av.) (E)	US

Unit (synonym, acronym)	Symbol	Physical quantity	Dimension	Conversion factor (SI equivalent unit)	Notes, definitions, other conversion factors	System
quarter (US, short)	–	mass	M	1 quarter (US, short) = 226.796185 kg (E)	Obsolete American unit of mass. 1 quarter (US, short) = 500 lb (av.) 1 quarter (US, short) = 5 sh. cwt (US) (E)	US
quarter section	–	surface, area	L^2	1 quarter section = 6.474970275×10^5 m^2	Obsolete British and American unit of area. 1 quarter section = 160 acres (E) 1 quarter section = 77 440 square yards (E)	UK, US
quarteron (de Paris)	–	mass	M	1 quarteron (de Paris) = $122.3764625 \times 10^{-3}$ kg	Obsolete French unit of mass employed before the French Revolution (1789). 1 quarteron (de Paris) = 1/4 livre (de Paris) (E)	French
quartillo (Portuguese)	–	capacity, volume	L^3	1 quartillo (Portuguese) = 0.34375×10^{-3} m^3	Obsolete Portuguese unit of capacity used for liquid substances. 1 quartillo (Portuguese) = 1/48 almude (E)	Portuguese
quarto (Portuguese, dry)	–	capacity, volume	L^3	1 quarto (Portuguese, dry) = 3.375×10^{-3} m^3	Obsolete Portuguese unit of capacity used for dry substances. 1 quarto (Portuguese, dry) = 1/16 fanga (E)	Portuguese
quei (Chinese)	–	capacity, volume	L^3	1 quei (Chinese) = 1.03544×10^{-7} m^3	Obsolete Chinese unit of capacity used in ancient times. 1 quei (Chinese) = 1/10 000 tcheng (E)	Chinese
quentchen (Austrian)	–	mass	M	1 quentchen (Austrian) = $4.375078125 \times 10^{-3}$ kg	Obsolete Austrian unit of mass for general uses. 1 quentchen (Austrian) = 1/128 pfund (E)	Austrian

quentchen (Prussian)	–	mass	M	1 quentchen (Prussian) $= 4.871989583 \times 10^{-3}$ kg	Obsolete German unit of mass. 1 quentchen (Prussian) = 1/96 pfund (E)	German
Quevenne degree	°Q	specific gravity index on lactometric scale	nil	$°Q = 1000 \times (1 - d)$	Hydrometer scale where the 15°Q and 40°Q correspond to solutions having the specific gravities 1.015 and 1.040 respectively.	
quicunx (Roman)	–	mass	M	1 quicunx (Roman) $= 136.25 \times 10^{-3}$ kg	Obsolete Roman unit of mass used in ancient times. 1 quicunx (Roman) = 5 unciae (Roman) (E)	Roman
quintal (Ancien Régime)	q	mass	M	1 quintal (Ancien Régime) = 489.505850 kg	Obsolete French unit of mass employed before the French Revolution (1789). 1 quintal (Ancien Régime) = 1000 livres (de Paris) (E)	French
quintal (metric)	q	mass	M	1 q = 100 kg (E)	Obsolete French metric unit of mass still used in a few countries.	French
quintal (Portuguese)	–	mass	M	1 quintal (Portuguese) = 58.752 kg	Obsolete Portuguese unit of mass. 1 quintal (Portuguese) = 128 libra (E)	Portuguese
quintal (Spanish)	–	mass	M	1 quintal (Spanish) = 46.0093 kg	Obsolete Spanish unit of mass. 1 quintal (Spanish) = 100 libra (E)	Spanish
quintal (US)	quint. (US)	mass	M	1 quintal (US) = 45.359237 kg (E)	Obsolete British and American unit of mass. 1 quintal (US) = 100 lb (av.) (E) 1 quintal (US) = 1 quintal UK (E)	UK, US
quintalmacho (Spanish)	–	mass	M	1 quintalmacho (Spanish) = 150.0460093 kg	Obsolete Spanish unit of mass. 1 quintalmacho (Spanish) = 150 libra (E)	Spanish

Unit (synonym, acronym)	Symbol	Physical quantity	Dimension	Conversion factor (SI equivalent unit)	Notes, definitions, other conversion factors	System
quinze-seize (de Paris)	15/16	length, distance	L	1 quinze-seize (de Paris) = 1.76301362 m	Obsolete French unit of length employed in textiles and clothing measurements. It was used in France during the Ancien Régime before the French Revolution (1789).	French
R-value	R-value	heat insulation coefficient (thermal resistance multiplied by area)	$M^{-1}T^3\Theta$	1 R-value = 0.1761101838 $W^{-1}.m^2.K$	Obsolete unit of heat-insulation coefficient employed in the British and American building engineering. One R-value corresponds to the heat insulation coefficient of a wall which conserves a temperature difference of 1°F between its faces when the heat flux is equal to one $Btu_{IT}.h^{-1}.ft^{-2}$. 1 R-value = 1 $(Btu_{IT}/h)^{-1}.ft^2.°F$ (E) 1 R-value = 0.1761101838 RSI 1 R-value = 1 $(U\text{-factor})^{-1}$ (E)	UK, US
rabah (Hebrew) [Sacred system]	–	mass	M	1 rabah (Hebrew) = 3.541667×10^{-3} kg	Obsolete Hebrew unit of mass used in ancient times. Sacred system. 1 rabah (Hebrew) = 1/240 mina (E)	Hebrew
racion (Spanish, dry)	–	capacity, volume	L^3	1 racion (Spanish, dry) = $2.89067708 \times 10^{-4}$ m^3	Obsolete Spanish unit of capacity used for dry substances. 1 racion (Spanish, dry) = 1/192 fanega (E)	Spanish
rad (radiation absorbed dose)	rd, rad	radiation absorbed dose, specific energy, kerma, index of absorbed dose	L^2T^{-2}	1 rad = 10^{-2} Gy (E)	Obsolete absorbed dose unit employed in ionizing radiation dosimetry. It was adopted in 1953 by the ICRU. The rad is the absorbed dose when the energy per unit mass imparted to matter by ionizing	

					1 rad = 10 J.kg (E) 1 rad = 10^{2} erg.g^{-1} (E) 1 rad = 2.388 × 10^{-6} cal.g^{-1} 1 rad = 6.242 × 10^{13} eV.g^{-1}	
radian	rad	plane angle	α	**SI Supplementary Unit**	The radian is an SI supplementary unit. One radian is a plane angle enclosed between two radii which intercept, on a circle, an arc of length equal to the radius.	SI
radian per minute	rad.min^{-1}	angular velocity, angular frequency, circular frequency	αT^{-1}	1 rad.min^{-1} = 0.0166667 rad.s^{-1}		
radian per second	rad.s^{-1}	angular velocity, angular frequency, circular frequency	αT^{-1}	**SI derived unit**	The radian per second is the unit of angular velocity of an object which is in uniform revolution around an axis and which is defined as a plane angle of one radian in one second.	SI
radian per square second	rad.s^{-2}	angular acceleration	αT^{-2}	**SI derived unit**	The radian per square second is the SI derived unit of angular acceleration. It is the acceleration of an object which is in revolution around an axis and which undergoes a variation of angular velocity of one radian per second in one second.	SI
radian square metre per kilogram	rad.m^{2}.kg^{-1}	specific optical rotatory power	$M^{-1}L^{2}\alpha$	**SI derived unit**		SI
radian square metre per mole	rad.m^{2}.mol^{-1}	molar optical rotatory power	$N^{-1}L^{2}\alpha$	**SI derived unit**		SI

Unit (synonym, acronym)	Symbol	Physical quantity	Dimension	Conversion factor (SI equivalent unit)	Notes, definitions, other conversion factors	System
Rankine degree	°R	temperature	Θ	T(K) = 5/9T(°R)	Usual absolute temperature scale in the English-speaking countries. T(°R) = T(°F) + 459.67 1 °F = 1°R (E)	UK, US
rayl (cgs)	rayl	specific acoustic impedance	$ML^{-2}T^{-1}$	1 rayl = 10 $kg.m^{-2}.s^{-1}$ (E)	Obsolete cgs unit of specific acoustic impedance. A specific acoustic impedance has a magnitude of 1 rayl when a sound pressure of 1 barye produces a linear velocity of 1 $cm.s^{-1}$. The unit is named after Lord Rayleigh (1842–1919).	cgs
rayl (MKSA)	rayl	specific acoustic impedance	$ML^{-2}T^{-1}$	1 rayl = 1 $kg.m^{-2}.s^{-1}$ (E)	Obsolete MKSA unit of specific acoustic impedance. A specific acoustic impedance has a magnitude of 1 rayl when a sound pressure of 1 Pa produces a linear velocity of 1 $m.s^{-1}$. The unit is named after Lord Rayleigh (1842–1919).	MKSA
rayleigh	R	photon fluence rate	$L^{-2}T^{-1}\Omega^{-1}$	1 rayleigh = $(1/4\pi) \times 10^{10}$ $m^{-2}.s^{-1}.sr^{-1}$	Obsolete unit used in photometry.	
Réaumur degree	°Ré	temperature	Θ	T (°Ré) = 0.8 T(°C)	Obsolete French temperature scale. 0°Ré = 0°C (melting ice) 80°Ré = 100°C (boilling water)	French
reciprocal joule per cubic metre	$J^{-1}.m^{-3}$	density of state	$M^{-1}L^{-5}T^2$	**SI derived unit**		SI
reciprocal metre	m^{-1}	wavenumber	L^{-1}	**SI derived unit**		SI

reciprocal ohm per centimetre	roc	electrical conductivity	$M^{-1}L^{-3}T^{3}I^{2}$	1 roc = 100 S.m^{-1} (E)	Name suggested in 1964 for the practical cgs electrical conductivity unit. 1 roc = 1 Ω^{-1}.cm^{-1} (E)	
reciprocal ohm per metre	rom	electrical conductivity	$M^{-1}L^{-3}T^{3}I^{2}$	1 rom = 1 S.m^{-1} (E)	Name suggested in 1964 for the practical MKS electrical conductivity unit. 1 rom = 1 Ω^{-1}.m^{-1} (E)	
reciprocal second	s^{-1}	frequency	T^{-1}	**SI derived unit** 1 s^{-1} = 1 Hz (E)		SI
reciprocal volt square metre second	$m^{2}.V^{-1}.s^{-1}$	electric mobility of ions and charged particles	$M^{-1}T^{2}I$	**SI derived unit**		SI
Redwood (second)	–	index of kinematic viscosity	nil	(see note)	Obsolete British empirical index of kinematic viscosity which corresponds to the flowing time expressed in seconds, of 50 cm^3 of liquid in a Redwood viscosimeter.	UK
ref (Swedish)	–	length, distance	L	1 ref (Swedish) = 296.90 m	Obsolete Swedish unit of length.	Swedish
réhoboam (rehoboam)	–	capacity, volume	L^{3}	1 réhoboam = 4.546092 × 10^{-3} m^3	Obsolete British unit which expressed the capacity of a wine container. Still employed in oenology, especially in France. 1 réhoboam = 1 gallon (UK) (E) 1 réhoboam = 6 bouteilles (E)	UK, French
rem (rad equivalent mammals or man)	rem	dose equivalent, index of dose equivalent	$L^{2}T^{-2}$	1 rem = 10^{-2} Sv (E)	Obsolete unit of dose equivalent employed in radioprotection. One rem was equal to the quantity of ionizing radiation which produces a specific absorbed energy on living substance which produces the same biological effect as one rad of X-rays generated by an electric voltage of 200–250 kilovolts.	

Unit (synonym, acronym)	Symbol	Physical quantity	Dimension	Conversion factor (SI equivalent unit)	Notes, definitions, other conversion factors	System
rema (Egyptian)	–	surface, area	L^2	1 rema (Egyptian) = 1370.25 m^2	Obsolete Egyptian unit of area used in ancient times. 1 rema (Egyptian) = 50 pekeis (E)	Egyptian
rep (rad equivalent physical)	rep	dose equivalent, index of dose equivalent	L^2T^{-2}	1 rep = 10^{-2} sievert (E)	Obsolete unit of dose equivalent employed in radioprotection. The rep is equal to the quantity of radiation which releases, in one gram of body tissue, the same energy as one röntgen of X or gamma rays in the same air mass.	
reputed quart (bouteille champenoise)	–	capacity, volume	L^3	1 reputed quart = $0.757682000 \times 10^{-3}$ m^3	Obsolete British unit which expressed the capacity of a wine container. Still employed in oenology, especially in France. 1 reputed quart = 1/6 gallon (UK) (E)	UK, French
retti (Indian)	–	mass	M	1 retti (Indian) = 1.468750×10^{-4} kg	Obsolete Indian unit of mass used in ancient times. 1 retti (Indian) = 320 pala (E)	Indian
revolutions per minute	rpm, rev. min^{-1}	angular velocity, angular frequency, circular frequency	αT^{-1}	1 rpm = 0.1047197551 $rad.s^{-1}$	Anglo-saxon unit of angular velocity. 1 rpm = $\pi/30$ $rad.s^{-1}$ (E)	UK, US
revolutions per second	rps, $rev.s^{-1}$	angular velocity, angular frequency, circular frequency	αT^{-1}	1 $rev.s^{-1}$ = 6.28319 $rad.s^{-1}$		
reynolds (reyns)	reyns (reyn)	dynamic viscosity, absolute viscosity	$ML^{-1}T^{-1}$	1 reyns = 1.488163944 Pa.s	Unit of dynamic viscosity or absolute viscosity in the FPS system. The unit is	FPS

					named after O. Reynolds (1842–1912). 1 reyn = $lb.ft^{-1}.s^{-1}$	
rhe	–	reciprocal of dynamic viscosity fluidity	$M^{-1}LT$	1 rhe = 10 $(Pa.s)^{-1}$ (E)	Obsolete cgs unit of fluidity. 1 rhe = 1 Po^{-1} (E)	cgs
ri (Japanese)	–	length, distance	L	1 ri (Japanese) = 3926.88 m	Obsolete Japanese unit of length. 1 ri = 12 960 shaku (E)	Japanese
right angle	⊥	plane angle	α	1 right angle = 1.570796327 rad (E)	1 right angle = $\pi/2$ rad (E) 1 plane angle = 90 degrees (E) 1 plane angle = 100 gon (E)	
rin (Japanese)	–	length, distance	L	1 rin (Japanese) = $3.030303030 \times 10^{-4}$ m	Obsolete Japanese unit of length. 1 rin = 1/1000 shaku (E)	Japanese
ringing equivalent number	REN	electric resistance	$ML^2T^{-3}I^{-2}$	1 REN = 1/4000 Ω (E)	Obsolete unit of resistance	UK, US
rod (perch, pole)	rd	length, distance	L	1 rod = 5.0292 m (E)	Obsolete American and British unit of length used in surveyor's measurements. 1 rod = 1 perch (E) 1 rod = 1 pole (E) 1 rod = 5.5 yards (E) 1 rod = 16.5 feet (E)	UK, US
roeden (Dutch)	–	length, distance	L	1 roeden (Dutch) = 3.679770 m	Obsolete unit of length used in Amsterdam (Netherlands)	Dutch
roentgen (röntgen)	R	exposure	ITM^{-1}	1 R = 2.58×10^{-4} $C.kg^{-1}$	Obsolete unit of exposure employed in ionizing radiation dosimetry. It was equal *(continued overleaf)*	

Unit (synonym, acronym)	Symbol	Physical quantity	Dimension	Conversion factor (SI equivalent unit)	Notes, definitions, other conversion factors	System
roentgen (röntgen) *(continued)*					to the quantity of gamma or X-rays which produce an electric charge of one franklin of two signs per ionization in one cubic centimetre of dry air ($\rho = 1.293$ g.cm^{-3}) with standard T and P (0°C, 760 torr). The unit is named after the German Physicist W.C. Roentgen (1845–1923) who discovered the X-ray in 1895.	
rood (UK)	–	surface, area	L^2	1 rood (UK) $= 1.011714106 \times 10^3$ m^2	Obsolete British unit of area used in surveyor's measurements. 1 rood (UK) = 1/4 acre (E) 1 rood (UK) = 40 square rods (UK) (E)	UK
rope (UK)	–	length, distance	L	1 rope (UK) = 6.096 m (E)	Obsolete British unit of length. 1 rope (UK) = 20 feet (E)	UK
roquille (de Paris)	–	capacity, volume	L^3	1 roquille (de Paris) $= 2.90625 \times 10\text{-}5$ m^3	Obsolete French unit of capacity employed before the French Revolution. It served to express the capacity of liquids and grains. It varied according to location and merchandise.	French
roquille (US)	–	capacity, volume	L^3	1 roquille (US) $= 1.182941183 \times 10^{-4}$ m^3	Obsolete American unit of capacity. 1 roquille (US) = 1/32 gallon (US) (E) 1 roquille (US) = 1 gill (US) (E) 1 roquille (US) = 1 noggin (US)	US
roquille (UK)	–	capacity, volume	L^3	1 roquille (UK) $= 1.420653750 \times 10^{-4}$ m^3	Obsolete British unit of capacity. 1 roquille (UK) = 1/32 gallon (UK) (E) 1 roquille (UK) = 1 gill (UK) (E) 1 roquille (UK) = 1 noggin (UK) (E)	UK

rotl (Arabic)	–	mass	M	1 rotl (Arabic) = 0.340 kg	Obsolete Arabic unit of mass used in ancient times. System of the Prophet.	Arabic
round	–	plane angle	α	1 round = 6.283185307 rad (E)	1 round = 2π rad (E) 1 round = 360° (E)	
rowland	–	length, distance	L	1 rowland $\approx 10^{-10}$ m	Obsolete unit of wavelength used in spectroscopy. The unit is named after H.A. Rowland (1848–1901). 1 rowland $\approx$ 0.1 nm	
RSI (metric R-value)	RSI	heat insulation coefficient (thermal resistance multiplied by area)	$M^{-1}T^3\Theta$	1 RSI = 1 $W^{-1}.m^2.K$ (E)	Unit of heat insulation coefficient employed in the British and American building engineering. One RSI corresponds to the heat insulation coefficient of a wall which conserves a temperature difference of $1^\circ C$ between its faces when the heat flux is equal to one $W.m^{-2}$. 1 RSI = 1 $W^{-1}.m^2.{}^\circ C$ (E) 1 RSI = 5.678285660 R-value 1 RSI = 1 $(USI)^{-1}$ (E)	UK, US
rubbo (Italian)	–	mass	M	1 rubbo (Italian) = 7.675 kg	Obsolete Italian unit of mass. 1 rubbo (Italian) = 25 libbra (E)	Italian
ruthe (Prussian)	–	length, distance	L	1 ruthe (Prussian) = 3.766284000 m	Obsolete German unit of length. 1 ruthe = 12 fuss (E)	German
rutherford	–	radioactivity	T^{-1}	1 rutherford = 10^6 Bq	Obsolete unit of radioactivity used in nuclear physics. One rutherford is defined as the quantity of radioactive material which undergoes one million spontaneous nuclear transitions per second. The unit is named after Lord Rutherford (1871–1937). 37 rutherfords = 1 mCi (E).	

Unit (synonym, acronym)	Symbol	Physical quantity	Dimension	Conversion factor (SI equivalent unit)	Notes, definitions, other conversion factors	System
rydberg	Ry	energy, work, heat	ML^2T^{-2}	1 Ry = $2.179874113 \times 10^{-18}$ J	Obsolete unit of energy used in atomic and quantum spectroscopy. 1 Ry = 13.605698150 eV	
sâa (Arabic)	–	capacity, volume	L^3	1 sâa (Arabic) = 2.72×10^{-3} m^3 (of water)	Obsolete Arabic unit of capacity used in ancient times. Measured by weight. 1 sâa (Arabic) = 1/12 cafiz (E)	Arabic
saashen (Russian)	–	length, distance	L	1 saashen (Russian) = 2.1336 m	Obsolete Russian unit of length used before 1917. 1 saashen (Russian) = 7 foute (E)	Russian
sabine (metric)	–	acoustical absorption area	L^2	1 sabine (metric) = 1 m^2 (E)	Obsolete metric unit of absorption of surface covering in room acoustics equal to that of 1 square metre of perfectly absorbing material. The unit is named after W.C. Sabine (1868–1919). 1 sabine (metric) = 1 m^2 (E)	MKSA
sabine (UK)	–	acoustical absorption area	L^2	1 sabine (UK) = 9.290304×10^{-3} m^2 (E)	Obsolete British and American unit of absorption of surface covering in room acoustics equal to that of 1 square foot of perfectly absorbing material. The unit is named after W.C. Sabine (1868–1919). 1 sabine (UK) = 1 ft^2 (E)	UK, US
sack (UK)	sk (UK)	capacity, volume	L^3	1 sack (UK) = $109.106208 \times 10^{-3}$ m^3	Obsolete British unit used for capacity measurements for all merchandise (solids, liquids, foodstuffs, etc). 1 sack (UK) = 3 bushels (UK) (E) 1 sack (UK) = 12 pecks (UK) (E) 1 sack (UK) = 24 gallons (UK) (E)	UK

					1 sack (UK) = 96 quarts (UK) (E) 1 sack (UK) = 1 bag (UK) (E)	
sack (UK, weight)	–	mass	M	1 sack (UK, weight) = 158.7573295 kg (E)	Obsolete British unit of mass used in the weighing of wool. 1 sack (UK, weight) = 350 lb (av.) (E) 1 sack (UK, weight) = 2 weys (UK, weight) (E) 1 sack (UK, weight) = 28 stones (UK, wool) (E)	UK
salmanazar (salmarazd)	–	capacity, volume	L^3	1 salmanazar $= 9.092184 \times 10^{-3}$ m^3 (E)	Obsolete British unit used to express the capacity of a wine container. Still employed in oenology, especially in France. 1 salmanazar = 2 gallons (UK) (E) 1 salmanazar = 12 bouteilles (E)	UK, French
saltus (Roman)	–	surface, area	L^2	1 saltus (Roman) $= 2.021401498 \times 10^6$ m^2	Obsolete Roman unit of area employed in ancient times. 1 saltus (Roman) = 23 040 000 quadratus pes (E)	Roman
sath (Hebrew)	–	capacity, volume	L^3	1 sath (Hebrew) $= 6.426 \times 10^{-3}$ m^3 (of water)	Obsolete Hebrew unit of capacity used in ancient times. Measured by weight. 1 sath (Hebrew) = 3/10 ephah (new)	Hebrew
saum (Austrian)	–	mass	M	1 saum (Austrian) = 154.002750 kg	Obsolete Austrian unit of mass for general use. 1 saum (Austrian) = 275 pfund (E)	Austrian
savart	–	logarithmic musical interval	nil	$I = 1000 \times \log_{10}(f_1/f_2)$ (see note)	The savart is the interval between two musical sounds having as a basic frequency ratio the 1000th root of two. *(continued overleaf)*	

Unit (synonym, acronym)	Symbol	Physical quantity	Dimension	Conversion factor (SI equivalent unit)	Notes, definitions, other conversion factors	System
savart *(continued)*					The number of cents between frequencies f_1 and f_2 is: $I = 1000 \times \log_{10}(f_1/f_2)$. The unit is named after the French physicist F. Savart (1791–1841). 1 savart = 1/301 octave (E) 1 savart = 1200/301 cent (E)	
Saybolt (universal second)	Saybolt, SUS	index of kinematic viscosity	nil	(see note)	Obsolete American empirical index of kinematic viscosity. The Saybolt universal second (SUS) corresponds to the flowing time expressed in seconds of 60 cm^3 of liquid in a Saybolt viscosimeter.	US
scheffel (Prussian, dry)	–	capacity, volume	L^3	1 scheffel (Prussian, dry) $= 54.974240 \times 10^{-3}\ m^3$	Obsolete German unit of capacity used for dry substances. 1 scheffel (Prussian, dry) = 16 metzel (E)	German
schepel (Dutch, dry)	–	capacity, volume	L^3	1 schepel (Dutch, dry) $= 27.26 \times 10^{-3}\ m^3$	Obsolete Dutch unit of capacity used for dry substances.	Dutch
schiffspfund (Prussian)	–	mass	M	1 schiffspfund (Prussian) = 154.34430 kg	Obsolete German unit of mass. 1 schiffspfund (Prussian) = 330 pfund (E)	German
schoeme (Persian)	–	length, distance	L	1 schoeme (Persian) = 6 912 m	Obsolete unit of length in the Assyrio-Chaldean-Persian system used in ancient times. 1 schoeme = 21 600 zereths (E)	Persian
scripulum (Roman) [Roman scruple]	–	mass	M	1 scripulum (Roman) $= 1.135416667 \times 10^{-3}$ kg	Obsolete Roman unit of mass used in ancient times. 1 scripulum (Roman) = 1/24 uncia (Roman) (E)	Roman

scruple (Austrian, apothecary)	–	mass	M	1 scruple (Austrian, apothecary) = $1.458368055 \times 10^{-3}$ kg	Obsolete Austrian unit of mass used in pharmacy. 1 scruple (Austrian, apothecary) = 1/288 pfund (Austrian, apothecary) (E)	Austrian
scruple (UK fluid)	–	capacity, volume	L^3	1 scruple (UK fl.) = $1.183878125 \times 10^{-6}$ m^3	Obsolete British unit of capacity. 1 scruple (UK fl.) = 20 minims (UK) (E)	UK
scruple (UK, apoth.)	s, scr (ap.)	mass	M	1 scruple (UK) = $1.295978200 \times 10^{-3}$ kg	Obsolete British unit of mass. Sometimes written scrupule. The unit is derived from the Latin *scripulum*. 1 scruple (UK) = 1/288 lb (apoth.) (E) 1 scruple (UK) = 20 grains (apoth.) (E)	UK
scruple (US, apoth.)	s, scr (ap.)	mass	M	1 scruple (US) = $1.295978200 \times 10^{-3}$ kg	Obsolete American unit of mass. Sometimes written scrupule. The unit is derived from the Latin *scripulum*. 1 scruple (US) = 1/288 lb (apoth.) (E) 1 scruple (US) = 20 grains (apoth.) (E)	US
scrupule (Russian)	–	mass	M	1 scrupule (Russian) = $1.244178328 \times 10^{-3}$ kg	Obsolete Russian unit of mass used before 1917 in pharmacy. 1 scrupule (Russian) = 28 doli (E)	Russian
se (Japanese)	–	surface, area	L^2	1 se = 99.17355372 m^2	Obsolete Japanese unit of area. 1 se = 30 bu (E)	Japanese
seam (UK)	–	capacity, volume	L^3	1 seam (UK) = $290.949888 \times 10^{-3}$ m^3	Obsolete British unit of capacity. 1 seam (UK) = 64 gallons (UK) (E)	UK
second	s	time, period, duration	T	**SI base unit**	The second is the duration of 9 192 631 770 periods of the radiation corresponding to the transition between the two hyperfine levels ($F=4$, $m_F=0$ to $F=3$, $m_F=0$) of the ground state of the caesium 133 atom [13th CGPM (1967)].	SI, MTS, MKSA, cgs, MKpS, FPS

Unit (synonym, acronym)	Symbol	Physical quantity	Dimension	Conversion factor (SI equivalent unit)	Notes, definitions, other conversion factors	System
second (sidereal)	–	time, period, duration	T	1 second (sidereal) = 0.9972696 s		@
second of angle	″	plane angle	α	1″ = $4.848136811 \times 10^{-4}$ rad	1″ = $\pi/(3600 \times 180)$ rad (E)	INT
second of angle (new)	c	plane angle	α	1 c = $1.570796327 \times 10^{-6}$ rad	1 c = 1/10 000 gon (grade) (E)	French
section (square statute mile)	sq.mi, mi^2 (stat.)	surface, area	L^2	1 section = 2.589988110×10^6 m^2	Obsolete British and American unit of area. 1 section = 1 square mile (stat.) (E) 1 section = 640 acres (E) 1 section = 2560 roods (UK) (E)	UK, US
sei (Chinese)	–	capacity, volume	L^3	1 sei (Chinese) = 103.544×10^{-3} m^3	Obsolete Chinese unit of capacity used in ancient times. 1 sei (Chinese) = 100 tcheng (E)	Chinese
seidel (Austrian)	–	capacity, volume	L^3	1 seidel (Austrian) = 3.537750×10^{-4} m^3	Obsolete Austrian unit of capacity used for liquid substances. 1 seidel (Austrian) = 1/4 mass (E)	Austrian
seir (Arabic) [Arabic stadion]	–	length, distance	L	1 seir (Arabic) = 192 m	Obsolete Arabic unit of length used in ancient times. 1 seir (Arabic) = 600 feet (Arabic) (E)	Arabic
semis (Roman)	–	mass	M	1 semis (Roman) = 163.50×10^{-3} kg	Obsolete Roman unit of mass used in ancient times. 1 semis (Roman) = 6 uncia (Roman) (E)	Roman
semiuncia (Roman)	–	mass	M	1 semiuncia (Roman) = 13.625×10^{-3} kg	Obsolete Roman unit of mass used in ancient times. 1 semiuncia (Roman) = 1/2 uncia (Roman) (E)	Roman

semodius (Roman)	–	capacity, volume	L	1 semodius (Roman) $= 4.394 \times 10^{-3}\ m^3$	Obsolete Roman unit of volume employed in ancient times. It was used for capacity measurements of dry substances. 1 semodius (Roman) = 8 sextarius (E)	Roman
sennus (Egyptian)	–	length, distance	L	1 sennus (Egyptian) = 52.35 m	Obsolete Egyptian unit of length used in ancient times. 1 sennus = 100 Royal cubit (E)	Egyptian
sep	–	mass	M	1 sep $= 1.365 \times 10^{-3}$ kg	Obsolete Egyptian unit of mass used in ancient times. 1 sep = 1/10 deben (E)	Egyptian
septunx (Roman)	–	mass	M	1 septunx (Roman) $= 190.75 \times 10^{-3}$ kg	Obsolete Roman unit of mass used in ancient times. 1 septunx (Roman) = 7 unciae (Roman) (E)	Roman
sesma (Spanish)	–	length, distance	L	1 sesma (Spanish) $= 1.393175000 \times 10^{-1}$ m	Obsolete Spanish unit of length. 1 sesma (Spanish) = 1/6 vara (E)	Spanish
sétier (de Paris, dry)	–	capacity, volume	L^3	1 sétier (de Paris, dry) $= 22.353360 \times 10^{-3}\ m^3$	Obsolete French unit of capacity employed before the French Revolution. It served to express the capacity of dry substances. It varied according to location and merchandise. The unit is named after Latin *sextarius*: sixth. 1 sétier (de Paris) = 12 boisseau (de Paris) (E)	French
setta (Egyptian)	–	surface, area	L^2	1 setta (Egyptian) = 27 405 m^2	Obsolete Egyptian unit of area used in ancient times. 1 setta (Egyptian) = 1000 pekeis (E)	Egyptian
sextans (Roman)	–	mass	M	1 sextans (Roman) $= 54.50 \times 10^{-3}$ kg	Obsolete Roman unit of mass used in ancient times. 1 sextans (Roman) = 2 unciae (Roman) (E)	Roman

Unit (synonym, acronym)	Symbol	Physical quantity	Dimension	Conversion factor (SI equivalent unit)	Notes, definitions, other conversion factors	System
sextarius (Roman)	–	capacity, volume	L^3	1 sextarius (Roman) $= 0.547 \times 10^{-3}$ m^3	Obsolete Roman unit of volume employed in ancient times. It was used for capacity measurements of liquids.	Roman
sextarius (Roman)	–	capacity, volume	L^3	1 sextarius (Roman) $= 0.547 \times 10^{-3}$ m^3	Obsolete Roman unit of volume employed in ancient times. It was used for capacity measurements of dry substances.	Roman
shacles	–	length, distance	L	1 shacles (UK) = 27.432 m (E)	Obsolete British unit of length employed in navigation. 1 shacles (UK) = 15 fathoms (E) 1 shacles (UK) = 30 yards (E) 1 shacles (UK) = 90 feet (E)	UK
shake	–	time, period, duration	T	1 shake $= 10^{-8}$ s (E)		US
shaku	–	capacity, volume	L^3	1 shaku $= 1.809068370 \times 10^{-5}$ m^3	Obsolete Japanese unit of capacity. 1 shaku = 1/100 shô (E)	Japanese
shaku (Japanese)	–	length, distance	L	1 shaku (Japanese) = 0.3030303030 m	Obsolete Japanese unit of length. 1 shaku = 10/33 m (E)	Japanese
shari (Indian)	–	capacity, volume	L^3	1 shari (Indian) $= 211.2 \times 10^{-3}$ m^3 (of water)	Obsolete Indian unit of capacity used in ancient times. Measured by weight. 1 shari (Indian) = 16 drona (E)	Indian
shed	–	surface, area	L^2	1 shed $= 10^{-52}$ m^2 (E)	Obsolete British unit of area used in nuclear physics to express the cross section of a nuclide. 1 shed $= 10^{-24}$ barn (E)	UK
shekel (Hebrew) [Sacred system]	–	mass	M	1 shekel (Hebrew) $= 14.166667 \times 10^{-3}$ kg	Obsolete Hebrew unit of mass used in ancient times. Sacred system.	Hebrew

shekel (Hebrew) [Talmudic system]	–	mass	M	1 shekel (Hebrew) $= 1.416800 \times 10^{-2}$ kg	Obsolete Hebrew unit of mass used in ancient times. Rabbinical or Talmudic system. 1 shekel (Hebrew) = 1/25 mina (E)	Hebrew
shi (Japanese)	–	length, distance	L	1 shi (Japanese) $= 3.030303030 \times 10^{-6}$ m	Obsolete Japanese unit of length. 1 shi = 1/100 000 shaku (E)	Japanese
shih (dan)	–	mass	M	1 shih = 30 kg	Obsolete Chinese unit of mass. 1 shih = 120 jin (E)	Chinese
shô	–	capacity, volume	L^3	1 shô $= 1.809068370 \times 10^{-3}$ m^3	Obsolete Japanese unit of capacity. 1 shô = 2401/1331 dm^3	Japanese
shoeme (Egyptian)	–	length, distance	L	1 shoeme (Egyptian) $= 6.282 \times 10^3$ m	Obsolete Egyptian unit of length used in ancient times. 1 shoeme = 12 000 Royal cubit (E)	Egyptian
short actus (Roman)	–	surface, area	L^2	1 short actus (Roman) = 35.093776 m^2	Obsolete Roman unit of area employed in ancient times. 1 short actus (Roman) = 400 quadratus pes (E)	Roman
short cumbha (Indian)	–	capacity, volume	L^3	1 short cumbha (Indian) $= 26.4 \times 10^{-3}$ m^3(of water)	Obsolete Indian unit of capacity used in the Ancien Times. Measured by weight. 1 short cumbha (Indian) = 2 drona (E)	Indian
short ton (US, ton)	sh. ton	mass	M	1 short ton (US) = 907.184740 kg (E)	American unit of mass. 1 short ton = 2000 lb (US) (E)	US
shower unit	s	length, distance	L	1 shower unit $= \ln 2/\mu$ (m^{-1})	Obsolete unit of length employed in astrophysics for cosmic-ray measurements. It was equal to the thickness of a medium which halves the *(continued overleaf)*	@

Unit (synonym, acronym)	Symbol	Physical quantity	Dimension	Conversion factor (SI equivalent unit)	Notes, definitions, other conversion factors	System
shower unit (continued)					intensity of an incident beam of charged particles. It corresponds to the half-thickness $X_{1/2}$ of the considered medium for a given charged particle of energy E.	
shtoff (Russian)	–	capacity, volume	L^3	1 shtoff (Russian) $= 1.537426250 \times 10^{-3}$ m^3	Obsolete Russian unit of capacity for liquid substances used before 1917. 1 shtoff (Russian) = 25/2 tcharka (E)	Russian
shu	–	mass	M	1 shu $= 104.166667 \times 10^{-6}$ kg	Obsolete Chinese unit of mass. 1 shu = 1/2400 jin (E)	Chinese
Siegbahn unit (CuK_{α_1}) (X unit)	UX, X	length, distance	L	1 UX (CuK_{α_1}) $= 1.00020778970 \times 10^{-13}$ m	Obsolete unit of length used in atomic spectroscopy and X-ray diffractometry measurements. It was introduced by K.N.G. Siegbahn (1886–1977) in 1925. It was equal to the distance between lattice planes with the Miller index (200) of a calcite crystal (Island Spath) measured by X-ray diffraction using a CuK_{α_1} spectral line. This interlattice distance is equal to 3029.45 UX.	
siemens (resistance)	–	electric resistance	$ML^2T^{-3}I^{-2}$	1 siemens = 0.96 Ω (E)	Obsolete German unit of electric resistance. It was defined in 1860. It was the resistance measured at 0°C of a mercury column of 100 cm length with a cross section of one square millimetre	German
siemens	S	electrical conductance	$M^{-1}L^{-2}T^3I^2$	**SI derived unit** 1 S = 1 $A.V^{-1}$ (E)	The siemens is the electrical conductance of a conductor in which a current of one ampere is produced by an electrical	SI

					CGPM, 1972). The unit is named after E.W. Siemens (1816–1892).	
siemens per metre	$S.m^{-1}$	electrical conductivity	$M^{-1}L^{-3}T^{3}I^{2}$	**SI derived unit** $1\ S.m^{-1} = 1A.V.m^{-1}$ (E)		
sievert	Sv	dose equivalent, index of dose equivalent	$L^{2}T^{-2}$	**SI derived unit** $1\ Sv = 1J.kg^{-1}$ (E)	The sievert is the SI derived unit of dose equivalent when the absorbed dose of ionizing radiation multiplied by the dimensionless factors *Q* (quality factor) and *N* (product of any other multiplying factors) stipulated by the ICRP is one joule per kilogram(16th CGPM, 1980). The unit is named after R.M. Sievert (1896–1960). The relationship between the absorbed dose of radiation *D* and the dose equivalent *H* is given by $H = Q \times N \times H$. *Q* is also known as the relative biological efficiency (RBE) which varies with the nature of the radiation. *N* is a factor which takes into account the distribution of energy throughout the dose. 1 Sv = 100 rem (E) RBE = 1 X-ray, gamma, beta particle RBE = 10 neutrons RBE = 20 alpha particles (ICRP)	SI
sigma	σ	length, distance	L	$1\ \sigma = 10^{-12}$m(E)	Obsolete submultiple of the metre. $1\ \sigma = 1$ pm	
sign	–	plane angle	α	$1\ \text{sign} = 5.235987756 \times 10^{-1}$ rad	$1\ \text{sign} = \pi/6$ rad (E) $1\ \text{sign} = 30^{\circ}$ (E)	

Unit (synonym, acronym)	Symbol	Physical quantity	Dimension	Conversion factor (SI equivalent unit)	Notes, definitions, other conversion factors	System
Sikes degree	–	percentage of alcohol in wines and spirits	nil	(see note)	Obsolete unit introduced in 1794	
siriometre	–	length, distance	L	1 siriometre = $1.49597870 \times 10^{15}$ m	Obsolete unit of distance employed in astronomy. 1 siriometre = 10^6 AU (E)	@
siriusweit	–	length, distance	L	1 siriusweit = $1.5428387847 \times 10^{17}$ m	Obsolete unit of distance employed in astronomy. 1 siriusweit = 5 parsecs (E)	@
skålpund (Swedish) [Swedish pound]	–	mass	M	1 skålpund = 0.4250797024 kg	Obsolete Swedish unit of weight. 1 kg = 2.3525 skålpunds (E)	Swedish
skein	–	length, distance	L	1 skein (UK) = 109.728000 m (E)	Obsolete British unit of length. 1 skein (UK) = 60 fathoms (E) 1 skein (UK) = 120 yards (E) 1 skein (UK) = 360 feet (E)	UK
skein	–	surface, area	L^2	1 skein (UK) = 1.09728×10^2 m^2	Obsolete British unit of area used in surveyor's measurements.	UK
skippund (Swedish)	–	mass	M	1 skippund = 170.031881 kg	Obsolete Swedish unit of weight. 1 skippund = 400 skålpunds (E)	Swedish
skot	–	illuminance	$JL^{-2}\Omega^{-1}$	1 skot = 10^{-3} lx (E)	Obsolete illuminance unit introduced in Germany during World War II for small illumination measurements.	German
slug (geepound)	–	mass	M	1 slug = 14.59390294 kg	Obsolete unit of mass. It was equal to the mass which under an acceleration of 1ft.s^{-2} gives a force of 1 lbf. The name	UK, US, FPS

					geepound sometimes used for the unit is derived from English: ***g pound***.	
slug-force	–	force, weight	MLT^{-2}	1 slug-force = 143.117305 N	Obsolete unit of force derived from German slug.	
solar mass	M⊙	mass	M	1 M⊙ = 1.989×10^{30} kg	Astronomical unit of mass.	@
solotnik (Russian) [Russian denier]	–	mass	M	1 solotnik = $4.265580395 \times 10^{-3}$ kg	Obsolete Russian unit of mass from before 1917 for general use. 1 solotnik = 1/96 funt (E)	Russian
sone	–	subjective loudness	nil	(see note)	If S is the loudness in sones and P the loudness level in phons: $\log_{10} S = 0.0301 \times P - 1.204$	
sotka (Russian)	–	length, distance	L	1 sotka (Russian) = 2.1336×10^{-2} m	Obsolete Russian unit of length used before 1917. 1 sotka (Russian) = 7/100 foute (E)	Russian
Soxhlet degree	°Sx	specific gravity index on lactometric scale	nil	$°Sx = 1000 \times (1 - d)$	Soxhlet's lactometer, for determining the density of milk, has a scale from 25°Sx (sp. gr. 1.025) to 35°Sx (sp. gr. 1.035) divided into suitable scale divisions.	
span	–	length, distance	L	1 span (UK) = 0.2286 m (E)	Obsolete British unit of length. 1 span (UK) = 9 inches (E) 1 span (UK) = 3/4 foot (E)	UK
spanna (Swedish, dry)	–	capacity, volume	L^3	1 spanna (Swedish, dry) = 73.276×10^{-3} m^3	Obsolete Swedish unit of capacity for dry substances. 1 spanna (Swedish, dry) = 28 kanna (E)	Swedish
spat	S	length, distance	L	1 spat = 10^{12} m (E)	Obsolete unit of distance employed in astronomy. 1 spat = 1 Tm (E)	@

Unit (synonym, acronym)	Symbol	Physical quantity	Dimension	Conversion factor (SI equivalent unit)	Notes, definitions, other conversion factors	System
spat	sp	solid angle	Ω	1 sp = 12.566370610 sr	The spat describes the solid angle which contains all the space around a point. 1 sp = 4π steradians (E)	
spithame (Egyptian) [Egyptian span]	–	length, distance	L	1 spithame (Egyptian) $= 4.653333 \times 10^{-1}$ m	Obsolete Egyptian unit of length used in ancient times. 1 spithame = 8/9 Royal cubit (E)	Egyptian
spithame (Greek, Attic) [Greek span]	–	length, distance	L	1 spithame (Greek, Attic) $= 2.3142 \times 10^{-1}$ m	Obsolete Greek unit of length employed in ancient times. 1 spthame (Attic) = 3/4 pous (E)	Attic
square (cubic metre) per mole	$m^6.mol^{-1}$	third virial coefficient	L^6N^{-1}	**SI derived unit**		SI
square ångström	$Å^2$	surface, area	L^2	1 $Å^2 = 10^{-20}$ m^2 (E)	Obsolete unit of area used in atomic physics to express cross sections of atoms.	
square attometre	am^2	surface, area	L^2	1 $am^2 = 10^{-36}$ m^2 (E)	Submultiple of the SI derived unit	SI
square bohr	$a_0{}^2$	surface, area	L^2	1 $a_0{}^2 = 2.800285611 \times 10^{-21}$ m^2	Obsolete unit of area used in atomic physics to express cross section of atoms.	a.u.
square centimetre	cm^2	surface, area	L^2	1 $cm^2 = 10^{-4}$ m^2 (E)	Submultiple of the SI derived unit	SI
square chain (engineer's)	sq. ch. (engin-eer's)	surface, area	L^2	1 sq. chain (engineer's) = 929.030400 m^2 (E)	American unit of area employed in surveyor's measurements. 1 sq. chain (engineer's) = 36 rd^2 (E) 1 sq. chain (engineer's) = 10 000 ft^2 (E)	US
square chain (Gunter's)	sq. ch. (Gunter's)	surface, area	L^2	1 sq. chain (Gunter's) = 404.6856422 m^2 (E)	British unit of area employed in surveyor's measurements. 1 sq. chain (Gunter's) = 16 rd^2 (E) 1 sq. chain (Gunter's) = 4356 ft^2 (E)	UK

square coulomb square metre per joule	$C^2.m^2.J^{-2}$	electric polarizability of a molecule	$M^{-1}L^{-2}T^4I^2$	**SI derived unit** $1\ C^2.m^2.J^{-2} = 1\ kg^{-2}.m^{-2}.s^6.A^2$ (E)		SI
square cubit (Egyptian)	–	surface, area	L^2	1 square cubit (Egyptian) $= 0.274050\ m^2$	Obsolete Egyptian unit of area used in ancient times. 1 square cubit (Egyptian) = 1/100 pekeis (E)	Egyptian
square decametre	dam^2	surface, area	L^2	$1\ dam^2 = 100\ m^2$ (E)	Multiple of the SI derived unit.	SI
square decimetre	dm^2	surface, area	L^2	$1\ dm^2 = 10^{-2}\ m^2$ (E)	Submultiple of the SI derived unit.	SI
square degree (old)	$\square^\circ$, $(^\circ)^2$	solid angle	Ω	$1\ (^\circ)^2 = 3.046174198 \times 10^{-4}$ sr	Obsolete unit of solid angle. $1\ (^\circ)^2 = (\pi/180)^2$ rad (E)	
square dekapode (Attic)	–	surface, area	L^2	1 square dekapode (Attic) $= 9.520927360\ m^2$	Obsolete Greek unit of area used in ancient times. 1 square dekapode (Attic) = 100 sq. pous (E)	Attic
square exametre	Em^2	surface, area	L^2	$1\ Em^2 = 10^{36}\ m^2$ (E)	Multiple of the SI derived unit.	SI
square femtometre	fm^2	surface, area	L^2	$1\ fm^2 = 10^{-30}\ m^2$ (E)	Submultiple of the SI derived unit.	SI
square fermi	F^2	surface, area	L^2	$1\ F^2 = 10^{-30}\ m^2$	Obsolete unit of area employed in nuclear physics to express the cross section of a nuclide.	
square foot	ft^2, sq.ft, $\square'$	surface, area	L^2	$1\ ft^2 = 9.290304000 \times 10^{-2}\ m^2$ (E)	Legal unit of area in the UK, US and FPS systems. $1\ ft^2 = 144\ in^2$ (E)	UK, US, FPS
square foot (Persian)	–	surface, area	L^2	1 square foot (Persian) $= 1.024 \times 10^{-1}\ m^2$	Obsolete Persian unit of area used in ancient times. 1 square foot (Persian) = 1/144 gar (E)	Persian

Unit (synonym, acronym)	Symbol	Physical quantity	Dimension	Conversion factor (SI equivalent unit)	Notes, definitions, other conversion factors	System
square foot (US, survey)	ft^2 (US, survey)	surface, area	L^2	1 ft^2 (US, survey) $= 9.290341161 \times 10^{-2}$ m^2 (E)	Obsolete American unit of area used in geodetic and surveyor's measurements.	US
square foot per hour	$ft^2.h^{-1}$, sq.ft/h	kinematic viscosity	$L^2.T^{-1}$	1 $ft^2.h^{-1} = 2.58064 \times 10^{-5}$ $m^2.s^{-1}$	British and American unit of kinematic viscosity.	UK, US
square fot (Swedish)	–	surface, area	L^2	1 square fot (Swedish) $= 8.814961 \times 10^{-2}$ m^2	Obsolete Swedish unit of area.	Swedish
square gigametre	Gm^2	surface, area	L^2	1 $Gm^2 = 10^{18}$ m^2 (E)	Multiple of SI derived unit.	SI
square gon (old)	$\square^g$, $(g)^2$	solid angle	Ω	1 $(g)^2 = 2.467401101 \times 10^{-4}$ sr	Obsolete unit of solid angle. 1 $(g)^2 = (\pi/200)^2$ rad (E)	French
square grade (old)	$\square^g$, $(g)^2$	solid angle	Ω	1 $(g)^2 = 2.467401101 \times 10^{-4}$ sr	Obsolete unit of solid angle. 1 $(g)^2 = (\pi/200)^2$ rad (E)	French
square hectometre	hm^2	surface, area	L^2	1 $hm^2 = 10^4$ m^2 (E)	Multiple of the SI derived unit.	SI
square inch	in^2, sq. in, $\square''$	surface, area	L^2	1 $in^2 = 6.4516 \times 10^{-4}$ m^2 (E)	Legal submultiple unit of area in the UK, US and FPS systems. 1 $ft^2 = 144$ in^2 (E)	UK, US, FPS
square inch per hour	$in^2.h^{-1}$, sq. in/h	kinematic viscosity	$L^2.T^{-1}$	1 $in^2.h^{-1}$ $= 1.792111111 \times 10^{-7}$ $m^2.s^{-1}$	British and American unit of kinematic viscosity.	UK, US
square kilometre	km^2	surface, area	L^2	1 $km^2 = 10^6$ m^2 (E)	Multiple of the SI derived unit.	SI
square league (Canadian)	sq. leag (Canada)	surface, area	L^2	1 sq. leag (Canada) $= 2.329999290 \times 10^7$ m^2	Obsolete Canadian unit of area. 1 sq. leag (Canada) = 9 sq. miles (Canada)	CAN
square league (int., naut.)	sq. leag (int., naut.)	surface, area	L^2	1 sq. leag (int., naut.) $= 3.086913600 \times 10^7$ m^2	Obsolete international unit of length employed in navigation. 1 sq. leag (int., naut.) = 9 sq. miles (int., naut.)	INT, UK, US

square league (statute, land)	sq. leag (statute, land)	surface, area	L^2	1 sq. leag (statute) $= 2.330989299 \times 10^7$ m^2	Obsolete British and American unit of length. 1 sq. leag (statute) = 9 sq. miles (statute)	UK, US
square league (US, naut.)	sq. leag (UK, naut.)	surface, area	L^2	1 sq. leag (US, naut.) $= 3.090861844 \times 10^7$ m^2	Obsolete American unit of length employed in navigation. 1 sq. leag (US, naut.) = 9 sq. miles (US, naut.)	US
square league (UK, naut.)	sq. leag (UK, naut.)	surface, area	L^2	1 sq. leag (UK, naut.) $= 3.090861844 \times 10^7$ m^2	Obsolete British unit of length used in navigation. 1 sq. leag (UK, naut.) = 9 sq. miles (UK, naut.)	UK
square megametre	Mm^2	surface, area	L^2	1 $Mm^2 = 10^{12}$ m^2 (E)	Multiple of the SI derived unit	SI
square metre	m^2	surface, area	L^2	**SI derived unit**	The square metre is the SI derived unit. It is equal to the area of a surface of the plane square with sides of one metre.	SI
square metre per hour	$m^2.h^{-1}$, sq. m/h	kinematic viscosity	$L^2.T^{-1}$	1 $m^2.h^{-1} = 2.777778 \times 10^{-4}$ $m^2.s^{-1}$	British and American unit of kinematic viscosity.	UK, US
square metre per mole	$m^2.mol^{-1}$	molar absorption coefficient	L^2N^{-1}	**SI derived unit**		SI
square metre per second	$m^2.s^{-1}$	kinematic viscosity, coefficient of diffusion, thermal diffusivity	L^2T^{-1}	**SI derived unit**	The square metre per second is the SI derived unit of kinematic viscosity of an homogenous fluid having a density of one kilogram per cubic metre and a dynamic viscosity of one pascal-second.	SI
square metre per steradian	$m^2.sr$	light gathering power, throughtput	$L^2\Omega$	**SI derived unit**		SI
square micrometre	μm^2	surface, area	L^2	1 $\mu m^2 = 10^{-12}$ m^2 (E)	Submultiple of the SI derived unit.	SI

Unit (synonym, acronym)	Symbol	Physical quantity	Dimension	Conversion factor (SI equivalent unit)	Notes, definitions, other conversion factors	System
square mile (geographical)	sq. mi. (Geogr.)	surface, area	L^2	1 sq. mi (geographical) $= 5.508001297 \times 10^7$ m^2	Obsolete American and British unit of area employed in geodetic measurements.	UK, US
square mile (int. naut.)	sq. mi (int. naut.)	surface, area	L^2	1 sq. mi (int. naut.) $= 3.429904000 \times 10^6$ m^2	International unit of area employed in navigation (mercantile marine, aviation).	INT
square mile (int.)	sq. mi (int.)	surface, area	L^2	1 sq. mi (int.) $= 2.589997766 \times 10^6$ m^2	Obsolete international unit of area employed in navigation (mercantile marine, aviation). 1 sq. mile (int.) = 640 acres (E)	INT, UK, US
square mile (statute, land)	sq. mi (stat.)	surface, area	L^2	1 sq. mi (statute) $= 2.589988110 \times 10^6$ m^2	Obsolete American and British unit of area employed in geodetic measurements.	UK, US
square mile (telegraph nautical)	sq. mi (teleg., naut.)	surface, area	L^2	1 sq. mi (teleg., naut.) $= 3.442203397 \times 10^6$ m^2	Obsolete American and British unit of area employed in navigation (mercantile marine).	UK, US
square mile (UK, naut.)	sq. mi (UK, naut.)	surface, area	L^2	1 sq. mi (UK, naut.) $= 3.434290938 \times 10^6$ m^2	Obsolete British unit of area employed in navigation (mercantile marine).	UK
square mile (US, naut.)	sq. mi (US, naut.)	surface, area	L^2	1 sq. mi (US, naut.) $= 3.434290938 \times 10^6$ m^2	Obsolete American unit of area employed in navigation (mercantile marine).	US
square mile (US, survey)	sq. mi (US, surv.)	surface, area	L^2	1 sq. mi (US, survey) $= 2.589998468 \times 10^6$ m^2	Obsolete American unit of area employed in geodetic and surveyors' measurements.	US
square millimetre	mm^2	surface, area	L^2	1 mm$^2 = 10^{-6}$ m^2 (E)	Submultiple of the SI derived unit.	SI
square nanometre	nm^2	surface, area	L^2	1 nm$^2 = 10^{-18}$ m^2 (E)	Submultiple of the SI derived unit.	SI

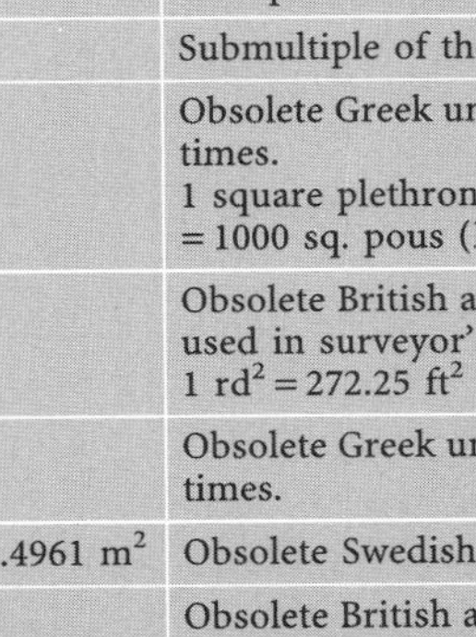

square perch	rd^2	surface, area	L^2	$1\ rd^2 = 25.293\ m^2$ (E)	Obsolete British and American unit of area used in surveyor's measurements. $1\ rd^2 = 272.25\ ft^2$ (E)	UK, US
square petametre	Pm^2	surface, area	L^2	$1\ Pm^2 = 10^{30}\ m^2$ (E)	Multiple of the SI derived unit.	SI
square picometre	pm^2	surface, area	L^2	$1\ pm^2 = 10^{-24}\ m^2$ (E)	Submultiple of the SI derived unit.	SI
square plethron (Attic)	–	surface, area	L^2	1 square plethron (Attic) $= 95.20927360\ m^2$	Obsolete Greek unit of area used in ancient times. 1 square plethron (Attic) = 1000 sq. pous (E)	Attic
square pole	rd^2	surface, area	L^2	$1\ rd^2 = 25.293\ m^2$ (E)	Obsolete British and American unit of area used in surveyor's measurements. $1\ rd^2 = 272.25\ ft^2$ (E)	UK, US
square pous (Attic)	–	surface, area	L^2	1 square pous (Attic) $= 9.520927360 \times 10^{-2}\ m^2$	Obsolete Greek unit of area used in ancient times.	Attic
square ref (Swedish)	–	surface, area	L^2	1 square ref (Swedish) $= 881.4961\ m^2$	Obsolete Swedish unit of area.	Swedish
square rod	rd^2	surface, area	L^2	$1\ rd^2 = 25.293\ m^2$ (E)	Obsolete British and American unit of area used in surveyor's measurements. $1\ rd^2 = 272.25\ ft^2$ (E)	UK, US
square terametre	Tm^2	surface, area	L^2	$1\ Tm^2 = 10^{24}\ m^2$ (E)	Multiple of the SI derived unit.	SI
square vara (Portuguese)	–	surface, area	L^2	1 square vara (Portuguese) $= 1.199025\ m^2$	Obsolete Portuguese unit of area.	Portuguese
square vara (Spanish)	–	surface, area	L^2	1 square vara (Spanish) $= 0.678737169\ m^2$	Obsolete Spanish unit of area.	Spanish
square volt per square kelvin	$V^2.K^{-2}$	Lorenz coefficient	$M^2L^4T^{-6}I^{-2}\Theta^{-2}$	**SI derived unit**		SI

Unit (synonym, acronym)	Symbol	Physical quantity	Dimension	Conversion factor (SI equivalent unit)	Notes, definitions, other conversion factors	System
square yard (old)	yd^2, sq. yd (old)	surface, area	L^2	1 yd^2 (old) = 0.836125897 m^2	Obsolete British unit of area.	UK
square yard (US)	yd^2, sq. yd (US)	surface, area	L^2	1 yd^2 (US) = 0.836307067 m^2	Obsolete American unit of area.	US
square yard (WMA, 1963)	yd^2, sq. yd	surface, area	L^2	1 yd^2= 0.83612736 m^2 (E)	Legal unit of area in the UK, US systems. 1 yd^2 = 9 ft^2 (E) 1 yd^2 = 144 in^2 (E)	UK, US
square yoctometre	ym^2	surface, area	L^2	1 $ym^2 = 10^{-48}$ m^2 (E)	Submultiple of the SI derived unit.	SI
square yottameter	Ym^2	surface, area	L^2	1 $Ym^2 = 10^{48}$ m^2 (E)	Multiple of the SI derived unit.	SI
square zeptometer	zm^2	surface, area	L^2	1 $zm^2 = 10^{-42}$ m^2 (E)	Submultiple of the SI derived unit.	SI
square zettameter	Zm^2	surface, area	L^2	1 $Zm^2 = 10^{42}$ m^2 (E)	Multiple of the SI derived unit.	SI
stade (Egyptian)	–	length, distance	L	1 stade (Egyptian) = 209.400 m	Obsolete Egyptian unit of length used in ancient times. 1 stade = 400 Royal cubit (E)	Egyptian
stadion (Greek, Attic)	–	length, distance	L	1 stadion (Greek, Attic) = 185.136 m	Obsolete Greek unit of length employed in ancient times. 1 stadion (Attic) = 600 pous (E)	Attic
standard (Petrograd)	–	capacity, volume	L^3	1 standard = 4.6722796878 m^3	1 standard = 165 cubic feet (E)	
standard atmosphere	atm	pressure, stress	$ML^{-1}T^{-2}$	1 atm = 101 325 Pa (E)	Obsolete unit of pressure and stress which should be disontinued. Unit of pressure equal to the air pressure measured at mean sea level.	INT

standard cubic foot	SCF, ft (60°F, 1 atm)	amount of substance	N	1 ft (60°F, 1 atm) = 1.195310318 mol	Obsolete American and British unit employed in physical chemistry to express the amount of substance of an ideal gas in the specified T and P conditions (60°F, 1 atm). The numerical value is obtained by the classical equation $PV = nRT$ applied to an ideal gas.	UK, US
standard cubic metre	SCM, m^3 (STP)	amount of substance	N	1 m^3 (STP) = 44.61590602 mol (273.15 K, 1 atm) 1 m^3 (STP) = 44.03247571 mol (273.15 K, 1 bar)	Obsolete metric unit employed in physical chemistry to express the amount of substance of an ideal gas in the standard T and P conditions. The numerical value is obtained by the classical equation $PV = nRT$ applied to an ideal gas.	
standard time	–	time, period, duration	T	(see note)	Time scale related to the law. It corresponds to the universal time adjusted by the fixed time period which depends on the country or regional location.	
stang (Swedish)	–	length, distance	L	1 stang (Swedish) = 4.7504 m	Obsolete Swedish unit of length.	Swedish
stat	–	radioactivity	T^{-1}	1 stat = 1.3431×10^{-16} Bq	Obsolete unit of radioactivity rarely used in nuclear sciences. 1 stat = 3.63×10^{-27} Ci	
statampere (esu of current)	statA	electric current intensity	I	1 statampere = 3.355641×10^{-10} A	Obsolete cgs unit of electric current in the esu subsystem.	cgs
statcoulomb (esu of charge)	statC	quantity of electricity, electric charge	IT	1 statcoulomb = 3.355641×10^{-10} C	Obsolete cgs unit of electric charge in the esu subsystem.	cgs
statfarad (esu of capacitance)	statC	electric capacitance	$M^{-1}L^{-2}T^4I^4$	1 statfarad = 1.112650×10^{-12} F	Obsolete unit of electric capacitance in the esu subsystem.	cgs
stathenry (esu of inductance)	statH	electric inductance	$ML^2T^{-2}I^{-2}$	1 stathenry = 8.987552×10^{11} H	Obsolete cgs unit of electric inductance in the esu subsystem.	cgs

Unit (synonym, acronym)	Symbol	Physical quantity	Dimension	Conversion factor (SI equivalent unit)	Notes, definitions, other conversion factors	System
statmho (esu of conductance)	statmho	electric conductance	$M^{-1}L^{-2}T^3I^2$	1 statmho = $1,112650 \times 10^{-12}$ S	Obsolete cgs unit of electric conductance in the esu subsystem.	cgs
statohm (esu of resistance)	statohm	electric resistance	$ML^2T^{-3}I^{-2}$	1 statohm = 8.987554×10^{11} Ω	Obsolete cgs unit of electric resistance in the esu subsystem.	cgs
statvolt (esu of electric potential)	statV	electric potential, electric potential difference, electromotive force	$ML^2T^{-3}I^{-1}$	1 statvolt = 2.99792458×10^2 V	Obsolete cgs unit of electric potential in the esu subsystem.	cgs
steekan (Dutch)	–	capacity, volume	L^3	1 steekan (Dutch) = 19.200×10^{-3} m^3	Obsolete Dutch unit of capacity used for liquid substances. 1 steekan (Dutch) = 16 mingelen (E)	Dutch
stein (Austrian)	–	mass	M	1 stein (Austrian) = 11.200200 kg	Obsolete Austrian unit of mass for general use. 1 stein (Austrian) = 20 pfund (E)	Austrian
stein (Prussian)	–	mass	M	1 stein (Prussian) = 10.289642 kg	Obsolete German unit of mass. 1 stein (Prussian) = 22 pfund (E)	German
stekar (Russian)	–	capacity, volume	L^3	1 stekar (Russian) = 18.449115×10^{-3} m^3	Obsolete Russian unit of capacity for liquid substances used before 1917. 1 stekar (Russian) = 3/2 vedro (E)	Russian
sten (Swedish)	–	mass	M	1 sten (Swedish) = 13.602550480 kg	Obsolete Swedish unit of weight. 1 sten (Swedish) = 32 skålpund (E)	Swedish
steradian	sr	solid angle	W	**SI supplementary unit**	The steradian is the solid angle which has its apex on the centre of a sphere, and cuts into this sphere surface an equivalent area of one square which has its side equal to the sphere radius. The space described by a solid angle could be the interior of a	SI

					conical or pyramidal surface.	
stère	st	capacity, volume	L^3	1 st = 1 m^3 (E)	Obsolete French unit of measure employed in the wood industry. One stère is equal to one cubic metre of stacked log wood for use as fuel.	French
sthène (Funal)	sthène, sn	force, weight	MLT^{-2}	1 sth = 10^3 N (E)	Obsolete MTS unit of force.	MTS
sthène per square metre	sthène. m^{-2} sn.m^{-2}	pressure, stress	$ML^{-1}T^{-2}$	1 sn.m^{-2} = 10^3 Pa (E)	Obsolete MTS pressure and stress unit. 1 sn.m^{-2} = 1 pz (E) 1 hpz = 1 bar (E) 1 atm = 1.01325 sn.m^{-2} (E)	MTS
stilb	sb	luminous luminance	JL^{-2}	1 stilb = 10^4 cd.m^{-2}(E)	Obsolete submultiple of the cgs unit of luminous luminance. The unit is named after Greek *stilbein*: shining.	cgs
stokes (lentor)	St	kinematic viscosity	L^2T^{-1}	1 St = 10^{-4} $m^2.s^{-1}$ (E)	Obsolete cgs unit of kinematic viscosity.	cgs
stone (UK)	st (UK)	mass	M	1 stone (UK) = 6.350293180 kg (E)	Obsolete British unit of mass. Sometimes used in the turf industry. 1 stone (UK) = 14 lb (av.) (E) 1 stone (UK) = 224 ounces (av.) (E)	UK
stone (UK, wool)	st (UK, wool)	mass	M	1 stone (UK, wool) = 5.669904625 kg (E)	Obsolete British unit of mass employed in the weighing of wool. 1 stone (UK, wool) = 12.5 lb (av.) (E)	UK
stoop (Dutch)	–	capacity, volume	L^3	1 stoop (Dutch) = 2.4×10^{-3} m^3	Obsolete Dutch unit of capacity used for liquid substances. 1 stoop (Dutch) = 2 mingelen (E)	Dutch

Unit (synonym, acronym)	Symbol	Physical quantity	Dimension	Conversion factor (SI equivalent unit)	Notes, definitions, other conversion factors	System
stop (Swedish)	–	capacity, volume	L^3	1 stop (Swedish) $= 1.308581 \times 10^{-3}$ m^3	Obsolete Swedish unit of capacity for liquid substances. 1 stop (Swedish) = 1/2 kanna (E)	Swedish
stop (Swedish, dry)	–	capacity, volume	L^3	1 stop (Swedish, dry) $= 1.3085 \times 10^{-3}$ m^3	Obsolete Swedish unit of capacity for dry substances. 1 stop (Swedish, dry) = 1/2 kanna (E)	Swedish
Stoppani degree	–	specific gravity of liquids, hydrometer index, hydrometer degree	nil	$°\text{Stoppani} = 166 - 166/d^{60°F}_{60°F}$		
sû (Chinese)	–	length, distance	L	1 sû (Chinese) $= 2.3 \times 10^{-6}$ m	Obsolete Chinese unit of length used in ancient times. 1 sû = 1/100 000 tchi (E)	
sû (Egyptian)	–	surface, area	L^2	1 sû (Egyptian) = 171.281250 m^2	Obsolete Egyptian unit of area used in ancient times. 1 sû (Egyptian) = 6.25 pekeis (E)	Egyptian
sun (Japanese)	–	length, distance	L	1 sun (Japanese) $= 3.030303030 \times 10^{-2}$ m	Obsolete Japanese unit of length. 1 sun = 1/10 shaku (E)	Japanese
svedberg (time)	S	time of sedimentation of molecules	T	1 S $= 10^{-13}$ s (E)		
svedberg (velocity)	Sv	velocity of sedimentation of molecules	LT^{-1}	1 Sv $= 10^{-15}$ $m.s^{-1}$ (E)	Obsolete unit of sedimentation velocity employed in biochemistry and biophysics. The unit is named after T. Svedberg (1884–1971) the Swedish pioneer in the use of the ultracentrifuge in biochemistry.	

					physical oceanography. The unit is named after H.U. Sverdup (1888–1957).	
tablespoon (metric)	–	capacity, volume	L^3	1 tablespoon (metric) = 1.5×10^{-5} m^3	Obsolete American unit of capacity employed in pharmacy. 1 tablespoon (metric) = 15 cm^3 1 tablespoon (metric) = 3 teaspoons (metric)	U.K, US
tablespoon (US)	–	capacity, volume	L^3	1 tablespoon (US) = $1.478676479 \times 10^{-5}$ m^3	Obsolete American unit of capacity employed in pharmacy. 1 tablespoon (US) = 1/2 fl oz (US) (E) 1 tablespoon (US) = 6 teaspoons (US) (E)	US
talent (Greek, Attic)	–	mass	M	1 talent (Greek, Attic) = 25.920 kg	Obsolete Greek unit of mass used in ancient times.	Attic
talent (Hebrew) [Talmudic system]	–	mass	M	1 talent (Hebrew) = 21.252 kg	Obsolete Hebrew unit of mass used in ancient times. Rabbinacal or Talmudic system. 1 talent of Moses = 60 mina (E)	Hebrew
talent of Moses (Hebrew) [Sacred system]	–	mass	M	1 talent of Moses = 425 kg	Obsolete Hebrew unit of mass used in ancient times. Sacred system. 1 talent of Moses = 50 mina (E)	Hebrew
tan (Japanese)	–	surface, area	L^2	1 tan = 991.7355372 m^2	Obsolete Japanese unit of area. 1 tan = 300 bu (E)	Japanese
tank-sala (Indian)	–	mass	M	1 tank-sala (Indian) = 2.643750×10^{-3} kg	Obsolete Indian unit of mass used in ancient times. 1 tank-sala (Indian) = 160/9 pala (E)	Indian
tavola (Italian)	–	surface, area	L^2	1 tavola (Italian) = 38 m^2	Obsolete Italian unit of area. 1 tavola (Italian) = 1/100 giornata (E)	Italian

Unit (synonym, acronym)	Symbol	Physical quantity	Dimension	Conversion factor (SI equivalent unit)	Notes, definitions, other conversion factors	System
tce (tonne coal equivalent)	tce	energy, work, heat	ML^2T^{-2}	1 tce = 29 GJ = 2.929×10^8 J	Large unit of energy employed in the oil industry and economics to express energy balances.	INT
tcharka (Russian)	–	capacity, volume	L^3	1 tcharka (Russian) = 1.229941×10^{-4} m^3	Obsolete Russian unit of capacity for liquid substances used before 1917.	Russian
tchast (Russian, dry)	–	capacity, volume	L^3	1 tchast (Russian, dry) = $1.09328067 \times 10^{-4}$ m^3	Obsolete Russian unit of capacity for dry substances used before 1917. tchast (Russian, dry) = 1/30 garnetz (E)	Russian
tchetverik (Russian, dry)	–	capacity, volume	L^3	1 tchetverik (Russian, dry) = 26.238736×10^{-3} m^3	Obsolete Russian unit of capacity for dry substances used before 1917. 1 tchetverik (Russian, dry) = 8 garnetz (E)	Russian
tchetvert (Russian, dry)	–	capacity, volume	L^3	1 tchetvert (Russian, dry) = $209.909888 \times 10^{-3}$ m^3	Obsolete Russian unit of capacity for dry substances used before 1917. 1 tchetvert (Russian, dry) = 64 garnetz (E)	Russian
tchevert (Russian)	–	length, distance	L	1 tchevert (Russian) = 0.1778 m	Obsolete Russian unit of length used before 1917. 1 tchevert = 7/12 foute (E)	Russian
tchi (Chinese)	–	length, distance	L	1 tchi (Chinese) = 0.23 m	Obsolete Chinese unit of length used in ancient times. Base unit of the Chinese system of length.	Chinese
tchin (jin)	–	mass	M	1 jin = 250×10^{-3} kg	Obsolete Chinese unit of mass.	Chinese
teaspoon (metric)	–	capacity, volume	L^3	1 teaspoon (metric) = 5×10^{-6} m^3	Obsolete American unit of capacity employed in pharmacy. 1 teaspoon (metric) = 5 cm^3 (E)	US

					1 teaspoon (metric) = 1/3 tablespoon (metric) (E)	
teaspoon (US)	–	capacity, volume	L^3	1 teaspoon (US) $= 4.928921597 \times 10^{-5}$ m^3	Obsolete American unit of capacity employed in pharmacy. 1 teaspoon (US) = 1/6 fl oz (US) (E) 1 teaspoon (US) = 1/3 tablespoon (US) (E)	US
techma (hyl, mug, metric slug, TME, par, techma)	techma	mass	M	1 techma = 9.80665 kg (E)	Obsolete technical metric unit of mass employed by mechanical engineers (base unit of the metric gravitational system). It was equal to the mass which under an acceleration of 1 $m.s^{-2}$ gives a force of 1 kgf. The name TME derived from the German acronym Technische Mass Einheit.	German
technical atmosphere	at	pressure, stress	$ML^{-1}T^{-2}$	1 at $= 9.80665 \times 10^4$ Pa (E)	Obsolete MKpS pressure and stress derived unit. Obsolete. 1 at = 1 $kgf.cm^{-2}$ (E) 1 atm = 1.033227453 at	MKpS
tempon (chronon)	–	time, period, duration	T	1 tempon $= 10^{-23}$ s (E)	Obsolete unit of time employed in atomic physics. It corresponds to the time needed by light to cover a distance equal to the electron radius.	
ten (Egyptian)	–	surface, area	L^2	1 ten (Egyptian) = 274.05 m^2	Obsolete Egyptian unit of area used in ancient times. 1 ten (Egyptian) = 10 pekeis (E)	Egyptian
ten (Persian)	–	surface, area	L^2	1 ten (Persian) = 147.456	Obsolete Persian unit of area used in ancient times. 1 ten (Persian) = 10 gar (E)	Persian

Unit (synonym, acronym)	Symbol	Physical quantity	Dimension	Conversion factor (SI equivalent unit)	Notes, definitions, other conversion factors	System
teragram	Tg	mass	M	1 Tg = 10^9 kg (E)	Multiple of the SI base unit. 1 Tg = 10^{12} g (E)	SI
terametre	Tm	length, distance	L	1 Tm = 10^{12} m (E)	Multiple of the SI base unit.	SI
tesla	T	magnetic induction field, magnetic flux density	$MI^{-1}T^{-2}$	**SI derived unit** 1 T = 1 W.m^{-2} = 1 kg.A^{-1}.s^{-2} (E)	The tesla is the magnetic flux density of a magnetic flux of one weber per square metre. The unit is named after N. Tesla (1857–1943).	SI
tex	tex	linear mass density	ML^{-1}	1 tex = 10^{-6} kg.m^{-1} (E)	Unit of linear mass density employed in the textile industry. It is temporarily maintained with the SI. 1 tex = 1 g.(1000 m)$^{-1}$ (E)	SI
thebs (Egyptian) [Egyptian finger]	–	length, distance	L	1 finger (Egyptian) = $1.871428571 \times 10^{-2}$ m	Obsolete Egyptian unit of length used in ancient times. 1 finger = 1/28 Royal cubit (E)	Egyptian
therm (EEG)	–	energy, work, heat	ML^2T^{-2}	1 therm (EEG) = $1.05505585262 \times 10^8$ J	Obsolete British and American unit of heat and energy employed in energy resource and reserve assessment. 1 therm = 10^5 Btu (IT) (E)	UK, US
therm (US)	–	energy, work, heat	ML^2T^{-2}	1 therm (US) = 1.054804×10^8 J	Obsolete American unit of heat and energy employed in energy resource and reserve assessment. 1 therm = 10^5 Btu (E)	US
thermie (15°C)	$th_{15°C}$	energy, work, heat	ML^2T^{-2}	1 $thermie_{15°C}$ = 4.1855×10^6 J	Obsolete unit of heat. 1 thermie = 10^6 cal (15°C)	
thou (mil)	thou, mil	length, distance	L	1 thou = 2.54×10^{-5} m (E)	Obsolete British and American submultiple of the inch	UK, US

					1 thou = 10[illegible] inch (E) 1 thou = 10^{-2} calibre (E) 1 thou = 1 mil (E)	
thousand mass unit (^{12}C)	TMU (^{12}C)	mass	M	1 TMU(^{12}C) = $1.492419113 \times 10^{-10}$ J.c^{-2}	Unit of mass employed in nuclear physics. It is defined by the Einstein equation $E = mc^2$, where m is expressed in atomic units of mass u(^{12}C). 1 TMU(^{12}C) = 9314943367 MeV.c^{-2}	
thousand mass unit (^{16}O)	TMU (^{16}O)	mass	M	1 TMU(^{16}O) = $1.49194477 \times 10^{-10}$ J.c^{-2}	Unit of mass employed in nuclear physics. It is defined by the Einstein equation $E = mc^2$, where m is expressed in atomic mass units u(^{16}O). 1 TMU(^{16}O) = 931.1982752 MeV.c^{-2}	
thousand mass unit (^{1}H)	TMU (^{1}H)	mass	M	1 TMU(^{1}H) = $1.504097345 \times 10^{-10}$ J.c^{-2}	Unit of mass employed in nuclear physics. It is defined by the Einstein equation $E = mc^2$, where m is expressed in atomic mass units u(^{1}H). 1 TMU(^{1}H) = 938.783 312 6 MeV.c^{-2}	
thsan (Chinese)	–	length, distance	L	1 thsan (Chinese) = 3.312×10^4 m	Obsolete Chinese unit of length used in ancient times. 1 thsan = 144 000 tchi (E)	Chinese
TME (hyl, mug, par, metric slug)	–	mass	M	1 TME = 9.80665 kg (E)	Obsolete technical metric unit of mass employed by mechanical engineers (base unit of the metric gravitational system). It was equal to the mass which under an acceleration of 1 m.s^{-2} gives a force of 1 kgf. The name TME derived from the German acronym Technische Mass Einheit.	German

Unit (synonym, acronym)	Symbol	Physical quantity	Dimension	Conversion factor (SI equivalent unit)	Notes, definitions, other conversion factors	System
to	–	capacity, volume	L^3	1 to = $18.09068370 \times 10^{-3}$ m^3	Obsolete Japanese unit of capacity. 1 to = 10 shô (E)	Japanese
to (Chinese)	–	capacity, volume	L^3	1 to (Chinese) = 10.3544×10^{-3} m^3	Obsolete Chinese unit of capacity used in ancient times. 1 to (Chinese) = 10 tcheng (E)	Chinese
tochka (Russian calibre)	–	length, distance	L	1 tochka (Russian) = 2.54×10^{-4} m	Obsolete Russian unit of length used before 1917. 1 tochka (Russian) = 1/1200 foute (E)	Russian
toe (tonne oil equivalent)	toe	energy, work, heat	ML^2T^{-2}	1 toe = 4.187×10^8 J	Large unit of energy employed in the oil industry and economics to express energy balances. 1 toe = 41.87 GJ (E)	INT
tog	tog	heat insulation coefficient (thermal resistance multiplied by area)	$M^{-1}T^3\Theta$	1 tog = 0.1 $W^{-1}.m^2.K$ (E)	Obsolete unit of heat insulation coefficient employed in the British textile industry. One tog corresponds to the heat-insulation coefficient of clothing which conserves a temperature difference of 0.1°C between its surfaces when the heat flux is equal to 1 $W.m^{-2}.h^{-1}$. 1 tog = 0.1 RSI (E) 1 tog = 0.116263888 $(kcal_{15}/h)^{-1}.m^2.°C$ 1 tog = 0.645910492 clo	UK
toise (de Pérou)	–	length, distance	L	1 toise (de Pérou) = 1.9490365 m	Obsolete French unit of length employed in France during the Ancien Régime before the French Revolution (1789). It was used in geodetic measurements. The name is	French

					derived from the latin *ex tensa* meaning distance. 1 toise (de Pérou) = 6 pieds (Paris)	
toise (metric)	–	length, distance	L	1 toise (metric) = 2 m (E)	Obsolete unit of length employed in France from 1812 to 1840.	French
tola (Indian)	–	mass	M	1 tola (Indian) = 11.750×10^{-3} kg	Obsolete Indian unit of mass used in ancient times. 1 tola (Indian) = 4 pala (E)	Indian
tomin (Spanish)	–	mass	M	1 tomin (Spanish) = $5.990794271 \times 10^{-4}$ kg	Obsolete Spanish unit of mass. 1 tomin (Spanish) = 1/768 libra (E)	Spanish
ton (American commercial of refrigeration)	CTR (US)	power	ML^2T^{-3}	1 CTR (US) = 3 516 800 W	Obsolete American unit of power employed in cryogenics. It was equal to the heat absorbed by the melting of one short ton (2000 lb) of ice at 0°C (32°F) in 24 hours. 1 CTR (US) = 12 000 Btu (refrigeration).h^{-1} (E) 1 CTR (US) = 11 376 kJ.h^{-1}	US
ton (assay UK)	(UK) AT	mass	M	1 ton (assay UK) = 32.66667×10^{-3} kg	Obsolete British unit of mass used by assayers and jewellers in the assaying of gold and silver. It was equal to the mass expressed in UK long tons [2240 lb (av.)] of the ore of a precious metal needed to give one troy ounce of pure precious metal. In other words, it was the mass of an ore which gives 30.59 mg of precious metal (gold or silver) by kilogram.	UK

Unit (synonym, acronym)	Symbol	Physical quantity	Dimension	Conversion factor (SI equivalent unit)	Notes, definitions, other conversion factors	System
ton (assay US)	(US) AT	mass	M	1 ton (assay US) $= 29.16667 \times 10^{-3}$ kg	Obsolete American unit of mass used by assayers and jewellers in the assaying of gold and silver. It was equal to the mass expressed in US short tons [2000 lb (av.)] of the ore of a precious metal needed to give one troy ounce of pure precious metal. In other words, it was the mass of an ore which gives 34.28 mg of precious metal (gold or silver) by kilogram.	US
ton (British commercial unit of refrigeration)	CTR (UK)	power	ML^2T^{-3}	1 CTR (UK) = 3922.696721 W	Obsolete British unit of power employed in cryogenics. It was equal to the heat absorbed by the melting of one long ton (2240 lb) of ice at 0°C (32°F) in 24 hours. 1 CTR (UK) = 13 384.8 Btu (refrigeration).h^{-1} (E) 1 CTR (UK) = 1.1154 CTR (US)	UK
ton (register)	–	capacity, volume	L^3	1 register ton = 2.83168466 m^3	Obsolete British unit of capacity. 1 register ton = 100 ft^3 (E)	UK
ton (UK, long, 2240 lb)	UK ton, lg ton	mass	M	1 long ton (UK) = 1016.0469088 kg (E)	British unit of mass in the avoirdupois system. 1 long ton (UK) = 2240 lb (av.) (E) 1 long ton (UK) = 20 cwt (E) 1 long ton (UK) = 80 quarters (UK, av.) (E) 1 long ton (UK) = 160 stone (UK, av.) (E)	UK

ton (UK, shipping)	–	mass	M	1 ton (UK, shipping) = 1169.698141 kg	Obsolete British unit of mass employed in navigation (mercantile marine) and which corresponds to the mass of 40.3 ft^3 of sea water.	UK
ton (US, shipping)	–	mass	M	1 ton (US, shipping) = 1219.040246 kg	Obsolete American unit of mass employed in navigation to describe the size of the vessels. It is equal to the mass which occupies 42 ft^3 of sea water.	US
ton (US, short, 2000 lb)	–	mass	M	1 short ton (US) = 907.184740 kg (E)	American unit of mass in the avoirdupois system. 1 short ton (US) = 2000 lb (US) (E) 1 short ton (US) = 20 short cwt (US) (E)	US
ton-force (metric)	–	force, weight	MLT^{-2}	1 ton-force (metric) = 9806.65 N (E)	Obsolete metric unit of force. 1 ton-force (metric)= 1000 kgf (E)	MTS
ton-force (short)	–	force, weight	MLT^{-2}	1 ton-force (short) = 8896.443231 N	American unit of force. 1 ton-force (short) = 2000 lbf (E)	US
ton-force (long)	–	force, weight	MLT^{-2}	1 ton-force (long) = 9964.016420 N	British unit of force. 1 ton-force (long) = 2240 lbf (E)	UK
tonelada (Spanish)	–	mass	M	1 tonelada (Spanish) = 920.186 kg	Obsolete Spanish unit of mass. 1 tonelada (Spanish) = 2000 libra (E)	Spanish
tonne (metric)	t	mass	M	1 t = 10^3 kg (E)	Obsolete MTS base unit of mass.	MTS
tonne coal equivalent	tce	energy, work, heat	ML^2T^{-2}	1 tce = 2.929×10^8 J	Large unit of energy employed in the oil industry and economics to express energy balances. 1 tce = 29.29 GJ (E)	INT
tonne oil equivalent	toe	energy, work, heat	ML^2T^{-2}	1 toe = 4.187×10^8 J	Large unit of energy employed in oil industry and economics to express energy balances. 1 toe = 41.87 GJ (E)	INT

Unit (synonym, acronym)	Symbol	Physical quantity	Dimension	Conversion factor (SI equivalent unit)	Notes, definitions, other conversion factors	System
tonneau de jauge (tonneau de mer)	–	capacity, volume	L^3	1 tonneau de jauge = 2.831680 m^3	Obsolete French unit of volume employed before the French Revolution in navigation.	French
tonnelada (Portuguese)	–	capacity, volume	L^3	1 tonnelada (Portuguese) = 858×10^{-3} m^3	Obsolete Portuguese unit of capacity used for liquid substances. 1 tonnelada (Portuguese) = 52 almude (E)	Portuguese
torr (mmHg at 0°C)	Torr, torr	pressure, stress	$ML^{-1}T^{-2}$	1 torr = 133.3223684 Pa	Obsolete unit of pressure employed in physics to measure small pressures. It is equal to the pressure exerted by a column of mercury one millimetre high, measured at 0°C (32°F). The unit is named after the Italian scientist E. Torricelli (1608–1647). 1 torr = 1 mmHg (0°C) 760 torr = 1 atm (E) 1 torr = (101325/760) Pa (E)	
tou (Chinese)	–	length, distance	L	1 tou (Chinese) = 1.035×10^5 m	Obsolete Chinese unit of length used in ancient times. 1 tou = 450 000 tchi (E)	Chinese
townsend	Td	gradient of electric field strength	$MLT^{-3}I^{-1}$	1 Townsend = 10^{21} $V.m^{-2}$ (E)	Obsolete cgs unit. The unit is named after J. Townsend (1868-1957). 1 townsend = 10^{17} $V.cm^{-2}$ (E)	cgs
township (US)	–	surface, area	L^2	1 township (US) = 93 239 571.96 m^2	Obsolete American unit of area 1 township (US) = 36 square miles (US, statute) (E)	US
trabucco (Italian)	–	length, distance	L	1 trabucco (Italian) = 3.082620 m	Obsolete Italian unit of length. 1 trabucco (Italian) = 6 piedi liprando (E)	Italian

triens (Roman)	–	mass	M	1 triens (Roman) = 109×10^{-3} kg	Obsolete Roman unit of mass used in ancient times. 1 triens (Roman) = 4 unciae (Roman) (E)	Roman
troland (luxon)	–	luminous luminance	$J.L^{-2}$	1 troland = 10^4 $cd.m^{-2}$ (E)	Obsolete unit of luminous luminance employed in ophtalmology. It was equal to the retinal luminous luminance received by eyes by one surface having a luminous luminance of one candela per square metre. The optical aperture of eyes is about one square millimetre.	
truss (UK)	–	mass	M	1 truss (UK) = 16.329325320 kg	Obsolete British unit of mass used by farmers. 1 truss (UK) = 36 pounds (UK, straw) (E)	UK
tsouen (Chinese) [cun]	–	length, distance	L	1 tsouen (Chinese) = 2.3×10^{-2} m	Obsolete Chinese unit of length used in ancient times. 1 cun = 1/10 tchi (E) 1 cun = 2.3 cm (E)	Chinese
tsubo (Japanese)	–	surface, area	L^2	1 tsubo = 3.305785124 m^2	Obsolete Japanese unit of area. 1 tsubo = 1 bu (E) 1 tsubo = 100/30.25 m^2	Japanese
tuba (Indian)	–	mass	M	1 tuba (Indian) = 4.70 kg	Obsolete Indian unit of mass used in ancient times. 1 tuba (Indian) = 100 pala (E)	Indian
tun (US, liq.)	–	capacity, volume	L^3	1 tun (US, liq.) = $953.923769600 \times 10^{-3}$ m^3	Obsolete American unit of capacity. 1 tun (US, liq.) = 252 gallons (US) (E) 1 tun (US, liq.) = 8 barrels (US) (E) 1 tun (US, liq.) = 4 hogsheads (US) (E)	US

Unit (synonym, acronym)	Symbol	Physical quantity	Dimension	Conversion factor (SI equivalent unit)	Notes, definitions, other conversion factors	System
tunland (Swedish)	–	surface, area	L^2	1 tunland (Swedish) = 4936.378160 m^2	Obsolete Swedish unit of area. 1 tunland (Swedish) = 56 000 square fot (E)	Swedish
tunna (Swedish, dry)	–	capacity, volume	L^3	1 tunna (Swedish, dry) = 146.552×10^{-3} m^3	Obsolete Swedish unit of capacity for dry substances. 1 tunna (Swedish, dry) = 56 kanna (E)	Swedish
turn	tr, rev	plane angle	α	1 tr = 6.283185307 rad (E)	1 tr = 2π rad (E) 1 tr = 360° (E)	
turn (Swedish) [Swedish inch]	–	length, distance	L	1 turn (Swedish) = 2.4741667×10^{-2} m	Obsolete Swedish unit of length.	Swedish
Twaddell degree	°Tw	specific gravity of liquids, hydrometer index, hydrometer degree	nil	$°\mathrm{Tw} = 200 \times \left[d_{60°\mathrm{F}}^{60°\mathrm{F}} - 1\right]$	Obsolete hydrometer unit introduced in 1830. The Twaddell hydrometer has the scale so arranged that the reading multiplied by 5 and added to 1000 gives the specific gravity with reference to water as 1000; it is always used for densities greater than water.	UK
twenty foot equivalent unit	TEU, TEQ	capacity, volume	L^3	1 TEU = 36.24556364 m^3	Obsolete American and British unit of capacity employed in marine applications. 1 TEU = $20 \times 8 \times 8$ ft^3	UK, US
typp	–	specific length	$M^{-1}L$	1 typp = 2015.906925 $m.kg^{-1}$	Obsolete American and British unit employed in the textile industry. The name of the unit is derived from the English acronym: thousand yard per pound. 1 typp = 1000 $yd.lb^{-1}$ (E)	UK, US

U-factor	U-factor	reciprocal of the heat insulation coefficient (thermal conductance divided by area)	$MT^{-3}\Theta\Omega^{-1}$	1 U factor = 5.678263337 $W.m^{-2}.K^{-1}$	Obsolete British and American unit: the reciprocal of heat insulation coefficient employed in civil and building engineering. 1 U-factor = 1 $(R\text{-value})^{-1}$ (E) 1 U-factor = 5.678263337 USI	UK
uncia (Roman ounce)	–	mass	M	1 uncia (Roman ounce) = 27.25×10^{-3} kg	Obsolete Roman unit of mass used in ancient times. 1 uncia (Roman ounce) = 1/12 libra (Roman pound) (E)	Roman
uncia (Roman inch)	–	length, distance	L	1 uncia (Roman inch) = 2.453333×10^{-2} m	Obsolete Roman unit of length employed in ancient times. 1 uncia (Roman inch) = 1/12 pes (E)	Roman
unit of entropy (entropy unit)	ue, eu	molar entropy	$ML^2T^{-2}\Theta^{-1}N^{-1}$	1 ue = 4.184 $J.K^{-1}.mol^{-1}$ (E)	1 ue = 1 $cal_{th}.K^{-1}.mol^{-1}$ (E)	
unit pole	–	magnetic induction flux	$ML^2T^{-2}I^{-1}$	1 unit pole = 1.256637×10^{-7} Wb	Obsolete British and American unit of magnetic flux.	UK, US
unité de masse	UdM	mass	M	1 UdM = 9.80665 kg (E)	Obsolete MKpS base unit of mass.	MKpS
universal time (Temps Universel)	UT	time, period, duration	T	(see note)	Universal time corresponds to GMT increased by 12 hours. It is described by the symbol UT_0. It is possible to calculate UT_1 introducing corrections for small displacements of the Earth around its axis of revolution. It is possible to calculate UT_2 introducing corrections for seasonal variations of the Earth's revolution.	

Unit (synonym, acronym)	Symbol	Physical quantity	Dimension	Conversion factor (SI equivalent unit)	Notes, definitions, other conversion factors	System
universal time coordinated (Temps Universel Coordonné)	UTC	time, period, duration	T	(see note)	Universal time coordinated replaced GMT on January 1st, 1972. It corresponds to the TAI fitted to one second to ensure approximate scale concordance with UT_1. Universal time coordinated is based on emitted coordinated time rate signals and standard frequencies. The international abbreviation UTC is employed in all languages. UTC = TAI – 10 seconds (SI)	INT
unze (Austrian)	–	mass	M	1 unze (Austrian) $= 3.5000625 \times 10^{-2}$ kg	Obsolete Austrian unit of mass for general use. 1 unze (Austrian) = 1/16 pfund (E)	Austrian
unze (Austrian, apothecary)	–	mass	M	1 unze (Austrian, apothecary) $= 3.500083 \times 10^{-2}$ kg	Obsolete Austrian unit of mass used in pharmacy. 1 unze (Austrian, apothecary) = 1/12 pfund (Austrian, apothecary) (E)	Austrian
unze (Dutch)	–	mass	M	1 unze (Dutch) $= 3.076048250 \times 10^{-2}$ kg	Obsolete Dutch unit of mass. 1 unze (Dutch) = 1/16 pond (Dutch) (E)	Dutch
urna (Roman)	–	capacity, volume	L^3	1 urna (Roman) $= 13.132 \times 10^{-3}$ m^3	Obsolete Roman unit of volume employed in ancient times. It was used for capacity measurements of liquids. 1 urna (Roman) = 24 sextarius (E)	Roman
USI	USI	reciprocal of heat insulation coefficient (thermal conductance divided by area)	$MT^{-3}\Theta^{-1}$	1 USI = 1 $W.m^{-2}.K^{-1}$ (E)	Unit of the reciprocal of the heat-insulation coefficient employed in the British and American building engineering. One USI corresponds to the	UK, US

					reciprocal of the heat insulation coefficient of a wall which conserves a temperature difference of 1°C between its faces when the heat flux is equal to one $W.m^{-2}$.	
vara (Portuguese)	–	length, distance	L	1 vara (Portuguese) = 1.095 m	Obsolete Portuguese unit of length. 1 vara (Portuguese) = 10/3 pe (E)	Portuguese
vara (Spanish)	–	length, distance	L	1 vara (Spanish) = 0.835905 m	Obsolete Spanish unit of length.	Spanish
vat (Dutch)	–	capacity, volume	L^3	1 vat (Dutch) = 921.600×10^{-3} m^3	Obsolete Dutch unit of capacity used for liquid substances. 1 vat (Dutch) = 768 mingelen (E)	Dutch
vedro (Russian)	–	capacity, volume	L^3	1 vedro (Russian) = 12.299410×10^{-3} m^3	Obsolete Russian unit of capacity for liquid substances used before 1917. 1 vedro (Russian) = 50 tcharka (E)	Russian
vedro (Russian, dry)	–	capacity, volume	L^3	1 vedro (Russian, dry) = 13.119368×10^{-3} m^3	Obsolete Russian unit of capacity for dry substances used before 1917. 1 vedro (Russian, dry) = 4 garnetz (E)	Russian
velte (de Paris)	–	capacity, volume	L^3	1 velte (de Paris) = $7.45111200 \times 10^{-3}$ m^3	Obsolete French unit of capacity employed before the French Revolution. It served to express the capacity of liquids and grains. It varied according to the location and merchandise. 1 velte (de Paris) = 8 pintes (de Paris)	French
vergée (de Paris)	–	surface, area	L^2	1 vergée (de Paris) = 1 276 m^2	Obsolete French unit of area used in surveyor's measurements before the French revolution (1789).	French
vergees (Guernsey)	–	surface, area	L^2	1 vergees (Guernsey) = 1.541659589×10^3 m^2	Obsolete British unit of area employed in surveyor's measurements in the Channel Islands. 1 vergees (Guernsey) = (1/2.625) acre (E)	UK

Unit (synonym, acronym)	Symbol	Physical quantity	Dimension	Conversion factor (SI equivalent unit)	Notes, definitions, other conversion factors	System
vergees (Jersey)	–	surface, area	L^2	1 vergees (Jersey) $= 1.798602854 \times 10^3$ m^2	Obsolete British unit of area employed in surveyor's measurements in the Channel Islands. 1 vergees (Jersey) = (1/2.25) acre (E)	UK
vershok (Russian)	–	length, distance	L	1 vershok (Russian) $= 4.444999 \times 10^{-2}$ m	Obsolete Russian unit of length used before 1917. 1 vershok (Russian) = 7/48 foute (E)	Russian
verst (Russian)	–	length, distance	L	1 verst (Russian) = 1066.8 m	Obsolete Russian unit of length used before 1917. 1 verst (Russian) = 3500 foute (E)	Russian
versum (Roman)	–	surface, area	L^2	1 versum (Roman) = 877.344400 m^2	Obsolete Roman unit of area employed in ancient times. 1 versum (Roman) = 10 000 quadratus pes (E)	Roman
vierd (Dutch, dry)	–	capacity, volume	L^3	1 vierd (Dutch, dry) $= 6.815 \times 10^{-3}$ m^3	Obsolete Dutch unit of capacity used for dry substances. 1 vierd (Dutch, dry) = 1/4 schepel (E)	Dutch
vierding (Austrian)	–	mass	M	1 vierding (Austrian) $= 1.400025 \times 10^{-1}$ kg	Obsolete Austrian unit of mass for general uses. 1 vierding (Austrian) = 1/4 pfund (E)	Austrian
vierling (Dutch)	–	mass	M	1 vierling (Dutch) $= 3.845060313 \times 10^{-4}$ kg	Obsolete Dutch unit of mass. 1 vierling (Dutch) = 1/1280 pond (Dutch) (E)	Dutch
viertel (Austrian)	–	capacity, volume	L^3	1 viertel (Austrian) $= 14.151 \times 10^{-3}$ m^3	Obsolete Austrian unit of capacity used for liquid substances. 1 viertel (Austrian) = 10 mass (E)	Austrian

dry)				$= 15.372250 \times 10^{-3}$ m^3	dry substances. 1 viertel (Austrian, dry) = 1/4 metzel (E)	
vieth degree	–	specific gravity index on lactometric scale	nil	(see note)	Hydrotimeter scale used in lactometry.	
violle	–	luminous intensity	J	1 violle = 20.4 cd (E)	Obsolete unit of luminous intensity. It was equal to the intensity of the blackbody at the temperature of platinum at melting point. It was used during the period 1884–1889.	
vistati	–	length, distance	L	1 vistati = 2.285×10^{-1} m	Obsolete Indian unit of length used in ancient times. 1 vistati = 1/2 hasta (E)	Indian
voeten (Dutch) [Dutch foot]	–	length, distance	L	1 voeten (Dutch) = 0.2830594 m	Obsolete unit of length used in Amsterdam (Netherlands)	Dutch
voie (de Paris)	–	capacity, volume	L^3	1 voie (de Paris) = 1.920 m^3	Obsolete French unit of capacity employed before the French Revolution.	French
volt	V	electric potential, electric potential difference, electromotive force	$ML^2T^{-3}I^{-1}$	**SI derived unit** 1 V = 1kg.m^2.s^{-3}.A^{-1} (E)	The volt is the electric potential difference between two points of a conductor carrying a constant current of one ampere, when the power dissipated between these points is equal to one watt. The unit is named after the Italian scientist A. Volta (1745–1827).	SI
volt (int., US)	V	electric potential, electric potential difference, electromotive force	$ML^2T^{-3}I^{-1}$	1 V (int., US) = 1.000330 V	Obsolete IEUS unit of electric potential.	US, IEUS

Unit (synonym, acronym)	Symbol	Physical quantity	Dimension	Conversion factor (SI equivalent unit)	Notes, definitions, other conversion factors	System
volt (int. mean)	V	electric potential, electric potential difference, electromotive force	$ML^2T^{-3}I^{-1}$	1 V (int. mean) = 1.00034 V	Obsolete IEUS unit of electric potential introduced in 1908. It is the electromotive force of a Weston electrochemical cell measured at 20°C. It is equal to 1.0183 int. volts.	IEUS, INT
volt ampere	VA	power	ML^2T^{-3}	1 VA = 1 W (E)		
volt ampere reactive	var	reactive power	ML^2T^{-3}	1 VA = 1 W (E)		
volt per kelvin	$V.K^{-1}$	Thomson coefficient	$ML^2T^{-3}I^{-1}\Theta^{-1}$	**SI derived unit**		SI
volt per metre	$V.m^{-1}$	electric field strength	$MLT^{-3}I^{-1}$	**SI derived unit**	The volt per metre is the electric field strength SI derived unit. It is equal to the electric potential difference of one volt between two points one metre apart.	SI
waag (Swedish)	–	mass	M	1 waag (Swedish) = 70.138150900 kg	Obsolete Swedish unit of weight. 1 waag (Swedish) =165 skålpund (E)	Swedish
watt	W	power	ML^2T^{-3}	**SI derived unit** 1 W = 1 $kg.m^2.s^{-3}$	The watt is the power which gives rise to the production of energy at the rate of one joule per second. The unit is named after J. Watt (1736–1819).	SI
watt (int. mean)	W	power	ML^2T^{-3}	1 W (int. mean) = 1.00019 W	Obsolete IEUS unit.	US
watt (int. US)	W	power	ML^2T^{-3}	1 W (int. US) = 1000165 W (E)	Obsolete IEUS unit.	US
watt per centimetre per celsius degree	$W.cm^{-1}.°C^{-1}$	thermal conductivity	$MLT^{-3}\Theta^{-1}$	1 $W.cm^{-1}.K^{-1}$ = 100 $W.m^{-1}.K^{-1}$	Obsolete cgs unit of thermal conductivity.	cgs

watt per kelvin	$W.K^{-1}$	thermal conductance	$ML^2T^{-3}\Theta^{-1}$	**SI derived unit** $1\ W.K^{-1} = 1\ kg.m^2.s^{-3}.K^{-1}$ (E)		SI
watt per metre per kelvin	$W.m^{-1}.K^{-1}$	thermal conductivity	$MLT^{-3}\Theta^{-1}$	**SI derived unit** $1\ W.m^{-1}.K^{-1} = 1\ kg.m.s^{-3}.K^{-1}$		SI
watt per square metre	$W.m^{-2}$	irradiance, radiant flux received, energy flux	MT^{-3}	**SI derived unit** $1\ W.m^{-2} = 1\ kg.s^{-3}$		SI
watt per square metre per steradian	$W.m^{-2}.sr^{-1}$	radiance	$ML^{-4}T^{-3}\Omega^{-1}$	**SI derived unit** $1\ W.m^{-2}.sr^{-1} = 1\ kg.m^{-4}.s^{-3}.sr^{-1}$ (E)		SI
watt per steradian	$W.sr^{-1}$	radiant intensity	$ML^{-2}T^{-3}\Omega^{-1}$	**SI derived unit** $1\ W.sr^{-1} = 1\ kg.m^{-2}.s^{-3}.sr^{-1}$ (E)		SI
weber	Wb	magnetic induction flux	$ML^2T^{-2}I^{-1}$	**SI derived unit** $1\ Wb = 1\ kg.m^2.s^{-2}.A^{-1}$ (E)	The weber is the magnetic flux which, linking a circuit of one turn, produces in it an electromotive force of one volt as it is reduced to zero at a uniform rate in one second (8th CGPM, 1948). The unit is named after W.E. Weber (1804–1891).	SI
weber per metre	$Wb.m^{-1}$	magnetic potential vector		**SI derived unit**		SI
week	w	time, period, duration	T	1 week = 604 800 s	1 week = 7 days (E)	
Weißkopf unit (Moszkowski)	–	probability of transition	nil	(see note)	Obsolete unit employed in nuclear physics to express the nuclear quantum state's transition probability.	
wey (UK, capacity)	–	capacity, volume	L^3	1 wey (UK, capacity) $= 1.45474884\ m^3$	Obsolete British unit of capacity. 1 wey (UK, capacity) = 40 bushels (UK) 1 wey (UK, capacity) = 80 buckets (UK) 1 wey (UK, capacity) = 320 gallons (UK)	UK

Unit (synonym, acronym)	Symbol	Physical quantity	Dimension	Conversion factor (SI equivalent unit)	Notes, definitions, other conversion factors	System
wey (UK, weight)	–	mass	M	1 wey (UK, weight) = 114.30527724 kg (E)	Obsolete British unit of mass employed in the weighing of wool. 1 wey (UK, weight) = 252 lb (av.) (E) 1 wey (UK, weight) = 14 stones (UK, wool) (E)	UK
wheathstone	–	electric resistance	$ML^2T^{-3}I^{-2}$	1 wheathstone = 0.0025 Ω (E)	Obsolete British unit of electric resistance. It was defined in 1843. It is equal to the electric resistance of a copper wire of one foot length and which has a mass of 100 grains.	UK
whizz	–	velocity, speed	LT^{-1}	1 whizz = $3.335640952 \times 10^{-9}$ $m.s^{-1}$	Unit of velocity suggested in 1981. Anecdotal. 1 whizz = (1/c) $m.s^{-1}$ (E)	
Winchester bushel (bushel (US))	bu (US, dry)	capacity, volume	L^3	1 bushel (US, dry) = 35.23907×10^{-3} m^3	Obsolete American unit of capacity used to measure the volume of powdered or divided solid materials (flour, sand, cement, ores, etc). 1 bushel (US, dry) = 4 pecks (US, dry) (E) 1 bushel (US, dry) = 32 quart (US, dry) (E) 1 bushel (US, dry) = 64 pint (US, dry) (E) 1 bushel (US, dry) = 0.968940 bushel (UK)	US
wink	–	time, period, duration	T	1 wink = $3.3333333 \times 10^{-10}$ s	Obsolete unit of time suggested in 1957. 1 wink = (1/3000) μs (E)	

woëbe (Persian)	–	capacity, volume	L^3	1 woëbe (Persian) $= 16.3 \times 10^{-3}$ m^3 (of water)	Obsolete unit of capacity of the Assyrio-Chaldean-Persian system used in ancient times. Measured by weight. 1 woëbe (Persian) = 1/2 amphora (E)	Persian
worsted	–	specific length	$M^{-1}L$	1 worsted = 1128.907878 $m.kg^{-1}$	Obsolete American and British unit employed in the textile industry. 1 worsted = 1 lb/560 yd (E)	UK, US
X unit (CuK_{α_1}) (Siegbahn unit)	UX, X	length, distance	L	1 UX (CuK_{α_1}) $= 1.00020778970 \times 10^{-13}$ m	Obsolete unit of length used in atomic spectroscopy and X-ray diffractommetry measurements. It was introduced by K.N.G. Siegbahn (1886–1977) in 1925. It was equal to the distance between lattice planes with the Miller index (200) of a calcite crystal (Island Spath) measured by X-ray diffraction using a CuK_{α_1} spectral line. This interlattice distance is accurately equal to 3029.45 UX.	
xilon (Egyptian)	–	length, distance	L	1 xilon (Egyptian) = 2.617500 m	Obsolete Egyptian unit of length used in ancient times. 1 xilon = 3 Royal cubit (E)	Egyptian
yabiki (Japanese)	–	length, distance	L	1 yabiki (Japanese) $= 7.575757575 \times 10^{-3}$ m	Obsolete Japanese unit of length. 1 yabiki = 2.5 shaku (E)	Japanese
yan (Chinese) [yin]	–	length, distance	L	1 yin (Chinese) = 23 m	Obsolete Chinese unit of length used in ancient times. 1 yin = 100 tchi (E)	Chinese
yard	yd	length, distance	L	1 yd = 0.9144 m (E)	Legal unit in the UK system (since the WMA, 1963) and in the US system (since the USMB, 1959). 1 yard = 3 feet (E) 1 yard = 36 inches (E)	UK, US

Unit (synonym, acronym)	Symbol	Physical quantity	Dimension	Conversion factor (SI equivalent unit)	Notes, definitions, other conversion factors	System
yard (obsolete)	yd	length, distance	L	1 yard (obsolete) = 0.9143992 m	Obsolete British unit of length. It was used in UK before the WMA of 1963. 1 yard (obsolete) = 3 feet (obsolete)	UK
yard (Prussian) (ruthe)	–	length, distance	L	1 ruthe (Prussian) = 3.7662 m	Obsolete German unit of length. 1 ruthe = 12 fuss (E)	German
yard (US)	yd (US)	length, distance	L	1 yd (US) = 0.91440183 m	Obsolete American unit of length. It was adopted by the Mendenhall Order of 1893. It was equal to exactly 3600/3937 metres. Now obsolete, it was replaced by the yard described by the USMB, 1959.	US
yard (WMA 1963, scientific)	yd	length, distance	L	1 yd (WMA, 1963) = 0.9144 m (E)	Legal British and American unit of length. It was adopted by the Weights and Measures Acts of 1963 and of 1985, and the US system (since the USMB, 1959). 1 yard = 3 feet (E) 1 yard = 36 inches (E)	UK, US
yava (Indian)	–	mass	M	1 yava (Indian) = 1.468750×10^{-5} kg	Obsolete Indian unit of mass used in ancient times. 1 yava (Indian) = 3200 pala (E)	Indian
year (365.25 days) (solar mean)	a, y	time, period, duration	T	1 year (solar mean) = 3.1557600×10^{7} s	Time period which contains 365.25 days. It corresponds conventionally to the duration of revolution of the Earth around the Sun. It could be extended to other planets of the solar system.	
year (anomalistic)	$a_{anom.}$	time, period, duration	T	1 year (anomalistic) = 3.15584329×10^{7} s	Obsolete unit of time employed in astronomy. It was equal to the time between two passages of the Sun at the	

					perigee. In 1900 it was equal to 365 days 6 hours 13 minutes and 53.0 seconds. 1 year (anomalistic) $=(365.256641 + 0.27 \times 10^{-6}T)$ days, where T is the time expressed in centuries since January 1st 1900.	
year (astronomical) (Bessel year, annus fictus)	$a_{astr.}$	time, period, duration	T	1 year (Bessel) $= 3.155787559 \times 10^7$ s	1 year (Bessel) = 365.2531897 days.	
year (calendar) (civil year)	a, y	time, period, duration	T	1 year (calendar) $= 3.1536000 \times 10^7$ s	Interval of time which contains 365 or 366 entire days. It approaches the duration of the tropical year by the use of bissextile years. The calendar year begins January 1st at 0.00 h and finishes December 31st at 24.00 h. 1 year (calendar) = 365 or 366 days	
year (Gaussian)	a_{gauss}	time, period, duration	T	1 year (Gaussian) $= 3.1558368787 \times 10^7$ s	Time period which corresponds to the duration of revolution of the Earth around the Sun calculated from Kepler's law. 1 year (Gaussian) = 365.258 898 days	
year (Gregorian)	–	time, period, duration	T	1 year (Gregorian) $= 3.1557816 \times 10^7$ s	1 year (Gregorian) = 365.25 days	
year (Julian)	–	time, period, duration	T	1 year (Julian) $= 3.1556952 \times 10^7$ s	1 year (Julian) = 365.2425 days	
year (sidereal)	–	time, period, duration	T	1 year (sidereal) $= 3.155814954 \times 10^7$ s	Obsolete unit of time employed in astronomy. It was equal to the time between two conjunctions of the Sun with the same star. In 1900 it was equal exactly to 365 days 6 hours 9 minutes 9.54 seconds. *(continued overleaf)*	

Unit (synonym, acronym)	Symbol	Physical quantity	Dimension	Conversion factor (SI equivalent unit)	Notes, definitions, other conversion factors	System
year (sidereal) *(continued)*					1 year (sidereal) = $(365.256360 + 10^{-7}T)$ days, where T is the time expressed in centuries since January 1st 1900.	
year (tropical)	a_{trop}	time, period, duration	T	1 year (tropical) = $3.15569299747 \times 10^{7}$ s	Obsolete unit of time employed in astronomy. It was equal to the period of one revolution of the Earth about the Sun measured between successive mean vernal equinoxes. This time interval corresponds to the mean difference of sun longitude which is time dependent in a non-linear form. Thus the tropical year is not constant and decreases by 0.530 seconds each century. Before the introduction in 1967 of atomic time, the tropical year 1900.00 served for the definition of the second and was exactly equal to 365 days 5 hours 48 minutes and 49.97 seconds. 1 year (tropical) = $(365.242199 - 0.53 \times 10^{-6}T)$ days, where T is time expressed in centuries since January 1st 1900.	
yin (Chinese) [yan]	–	length, distance	L	1 yin (Chinese) = 23 m	Obsolete Chinese unit of length used in ancient times. 1 yin = 100 tchi (E)	Chinese
yn (Chinese)	–	length, distance	L	1 yn (Chinese) = 23 m	Obsolete Chinese unit of length used in ancient times. 1 yn = 100 chi (E)	Chinese

yo (Chinese)	–	capacity, volume	L[illegible]	1 yo (Chinese) = 3.177200 × 10[illegible] m	Obsolete Chinese unit of capacity used in ancient times. 1 yo (Chinese) = 1/2 tcheng (E)	Chinese
yoctogram	yg	mass	M	1 yg = 10^{-27} kg (E)	Submultiple of the SI base unit. 1 yg = 10^{-24} g (E)	SI
yoctometre	ym	length, distance	L	1 ym = 10^{-24} m (E)	Submultiple of the SI base unit.	SI
yodjana	–	length, distance	L	1 yodjana = 1.462400 × 10^4 m	Obsolete Indian unit of length used in ancient times. 1 yodjana = 32 000 hasta (E)	Indian
yottagram	Yg	mass	M	1 Yg = 10^{21} kg (E)	Multiple of the SI base unit. 1 Yg = 10^{24} g (E)	SI
yottametre	Ym	length, distance	L	1 Ym = 10^{24} m (E)	Multiple of the SI base unit.	SI
yugada (Spanish)	–	surface, area	L^2	1 yugada (Spanish) = 3.219780874 × 10^5 m^2	Obsolete Spanish unit of area. 1 yugada (Spanish) = 460 800 sq. vara (E)	Spanish
zak (Dutch, dry)	–	capacity, volume	L^3	1 zak (Dutch, dry) = 81.780 × 10^{-3} m^3	Obsolete Dutch unit of capacity used for dry substances. 1 zak (Dutch, dry) = 3 schepel (E)	Dutch
zentner (Austrian)	–	mass	M	1 zentner (Austrian) = 56.001 kg	Obsolete Austrian unit of mass for general use. 1 zentner (Austrian) = 100 pfund (E)	Austrian
zentner (Prussian)	–	mass	M	1 zentner = 51.448210 kg (E)	Obsolete German unit of mass. 1 zentner = 110 pfund (E)	German
zeptogram	zg	mass	M	1 zg = 10^{-24} kg (E)	Submultiple of the SI base unit. 1 zg = 10^{-21} g (E)	SI
zeptometre	zm	length, distance	L	1 zm = 10^{-21} m (E)	Submultiple of the SI base unit.	SI

Unit (synonym, acronym)	Symbol	Physical quantity	Dimension	Conversion factor (SI equivalent unit)	Notes, definitions, other conversion factors	System
zereth (Egyptian) [Royal Egyptian foot]	–	length, distance	L	1 zereth (Egyptian) = 3.490×10^{-1} m	Obsolete Egyptian unit of length used in ancient times. 1 zereth = 2/3 Royal cubit (E)	Egyptian
zereth (Hebrew)	–	length, distance	L	1 zereth (Hebrew) = 2.775×10^{-1} m	Obsolete Hebrew unit of length used in ancient times. 1 zereth = 1/2 cubit (E)	Hebrew
zerteh (Persian) [Babylonian foot]	–	length, distance	L	1 zerteh (Persian) = 0.320 m	Obsolete unit of length in the Assyrio-Chaldean-Persian system used in ancient times. Base unit of the system of length.	Persian
zettagram	Zg	mass	M	1 Zg = 10^{18} kg (E)	Multiple of the SI base unit. 1 Zg = 10^{21} g (E)	SI
zettametre	Zm	length, distance	L	1 Zm = 10^{21} m (E)	Multiple of the SI base unit.	SI
zhang (Chinese)	–	length, distance	L	1 zhang (Chinese) = 2.3 m	Obsolete Chinese unit of length used in ancient times. 1 zhang = 10 chi (E)	Chinese
zhu	–	mass	M	1 zhu = $1.041666667 \times 10^{-3}$ kg	Obsolete Chinese unit of mass. 1 zhu = 1/24 jin (E)	Chinese
zoll (Austrian) [Austrian inch]	–	length, distance	L	1 zoll (Austrian) = 2.634000×10^{-2} m	Obsolete Austrian unit of length. 1 zoll (Austrian) = 1/12 fuss (E)	Austrian
zoll (German) [Prussian inch]	–	length, distance	L	1 zoll (Prussian) = $2.615475000 \times 10^{-2}$ m	Obsolete German unit of length. 1 zoll = 1/12 fuss (E)	German

zollpfund (Austrian)	–	mass	M	1 zollpfund (Austrian) $= 560.00448 \times 10^{-3}$ kg	Obsolete Austrian unit of weight. 1 kg = 1.7857 zollpund (E)	Austrian
zollpfund (Prussian)	–	mass	M	1 zollpfund (Prussian) $= 514.482100 \times 10^{-3}$ kg	Obsolete German unit of weight. 1 zollpund = 2 pfund (E)	German
zollpfund (Swiss)	–	mass	M	1 zollpfund (Swiss) $= 500 \times 10^{-3}$ kg	Obsolete Swiss unit of weight. 1 kg = 2 zollpund (E)	Swiss
zuzah (Hebrew) [Talmudic system]	–	mass	M	1 zuzah (Hebrew) $= 3.542000 \times 10^{-3}$ kg	Obsolete Hebrew unit of mass used in ancient times. Rabbinacal or Talmudic system. 1 zuzah (Hebrew) = 1/100 mina (E)	Hebrew

4.2 Conversion Tables Listed by Physical Quantities

Note: Conversion factors shown in **bold type** are exact values.

Table 4.2.1.1 Conversion table: units of mass: multiples and submultiples (SI)

Unit	Symbol	yg	zg	ag	fg	pg	ng	µg	mg	cg
yottagram	Yg	**1E+48**	**1E+45**	**1E+42**	**1E+39**	**1E+36**	**1E+33**	**1E+30**	**1E+27**	**1E+26**
zettagram	Zg	**1E+45**	**1E+42**	**1E+39**	**1E+36**	**1E+33**	**1E+30**	**1E+27**	**1E+24**	**1E+23**
exagram	Eg	**1E+42**	**1E+39**	**1E+36**	**1E+33**	**1E+30**	**1E+27**	**1E+24**	**1E+21**	**1E+20**
petagram	Pg	**1E+39**	**1E+36**	**1E+33**	**1E+30**	**1E+27**	**1E+24**	**1E+21**	**1E+18**	**1E+17**
teragram	Tg	**1E+36**	**1E+33**	**1E+30**	**1E+27**	**1E+24**	**1E+21**	**1E+18**	**1E+15**	**1E+14**
gigagram	Gg	**1E+33**	**1E+30**	**1E+27**	**1E+24**	**1E+21**	**1E+18**	**1E+15**	**1E+12**	**1E+11**
megagram	Mg	**1E+30**	**1E+27**	**1E+24**	**1E+21**	**1E+18**	**1E+15**	**1E+12**	**1E+09**	**1E+08**
kilogram	kg	**1E+27**	**1E+24**	**1E+21**	**1E+18**	**1E+15**	**1E+12**	**1E+09**	**1E+06**	**1E+05**
hectogram	hg	**1E+26**	**1E+23**	**1E+20**	**1E+17**	**1E+14**	**1E+11**	**1E+08**	**1E+05**	**1E+04**
decagram	dag	**1E+25**	**1E+22**	**1E+19**	**1E+16**	**1E+13**	**1E+10**	**1E+07**	**1E+04**	**1E+03**
gram	g	**1E+24**	**1E+21**	**1E+18**	**1E+15**	**1E+12**	**1E+09**	**1E+06**	**1E+03**	**1E+02**
decigram	dg	**1E+23**	**1E+20**	**1E+17**	**1E+14**	**1E+11**	**1E+08**	**1E+05**	**1E+02**	**1E+01**
centigram	cg	**1E+22**	**1E+19**	**1E+16**	**1E+13**	**1E+10**	**1E+07**	**1E+04**	**1E+01**	**1**
milligram	mg	**1E+21**	**1E+18**	**1E+15**	**1E+12**	**1E+09**	**1E+06**	**1E+03**	**1**	**1E-01**
microgram	µg	**1E+18**	**1E+15**	**1E+12**	**1E+09**	**1E+06**	**1E+03**	**1**	**1E-03**	**1E-04**
nanogram	ng	**1E+15**	**1E+12**	**1E+09**	**1E+06**	**1E+03**	**1**	**1E-03**	**1E-06**	**1E-07**
picogram	pg	**1E+12**	**1E+09**	**1E+06**	**1E+03**	**1**	**1E-03**	**1E-06**	**1E-09**	**1E-10**
femtogram	fg	**1E+09**	**1E+06**	**1E+03**	**1**	**1E-03**	**1E-06**	**1E-09**	**1E-12**	**1E-13**
attogram	ag	**1E+06**	**1E+03**	**1**	**1E-03**	**1E-06**	**1E-09**	**1E-12**	**1E-15**	**1E-16**
zeptogram	zg	**1E+03**	**1**	**1E-03**	**1E-06**	**1E-09**	**1E-12**	**1E-15**	**1E-18**	**1E-19**
yoctogram	yg	**1**	**1E-03**	**1E-06**	**1E-09**	**1E-12**	**1E-15**	**1E-18**	**1E-21**	**1E-22**

dg	g	dag	hg	kg	Mg	Gg	Tg	Pg	Eg	Zg	Yg
E+25	1E+24	1E+23	1E+22	1E+21	1E+18	1E+15	1E+12	1E+09	1E+06	1E+03	1
E+22	1E+21	1E+20	1E+19	1E+18	1E+15	1E+12	1E+09	1E+06	1E+03	1	1E-03
E+19	1E+18	1E+17	1E+16	1E+15	1E+12	1E+09	1E+06	1E+03	1	1E-03	1E-06
E+16	1E+15	1E+14	1E+13	1E+12	1E+09	1E+06	1E+03	1	1E-03	1E-06	1E-09
E+13	1E+12	1E+11	1E+10	1E+09	1E+06	1E+03	1	1E-03	1E-06	1E-09	1E-12
E+10	1E+09	1E+08	1E+07	1E+06	1E+03	1	1E-03	1E-06	1E-09	1E-12	1E-15
E+07	1E+06	1E+05	1E+04	1E+03	1	1E-03	1E-06	1E-09	1E-12	1E-15	1E-18
E+04	1E+03	1E+02	1E+01	1	1E-03	1E-06	1E-09	1E-12	1E-15	1E-18	1E-21
E+03	1E+02	1E+01	1	1E-01	1E-04	1E-07	1E-10	1E-13	1E-16	1E-19	1E-22
E+02	1E+01	1	1E-01	1E-02	1E-05	1E-08	1E-11	1E-14	1E-17	1E-20	1E-23
E+01	1	1E-01	1E-02	1E-03	1E-06	1E-09	1E-12	1E-15	1E-18	1E-21	1E-24
	1E-01	1E-02	1E-03	1E-04	1E-07	1E-10	1E-13	1E-16	1E-19	1E-22	1E-25
E-01	1E-02	1E-03	1E-04	1E-05	1E-08	1E-11	1E-14	1E-17	1E-20	1E-23	1E-26
E-02	1E-03	1E-04	1E-05	1E-06	1E-09	1E-12	1E-15	1E-18	1E-21	1E-24	1E-27
E-05	1E-06	1E-07	1E-08	1E-09	1E-12	1E-15	1E-18	1E-21	1E-24	1E-27	1E-30
E-08	1E-09	1E-10	1E-11	1E-12	1E-15	1E-18	1E-21	1E-24	1E-27	1E-30	1E-33
E-11	1E-12	1E-13	1E-14	1E-15	1E-18	1E-21	1E-24	1E-27	1E-30	1E-33	1E-36
E-14	1E-15	1E-16	1E-17	1E-18	1E-21	1E-24	1E-27	1E-30	1E-33	1E-36	1E-39
E-17	1E-18	1E-19	1E-20	1E-21	1E-24	1E-27	1E-30	1E-33	1E-36	1E-39	1E-42
E-20	1E-21	1E-22	1E-23	1E-24	1E-27	1E-30	1E-33	1E-36	1E-39	1E-42	1E-45
E-23	1E-24	1E-25	1E-26	1E-27	1E-30	1E-33	1E-36	1E-39	1E-42	1E-45	1E-48

Table 4.2.1.2 Conversion table: units of mass (>1 kg)

Unit	Symbol	ton (UK)	lg ton (US)	ton (metric)	sh. ton (US)	kip	quintal (metric)	cwt	cental (cH)
ton (UK, shipping)	ton (UK, ship.)	1.151224545	1.151224545	1.169698141	1.289371491	2.578742982	11.696981410	23.024488847	25.787429824
ton (US, long)	lg ton (US)	1	1	1.016046909	1.12	2.24	10.160469088	19.999998205	22.4
ton (UK, short)	ton (UK)	1	1	1.016046909	1.12	2.24	10.160469088	19.999998205	22.4
ton (metric)	t	9.84206527E-01	9.84206527E-01	1	1.102311311	2.204622622	10	19.684128785	22.046226218
ton (US, short)	sh. ton (US)	8.92857143E-01	8.92857143E-01	9.07184740E-01	1	2	9.071847400	17.857141254	20
load (UK)	load (UK)	5.78571428E-01	5.78571428E-01	5.87855712E-01	81/125	162/125	5.878557115	11.571427533	12.960000000
kip (kilopound)	kip	4.46428571E-01	4.46428571E-01	0.45359237	1/2	1	4.535923700	8.928570627	10
quarter (US, long)	lg qt (US)	1/4	1/4	95/374	7/25	14/25	2.540120000	5.000004921	5.600006014
quarter (US, short)	sh. qt (US)	2.23214104E-01	2.23214104E-01	22/97	1/4	1/2	2.267960000	4.464281672	4.999995921
barrel (UK, cement)	bbl (UK, cem.)	1.67857143E-01	1.67857143E-01	1.70550731E-01	1.880E-01	3.76E-01	1.705507311	3.357142555	3 19/25
barrel (UK, salt)	bbl (UK, salt)	1.24994806E-01	1.24994806E-01	8/63	1.39994183E-01	2.79988366E-01	1.270005864	2.499895898	2.799883657
wey (UK)	wy (UK)	9/80	9/80	1.14305277E-01	1.260E-01	2.52E-01	1.143052772	2.249999798	2 13/25
quintal (metric)	q, dt	9.84206527E-02	9.84206527E-02	1E-01	1.10231131E-01	2.20462262E-01	1	1.968412879	2.204622622
hundredweight (gross or long)	cwt, lg cwt	5E-02	5E-02	5.08023500E-02	7/125	1.12000010E-01	5.08023500E-01	1	1.120000101
cental (centner, kintal)	sh. cwt, cH	4.46428571E-02	4.46428571E-02	4.5359237E-02	5E-02	1E-01	4.5359237E-01	8.92857063E-01	1
hundredweight (net or short)	sh. cwt	4.46428571E-02	4.46428571E-02	4.5359237E-02	5E-02	1E-01	4.5359237E-01	8.92857063E-01	1
quintal (US)	quint (US)	4.46428571E-02	4.46428571E-02	4.5359237E-02	5E-02	1E-01	4.5359237E-01	8.92857063E-01	1
bag (cement)	bg (UK, cem.)	4.19642857E-02	4.19642857E-02	4.26376828E-02	4/85	3/32	4.26376828E-01	8.39285639E-01	9.40E-01
pood (Russian)	pood	1.61217950E-02	1.61217950E-02	1/61	1.80564104E-02	3.61128209E-02	10/61	3.22435872E-01	3.61128209E-01
truss (UK)	truss (UK)	1.60714286E-02	1.60714286E-02	1.63293253E-02	9/500	9/250	1.632932532E-01	3.21428543E-01	9/25
slug (geepound)	slug	1.43634116E-02	1.43634116E-02	1/69	1.60870210E-02	3.21740421E-02	7/48	2.87268207E-01	3.21740421E-01
quarter (UK)	qt (UK)	1.25E-02	1.25E-02	1.27005864E-02	1/71	1/36	1.27005864E-01	2.49999978E-01	2.8E-01
myriagram	myg	9.84206527E-03	9.84206527E-03	1E-02	1.10231131E-02	2.20462262E-02	1E-01	1.96841288E-01	2.20462262E-01
hyl (techma, UdM, TME, mug, par)	hyl, TME	9.65176894E-03	9.65176894E-03	9.80665E-03	1.08099812E-02	2.16199624E-02	9.80665E-02	1.93035362E-01	2.16199624E-01
stone (UK)	st (UK)	6.25E-03	6.25E-03	6.35029318E-03	7E-03	1/71	6.35029318E-02	1.24999989E-01	1.40E-01
stone (UK, wool)	st (UK, wool)	5.58035714E-03	5.58035714E-03	5.66990463E-03	1/160	1/80	5.669904625E-02	1.11607133E-01	1/8
clove (UK)	cl (UK)	3.57142857E-03	3.57142857E-03	3.62873896E-03	4E-03	8E-03	3.62873896E-02	7.14285650E-02	8E-02
kilogram	kg	[illegible]	[illegible]	[illegible]	[illegible]	[illegible]	[illegible]	[illegible]	[illegible]

Table 4.2.1.2 (continued)

Unit	Symbol	cental (cH)	pood	slug (geepound)	quarter (UK)	hyl (techma)	stone (UK)	kg	lb (1963)
ton (UK, shipping)	**ton (UK, ship.)**	25.787429824	71.407963188	80.149798272	92.09796366	119.276015867	184.195927313	1169.698141000	2578.742982383
ton (US, long)	**lg ton (US)**	**22.4**	62.027832410	69.621342396	**80**	103.607950605	**160**	1016.046908800	**2240**
ton (UK, short)	**ton (UK)**	**22.4**	62.027832410	69.621342396	**80**	103.607950605	**160**	1016.046908800	**2240**
ton (metric)	**t**	**22.046226218**	61.048197552	68.521779648	78.736522209	101.971621298	157.473044418	**1000**	2204.622621849
ton (US, short)	**sh. ton (US)**	**20**	55.381993224	62.161912854	71.428571429	92.507098754	142.857142857	907.184740000	**2000**
load (UK)	**load (UK)**	12.960000000	35.887531609	40.280919529	46.285714286	59.944599993	92.571428571	587.855711520	**1296**
kip (kilopound)	**kip**	**10**	27.690996612	31.080956427	35.714285714	46.253549377	71.428571429	453.59237	**1000**
quarter (US, long)	**lg qt (US)**	5.600006014	15.506974757	17.405354292	20.000021479	25.902015469	**40**	254.011727200	**560**
quarter (US, short)	**sh. qt (US)**	4.999995921	13.845487012	15.540465537	17.857128291	23.126755824	35.714285714	226.796185000	**500**
barrel (UK, cement)	**bbl (UK, cem.)**	3.76	10.411814725	11.686439615	13.428571427	17.391334564	26.857142854	170.550731100	**376**
barrel (UK, salt)	**bbl (UK, salt)**	2.799883657	7.753156885	8.702306194	9.999584488	12.950455697	19.999168977	127.000586360	**280**
wey (UK)	**wy (UK)**	2.52	6.978131146	7.832401020	**9**	11.655894443	**18**	114.305277240	**252**
quintal (metric)	**q, dt**	2.204622622	6.104819755	6.852177965	7.873652221	10.197162130	15.747304442	**100**	220.462262185
hundredweight (gross or long)	**cwt, lg cwt**	1.120000101	3.101391899	3.481067432	4.000000359	5.180397995	8.000000718	**50.80235**	**112**
cental (centner, kintal)	**sh. cwt, cH**	**1**	2.769099661	3.108095643	3.571428571	4.625354938	7.142857143	**45.359237**	**100**
hundredweight (net or short)	**sh. cwt**	**1**	2.769099661	3.108095643	3.571428571	4.625354938	7.142857143	**45.359237**	**100**
quintal (US)	**quint (US)**	**1**	2.769099661	3.108095643	3.571428571	4.625354938	7.142857143	**45.359237**	**100**
bag (cement)	**bg (UK, cem.)**	9.40E-01	2.602953682	2.921609904	3.357142857	4.347833641	6.714285714	42.637682780	**94**
pood (Russian)	**pood**	3.61128209E-01	**1**	1.122421012	1.289743602	1.670346143	2.579487204	16.380500000	36.112820857
truss (UK)	**truss (UK)**	9/25	9.96875878E-01	1.118914431	1.285714286	1.665127778	2.571428571	16.329325320	**36**
slug (geepound)	**slug**	3.21740421E-01	8.90931290E-01	**1**	1.149072931	1.488163644	2.298146326	14.593902940	32.174042081
quarter (UK)	**qt (UK)**	**2.8E-01**	7.75347905E-01	8.70266780E-01	**1**	1.295099383	2	12.700586360	**28**
myriagram	**myg**	2.20462262E-01	6.10481976E-01	6.85217796E-01	7.87365222E-01	1.019716213	1.574730444	**10**	22.046226218
hyl (techma, UdM, TME, mug, par)	**hyl, TME**	2.16199624E-01	5.98678307E-01	6.71969110E-01	7.72141516E-01	1	1.544283031	**9.80665**	21.619962435
stone (UK)	**st (UK)**	**1.40E-01**	3.87673953E-01	4.35133390E-01	**5E-01**	6.47549691E-01	1	6.350293180	**14**
stone (UK, wool)	**st (UK, wool)**	**1/8**	3.46137458E-01	3.88511955E-01	4.46428571E-01	5.78169367E-01	8.92857143E-01	5.669904625	**12.5**
clove (UK)	**cl (UK)**	**8E-02**	2.21527973E-01	2.48647651E-01	2.85714286E-01	3.70028395E-01	5.71428571E-01	3.62873896	**8**
kilogram	**kg**	2.20462262E-02	6.10481976E-02	6.85217796E-02	7.87365222E-02	1.01971621E-01	1.57473044E-01	**1**	2.204622622

Table 4.2.1.3 Conversion table: units of mass (<1 kg)

Unit	Symbol	kg	lb (av., old)	lb (av., 1963)
kilogram	kg	1	2.2046223414	2.2046226218
livre (de Paris)	livre	4.89510000E-01	1.0791846823	1.0791848196
pound (avoirdupois)	lb (av.)	4.53592428E-01	1	1.0000001272
pound (WMA, 1963)	lb (1963)	0.45359237	9.99999873E-01	1
pound (imperial standard)	lb (IS)	4.53592338E-01	9.99999802E-01	9.99999929E-01
pound (US)	lb (US)	4.53592243E-01	9.99999592E-01	9.99999720E-01
pound (apothecary, troy)	lb (troy)	3.73241722E-01	8.22856991E-01	8.22857095E-01
marc (de Paris)	marc	2.44755000E-01	5.39592341E-01	5.39592410E-01
quarteron (de Paris)	quarteron	1.22377500E-01	2.69796171E-01	2.69796205E-01
hectogram	hg	1E-01	2.20462234E-01	2.20462262E-01
ton (assay UK)	AT (UK)	3.26666700E-02	7.20176705E-02	7.20176797E-02
ounce (apothecary, troy)	oz (troy)	3.11034768E-02	6.85714198E-02	6.85714286E-02
once (de Paris)	once	3.05943750E-02	6.74490426E-02	6.74490512E-02
ton (assay US)	AT (US)	2.91666700E-02	6.43014923E-02	6.43015005E-02
ounce (avoirdupois)	oz (av.)	2.83495231E-02	6.24999920E-02	1/16
lot (de Paris)	lot	1.52971875E-02	3.37245213E-02	3.37245256E-02
dram (troy)	dr (troy)	3.88793460E-03	8.57142748E-03	3/350
drachm (apothecary)	dr (ap.)	3.88793460E-03	8.57142748E-03	3/350
gros (de Paris)	gros	3.82429688E-03	8.43113033E-03	8.43113140E-03
momme (Japan, pearl)	momme	3.75E-03	8.26733378E-03	8.26733483E-03
dram (avoirdupois)	dr (av.)	1.77184520E-03	3.90624950E-03	1/256
scruple (av.)	scr (av.)	1.57497351E-03	3.47222178E-03	1/288
pennyweight (troy)	dwt (troy)	1.55517384E-03	3.42857099E-03	3.42857143E-03
denier (de Paris)	denier	1.27476563E-03	2.81037678E-03	2.81037713E-03
gram	g	1E-03	2.20462234E-03	2.20462262E-03
glug	glug	1E-03	2.20462234E-03	2.20462262E-03
carat (old)	ct (old)	2.05E-04	4.51947580E-04	4.51947637E-04
carat (metric)	ct	2E-04	4.40924468E-04	4.40924524E-04
crith	crith	8.92295583E-05	1.96717478E-04	1.96717503E-04
grain (avdp, troy, apoth.)	gr (av.)	6.47989100E-05	1.42857125E-04	1/7000
grain (de Paris)	grain	5.31152344E-05	1.17099032E-04	1.17099047E-04
grain (jeweller's)	gr (jew.)	5E-05	1.10231117E-04	1.10231131E-04
centigram	cg	1E-05	2.20462234E-05	2.20462262E-05
milligram	mg	1E-06	2.20462234E-06	2.20462262E-06
microgram	μg	1E-09	2.20462234E-09	2.20462262E-09
gamma	γ	1E-09	2.20462234E-09	2.20462262E-09
milligamma	γγ	1E-12	2.20462234E-12	2.20462262E-12
microgamma	γγγ	1E-15	2.20462234E-15	2.20462262E-15
dalton	Da	1.66057000E-27	3.66092972E-27	3.66093019E-27
atomic mass unit (^{12}C)	u(^{12}C)	1.66054018E-27	3.66086399E-27	3.66086445E-27
atomic mass unit (^{16}O)	u(^{16}O)	1.66001241E-27	3.65970044E-27	3.65970090E-27
atomic mass unit (^{1}H)	u(^{1}H)	1.67353397E-27	3.68951038E-27	3.68951085E-27
a.u. of mass	m_e	9.10938970E-31	9.10938970E-31	9.10938970E-31

lb (US)	lb (troy)	ton (UK, assay)	oz (troy)	ton (US, assay)
.2046232402	2.6792290358	30.6122417743	32.1507465686	34.2857103673
.0791851223	1.3115094053	14.9849984709	15.7381119528	16.7831980819
.0000004077	1.2152780027	13.8854810637	14.5833351884	15.5517386009
.0000002805	1.2152778481	13.8854792974	14.5833333333	15.5517366227
.0000002099	1.2152777624	13.8854783178	14.5833323045	15.5517355255
	1.2152775072	13.8854754026	14.5833292428	15.5517322605
.22857326E-01	1	11.4257658219	12	12.7968575638
.39592561E-01	6.55754703E-01	7.4924992355	7.8690559764	8.3915990410
.69796281E-01	3.27877351E-01	3.7462496177	3.9345279882	4.1957995205
.20462324E-01	2.67922904E-01	3.0612241774	3.2150746569	3.4285710367
20176999E-02	8.75214908E-02	1	1.0502578284	1.1199999863
85714478E-02	**1/12**	9.521471518E-01	1	1.0664047970
74490701E-02	8.19693378E-02	9.365624044E-01	9.83631997E-01	1.0489498801
43015185E-02	7.81441891E-02	8.928571538E-01	9.37730215E-01	1
25000175E-02	7.59548655E-02	8.678424561E-01	9.11458333E-01	9.71983539E-01
.37245351E-02	4.09846689E-02	4.682812022E-01	4.91815999E-01	5.24474940E-01
.57143098E-03	**1/96**	1.190183940E-01	**1/8**	1.33300600E-01
.57143098E-03	**1/96**	1.190183940E-01	**1/8**	1.33300600E-01
.43113377E-03	1.02461672E-02	1.170703006E-01	1.22954000E-01	1.31118735E-01
.26733715E-03	1.00471089E-02	1.147959067E-01	1.20565300E-01	1.28571414E-01
90625110E-03	4.74717909E-03	5.424015351E-02	5.69661458E-02	6.07489712E-02
.47222320E-03	4.21971475E-03	4.821346978E-02	5.06365741E-02	5.39990855E-02
.42857239E-03	**1/240**	4.760735759E-02	**1/20**	5.33202398E-02
.81037792E-03	3.41538908E-03	3.902343352E-02	4.09846665E-02	4.37062450E-02
20462324E-03	2.67922904E-03	3.061224177E-02	3.21507466E-02	3.42857104E-02
20462324E-03	2.67922904E-03	3.061224177E-02	3.21507466E-02	3.42857104E-02
51947764E-04	5.49241952E-04	6.275509564E-03	6.59090305E-03	7.02857063E-03
40924648E-04	5.35845807E-04	6.122448355E-03	6.43014931E-03	6.85714207E-03
96717558E-04	2.39066423E-04	2.731516813E-03	2.86879692E-03	3.05929879E-03
42857183E-04	**1/5760**	1.983639900E-03	**1/480**	2.22167666E-03
.17099080E-04	1.42307878E-04	1.625976397E-03	1.70769444E-03	1.82109354E-03
.10231162E-04	1.33961452E-04	1.530612089E-03	1.60753733E-03	1.71428552E-03
.20462324E-05	2.67922904E-05	3.061224177E-04	3.21507466E-04	3.42857104E-04
20462324E-06	2.67922904E-06	3.061224177E-05	3.21507466E-05	3.42857104E-05
20462324E-09	2.67922904E-09	3.061224177E-08	3.21507466E-08	3.42857104E-08
20462324E-09	2.67922904E-09	3.061224177E-08	3.21507466E-08	3.42857104E-08
20462324E-12	2.67922904E-12	3.061224177E-11	3.21507466E-11	3.42857104E-11
20462324E-15	2.67922904E-15	3.061224177E-14	3.21507466E-14	3.42857104E-14
66093121E-27	4.44904736E-27	5.083377032E-26	5.33885652E-26	5.69338221E-26
66086548E-27	4.44896748E-27	5.083285759E-26	5.33876066E-26	5.69327998E-26
65970193E-27	4.44755344E-27	5.081670111E-26	5.33706382E-26	5.69147045E-26
68951188E-27	4.48378080E-27	5.123062648E-26	5.38053665E-26	5.73783010E-26
10938970E-31	9.10938970E-31	9.109389700E-31	9.10938970E-31	9.10938970E-31

(Continued overleaf)

Table 4.2.1.3 *(continued)* Conversion table: units of mass (<1 kg)

Unit	Symbol	ton (US, assay)	oz (av.)	dr (troy)
kilogram	kg	34.2857103673	35.2739619496	257.2059725490
livre (de Paris)	livre	16.7831980819	17.2669571139	125.9048956225
pound (avoirdupois)	lb (av.)	15.5517386009	16.0000020353	116.6666815075
pound (WMA, 1963)	lb (1963)	15.5517366227	16	116.6666666667
pound (imperial standard)	lb (IS)	15.5517355255	15.9999988712	116.6666584361
pound (US)	lb (US)	15.5517322605	15.9999955121	116.6666339424
pound (apothecary, troy)	lb (troy)	12.7968575638	13.1657135238	96
marc (de Paris)	marc	8.3915990410	8.6334785570	62.9524478112
quarteron (de Paris)	quarteron	4.1957995205	4.3167392785	31.4762239056
hectogram	hg	3.4285710367	3.5273961950	25.7205972549
ton (assay UK)	AT (UK)	1.1199999863	1.1522828746	8.4020626273
ounce (apothecary, troy)	oz (troy)	1.0664047970	1.0971428571	8
once (de Paris)	once	1.0489498801	1.0791848196	7.8690559764
ton (assay US)	AT (US)	1	1.0288240078	7.5018417234
ounce (avoirdupois)	oz (av.)	9.71983539E-01	1	175/24
lot (de Paris)	lot	5.24474940E-01	5.39592410E-01	3.9345279882
dram (troy)	dr (troy)	1.33300600E-01	1.37142857E-01	1
drachm (apothecary)	dr (ap.)	1.33300600E-01	1.37142857E-01	1
gros (de Paris)	gros	1.31118735E-01	1.34898102E-01	9.83631997E-01
momme (Japan, pearl)	momme	1.28571414E-01	1.32277357E-01	9.64522397E-01
dram (avoirdupois)	dr (av.)	6.07489712E-02	1/16	4.55729167E-01
scruple (av.)	scr (av.)	5.39990855E-02	1/18	4.05092593E-01
pennyweight (troy)	dwt (troy)	5.33202398E-02	5.48571429E-02	2/5
denier (de Paris)	denier	4.37062450E-02	4.49660342E-02	3.27877332E-01
gram	g	3.42857104E-02	3.52739619E-02	2.57205973E-01
glug	glug	3.42857104E-02	3.52739619E-02	2.57205973E-01
carat (old)	ct (old)	7.02857063E-03	7.23116220E-03	5.27272244E-02
carat (metric)	ct	6.85714207E-03	7.05479239E-03	5.14411945E-02
crith	crith	3.05929879E-03	3.14748004E-03	2.29503753E-02
grain (avdp, troy, apoth.)	gr (av.)	2.22167666E-03	2/875	1/60
grain (de Paris)	grain	1.82109354E-03	1.87358476E-03	1.36615555E-02
grain (jeweller's)	gr (jew.)	1.71428552E-03	1.76369810E-03	1.28602986E-02
centigram	cg	3.42857104E-04	3.52739619E-04	2.57205973E-03
milligram	mg	3.42857104E-05	3.52739619E-05	2.57205973E-04
microgram	μg	3.42857104E-08	3.52739619E-08	2.57205973E-07
gamma	γ	3.42857104E-08	3.52739619E-08	2.57205973E-07
milligamma	γγ	3.42857104E-11	3.52739619E-11	2.57205973E-10
microgamma	γγγ	3.42857104E-14	3.52739619E-14	2.57205973E-13
dalton	Da	5.69338221E-26	5.85748830E-26	4.27108522E-25
atomic mass unit (^{12}C)	u(^{12}C)	5.69327998E-26	5.85738313E-26	4.27107335E-25
atomic mass unit (^{16}O)	u(^{16}O)	5.69147045E-26	5.85552144E-26	4.26971586E-25
atomic mass unit (^{1}H)	u(^{1}H)	5.73783010E-26	5.90321735E-26	4.30449465E-25
a.u. of mass	m_e	9.10938970E-31	9.10938970E-31	9.10938970E-31

dr (av.)	dwt (troy)	scruple (UK)	gram	grain (av.)
64.3833911933	643.0149313726	634.9313150924	**1000**	15432.358352941
76.2713138230	314.7622390562	310.8052280509	489.5100000000	7554.2937373484
56.0000325649	291.6667037686	288.0000366355	453.5924277000	7000.0008904471
56	291.6666666667	**288**	453.59237	**7000**
55.9999819397	291.6666460902	287.9999796822	453.5923380000	6999.9995061645
55.9999281935	291.6665848559	287.9999192177	453.5922427700	6999.9980365411
10.6514163807	**240**	236.9828434283	373.2417000000	**5760**
38.1356569115	157.3811195281	155.4026140255	244.7550000000	3777.1468686742
9.0678284558	78.6905597640	77.7013070127	122.3775000000	1888.5734343371
6.4383391193	64.3014931373	63.4931315092	**100**	1543.2358352941
8.4365259936	21.0051565682	20.7410917428	32.6666700000	504.1237576373
7.5542857143	**20**	**20**	31.1034768000	**480**
7.2669571139	19.6726399410	19.4253267532	30.5943750000	4.721433586E+02
6.4611841244	18.7546043084	18.5188321400	29.1666700000	450.1105034020
6	18.2291666667	18	28.3495231250	**437.5**
.6334785570	9.8363199705	9.7126633766	15.2971875000	236.0716792921
.1942857143	**5/2**	2	3.8879346000	**60**
.1942857143	**5/2**	2	3.8879346000	**60**
.1583696392	2.4590799926	2.4281658441	3.8242968750	59.0179198230
.1164377170	2.4113059926	2.3809924316	3.75	5.787134382E+01
[illegible]	1.1393229167	**9/8**	1.7718451953	**875/32**
3/9	1.0127314815	1	1.5749735069	24.305555556
3.77714286E-01	1	9.87428571E-01	1.5551738400	**24**
7.19456546E-01	8.19693331E-01	8.09388615E-01	1.2747656250	19.6726399410
5.64383391E-01	6.43014931E-01	6.34931315E-01	1	15.4323583529
5.64383391E-01	6.43014931E-01	6.34931315E-01	1	15.4323583529
.15698595E-01	1.31818061E-01	1.30160920E-01	2.05E-01	3.1636334624
.12876678E-01	1.28602986E-01	1.26986263E-01	**2E-01**	3.0864716706
5.03596807E-02	5.73759383E-02	5.66546408E-02	8.92295583E-02	1.377022520E+00
32/875	**1/24**	**3/73**	6.47989100E-02	1
2.99773561E-02	3.41538888E-02	3.37245256E-02	5.31152344E-02	8.196933309E-01
2.82191696E-02	3.21507466E-02	3.17465658E-02	**5E-02**	7.716179176E-01
5.64383391E-03	6.43014931E-03	6.34931315E-03	**1E-02**	1.543235835E-01
5.64383391E-04	6.43014931E-04	6.34931315E-04	**1E-03**	1.543235835E-02
5.64383391E-07	6.43014931E-07	6.34931315E-07	**1E-06**	1.543235835E-05
5.64383391E-07	6.43014931E-07	6.34931315E-07	**1E-06**	1.543235835E-05
5.64383391E-10	6.43014931E-10	6.34931315E-10	**1E-09**	1.543235835E-08
5.64383391E-13	6.43014931E-13	6.34931315E-13	**1E-12**	1.543235835E-11
9.37198128E-25	1.06777130E-24	1.05434789E-24	1.66057000E-24	2.562651131E-23
9.37195524E-25	1.06776834E-24	1.05434496E-24	1.66056539E-24	2.562644012E-23
9.36897651E-25	1.06742896E-24	1.05400986E-24	1.66003760E-24	2.561829513E-23
9.44529112E-25	1.07612366E-24	1.06259525E-24	1.67355937E-24	2.582696791E-23
9.10938970E-31	9.10938970E-31	9.10938970E-31	9.10938970E-31	9.109389700E-31

Table 4.2.1.4 Conversion table for atomic units of mass

The atomic mass unit is defined as follows: it is the *A*th part of the mass of a atom $^{A}_{Z}X$ of which the nuclide has a mass number *A*. It is defined by *u* (the ol symbol u.m.a. is definitely obsolete even if it is still often employed). Fo historical reasons, it was first defined from the nuclide isotope 1 of th hydrogen atom ($^{1}_{1}H$), and later from the isotope 16 of the oxygen ato ($^{16}_{8}O$). It was the chemical system of microscopic mass. Finally, it wa definitively set up from the mass of the isotope 12 of the carbon ato ($^{12}_{6}C$). It is the physical system of mass.

It should be noticed that the choice of the isotope of an element as th atomic mass unit has a direct impact on the numerical values of the atom masses of chemical elements and above all the numerical values of funda mental physical constants (e.g. Avogadro's number, Faraday's constant, Boltz man's constant, etc). The deviations between the atomic masses depending o the choice of the standard isotope are negligible when dealing with elementar chemical computations (e.g. molecular mass calculus, etc). However, it i completely different when dealing with accurate computation as in nuclea physics (e.g. packing fraction, mass excess and binding energy computations Generally speaking, atomic mass unit tables that one can find in old textbook often indicate which isotope is taken as the standard reference. In case c doubt, one should know either at which date the table was compiled or kno the accurate value of the atomic mass of one element in the different system

Conversion factors between atomic mass units (u)			
	$u(^{1}H)$	$u(^{12}C)$	$u(^{16}O)$
$u(^{1}H)$	1	1.0078250351	1.0081454593
$u(^{12}C)$	0.9922357206	1	1.0003179364
$u(^{16}O)$	0.9919203531	0.9996821647	1

Atomic mass of the three standard elements in the three systems			
	$u(^{1}H)$	$u(^{12}C)$	$u(^{16}O)$
$m(^{1}H)$	1	1.0078250351	1.0081454593
$m(^{12}C)$	11.9068286477	12	12.0038152364
$m(^{16}O)$	15.8707256494	15.9949146350	16

Table 4.2.2.1 Conversion table: units of length: multiples and submultiples (SI)

Unit	Symbol	Ym	Zm	Em	Pm	Tm	Gm	Mm	km	hm	dam	m	dm	cm	mm	μm	nm	pm	fm	am	zm	ym
yottametre	Ym	1	1E+03	1E+06	1E+09	1E+12	1E+15	1E+18	1E+21	1E+22	1E+23	1E+24	1E+25	1E+26	1E+27	1E+30	1E+33	1E+36	1E+39	1E+42	1E+45	1E+48
zettametre	Zm	1E-03	1	1E+03	1E+06	1E+09	1E+12	1E+15	1E+18	1E+19	1E+20	1E+21	1E+22	1E+23	1E+24	1E+27	1E+30	1E+33	1E+36	1E+39	1E+42	1E+45
exametre	Em	1E-06	1E-03	1	1E+03	1E+06	1E+09	1E+12	1E+15	1E+16	1E+17	1E+18	1E+19	1E+20	1E+21	1E+24	1E+27	1E+30	1E+33	1E+36	1E+39	1E+42
petametre	Pm	1E-09	1E-06	1E-03	1	1E+03	1E+06	1E+09	1E+12	1E+13	1E+14	1E+15	1E+16	1E+17	1E+18	1E+21	1E+24	1E+27	1E+30	1E+33	1E+36	1E+39
terametre	Tm	1E-12	1E-09	1E-06	1E-03	1	1E+03	1E+06	1E+09	1E+10	1E+11	1E+12	1E+13	1E+14	1E+15	1E+18	1E+21	1E+24	1E+27	1E+30	1E+33	1E+36
gigametre	Gm	1E-15	1E-12	1E-09	1E-06	1E-03	1	1E+03	1E+06	1E+07	1E+08	1E+09	1E+10	1E+11	1E+12	1E+15	1E+18	1E+21	1E+24	1E+27	1E+30	1E+33
megametre	Mm	1E-18	1E-15	1E-12	1E-09	1E-06	1E-03	1	1E+03	1E+04	1E+05	1E+06	1E+07	1E+08	1E+09	1E+12	1E+15	1E+18	1E+21	1E+24	1E+27	1E+30
kilometre	km	1E-21	1E-18	1E-15	1E-12	1E-09	1E-06	1E-03	1	1E+01	1E+02	1E+03	1E+04	1E+05	1E+06	1E+09	1E+12	1E+15	1E+18	1E+21	1E+24	1E+27
hectometre	hm	1E-22	1E-19	1E-16	1E-13	1E-10	1E-07	1E-04	1E-01	1	1E+01	1E+02	1E+03	1E+04	1E+05	1E+08	1E+11	1E+14	1E+17	1E+20	1E+23	1E+26
decametre	dam	1E-23	1E-20	1E-17	1E-14	1E-11	1E-08	1E-05	1E-02	1E-01	1	1E+01	1E+02	1E+03	1E+04	1E+07	1E+10	1E+13	1E+16	1E+19	1E+22	1E+25
metre	m	1E-24	1E-21	1E-18	1E-15	1E-12	1E-09	1E-06	1E-03	1E-02	1E-01	1	1E+01	1E+02	1E+03	1E+06	1E+09	1E+12	1E+15	1E+18	1E+21	1E+24
decimetre	dm	1E-25	1E-22	1E-19	1E-16	1E-13	1E-10	1E-07	1E-04	1E-03	1E-02	1E-01	1	1E+01	1E+02	1E+05	1E+08	1E+11	1E+14	1E+17	1E+20	1E+23
centimetre	cm	1E-26	1E-23	1E-20	1E-17	1E-14	1E-11	1E-08	1E-05	1E-04	1E-03	1E-02	1E-01	1	1E+01	1E+04	1E+07	1E+10	1E+13	1E+16	1E+19	1E+22
millimetre	mm	1E-27	1E-24	1E-21	1E-18	1E-15	1E-12	1E-09	1E-06	1E-05	1E-04	1E-03	1E-02	1E-01	1	1E+03	1E+06	1E+09	1E+12	1E+15	1E+18	1E+21
micrometre	μm	1E-30	1E-27	1E-24	1E-21	1E-18	1E-15	1E-12	1E-09	1E-08	1E-07	1E-06	1E-05	1E-04	1E-03	1	1E+03	1E+06	1E+09	1E+12	1E+15	1E+18
nanometre	nm	1E-33	1E-30	1E-27	1E-24	1E-21	1E-18	1E-15	1E-12	1E-11	1E-10	1E-09	1E-08	1E-07	1E-06	1E-03	1	1E+03	1E+06	1E+09	1E+12	1E+15
picometre	pm	1E-36	1E-33	1E-30	1E-27	1E-24	1E-21	1E-18	1E-15	1E-14	1E-13	1E-12	1E-11	1E-10	1E-09	1E-06	1E-03	1	1E+03	1E+06	1E+09	1E+12
femtometre	fm	1E-39	1E-36	1E-33	1E-30	1E-27	1E-24	1E-21	1E-18	1E-17	1E-16	1E-15	1E-14	1E-13	1E-12	1E-09	1E-06	1E-03	1	1E+03	1E+06	1E+09
attometre	am	1E-42	1E-39	1E-36	1E-33	1E-30	1E-27	1E-24	1E-21	1E-20	1E-19	1E-18	1E-17	1E-16	1E-15	1E-12	1E-09	1E-06	1E-03	1	1E+03	1E+06
zeptometre	zm	1E-45	1E-42	1E-39	1E-36	1E-33	1E-30	1E-27	1E-24	1E-23	1E-22	1E-21	1E-20	1E-19	1E-18	1E-15	1E-12	1E-09	1E-06	1E-03	1	1E+03
yoctometre	ym	1E-48	1E-45	1E-42	1E-39	1E-36	1E-33	1E-30	1E-27	1E-26	1E-25	1E-24	1E-23	1E-22	1E-21	1E-18	1E-15	1E-12	1E-09	1E-06	1E-03	1

Table 4.2.2.2 Conversion table: units of length (>1 m)

Unit	Symbol	mile (geograph.)	league (UK, naut.)	league (int., naut.)
hubble	hubble	1.274730715E+21	1.701671286E+21	1.702759179E+21
siriusweit	-	2.078852095E+13	2.775113894E+13	2.776888049E+13
siriometre	srm	2.015711591E+13	2.690825987E+13	2.692546256E+13
parsec	pc	4.157704190E+12	5.550227788E+12	5.553776098E+12
spat	spat	1.347419980E+08	1.798706083E+08	1.799856012E+08
astronomic unit	AU	2.015711591E+07	2.690825987E+07	2.692546256E+07
myriametre	mym	1.34741998043	1.79870608279	1.79985601152
mile (geographical)	mi (geogr.)	1	1.33492608757	1.33577951764
league (UK, naut.)	lg (UK, naut.)	7.491051447E-01	1	1.00063930886
league (int., naut.)	lg (int., naut.)	7.486265411E-01	3.452338344E-03	1
league (US, stat.)	lg (US, stat.)	6.505386783E-01	8.684210526E-01	8.689762419E-01
league (Canadian)	lg (Can.)	6.503996246E-01	8.682354262E-01	8.687904968E-01
lieue (metric)	lieue (metric)	5.389679922E-01	7.194824331E-01	7.199424046E-01
lieue (de Poste)	lieue (de Poste)	5.252243084E-01	7.011356311E-01	7.015838733E-01
mile (telegr., naut.)	mi (tel., naut.)	2.499892004E-01	3.337171053E-01	3.339304536E-01
mile (UK, naut.)	mi (UK, naut.)	2.497017149E-01	1.151515152E-03	3.335464363E-01
mile (US, naut.)	mi (US, naut.)	2.497017149E-01	1.151515152E-03	3.335464363E-01
mile (int., naut.)	mi (int., naut.)	2.495421804E-01	1.150779448E-03	**1/3**
mile (US, survey)	mi (US, surv.)	2.168466598E-01	1.000002000E-03	2.896593266E-01
mile (int., statute)	mi (int., stat.)	2.168466303E-01	2.894742238E-01	2.896592873E-01
mile (US, statute)	mi (US, stat.)	2.168462261E-01	**1E-03**	2.896587473E-01
kilometre	km	1.347419980E-01	6.213711922E-04	1.799856012E-01
encablure	encablure	2.964323957E-02	3.957153382E-02	3.959683225E-02
furlong (UK)	fur	2.710577826E-02	**1/8000**	3.620734341E-02
cable length (UK)	cbl (UK, naut.)	2.497017149E-02	1/30	3.335464363E-02
cable length (int.)	cbl (int.)	2.495421804E-02	1.150779448E-04	**1/30**
skein (UK)	–	1.478496996E-02	1.973684211E-02	1.974946004E-02
arpent (Québec)	–	7.878541138E-03	1.051727010E-02	1.052399388E-02
bolt (US, cloth)	–	4.928323320E-03	2.272727273E-05	6.583153348E-03
chain (engineer's) or (Ramsden's)	ch (eng.), ch (Ramsden)	4.106936100E-03	1.893939394E-05	5.485961123E-03
shacles (UK)	–	3.696242490E-03	4.934210526E-03	4.937365011E-03
chain (surveyor's)	ch (surv.)	2.710577826E-03	**1.25E-05**	3.620734341E-03
chain (Gunter's)	ch (Gunt.)	2.710577826E-03	3.618421053E-03	3.620734341E-03
rope (UK)	rope	8.213872201E-04	3.787878788E-06	1.097192225E-03
perche (de Paris)	perche	7.878364626E-04	1.051703447E-03	1.052375810E-03
perch (pole, rod)	rod	6.776444566E-04	**3.125E-06**	9.051835853E-04
toise (metric)	toise (metric)	2.694839961E-04	3.597412166E-04	3.599712023E-04
toise (de Paris)	toise	2.626170857E-04	3.505743988E-04	3.507985241E-04
aune (de Paris)	aune	2.533887759E-04	3.382552873E-04	3.384715369E-04
fathom	fath	2.464161660E-04	1.136363636E-06	3.291576674E-04
ell	ell	1.540101038E-04	7.102272727E-07	2.057235421E-04
metre	m	1.347419980E-04	1.798706083E-04	1.799856012E-04
yard	yd	1.232080830E-04	1.644736842E-04	1.645788337E-04
pied (metric)	pied (metric)	4.446485935E-05	5.93573007E-05	5.939524838E-05
foot	ft	4.106936100E-05	5.482456140E-05	5.485961123E-05
link (engineer's)	li (eng.)	4.106936100E-05	5.482456140E-05	5.485961123E-05
span	span	3.080202075E-05	4.111842105E-05	4.114470842E-05
link (surveyor's)	li (surv.)	2.710577826E-05	3.618421053E-05	3.620734341E-05
pouce (metric)	pouce (metric)	3.705404946E-06	4.946441728E-06	4.949604032E-06
inch	in	3.422446750E-06	4.568713450E-06	4.571634269E-06

ague (US, stat.)	mile (UK, naut.)	mile (US, stat.)	kilometre	cable length (UK)
59500268E+21	5.108277538E+21	5.878500805E+21	9.460530E+21	5.105013857E+22
95585696E+13	8.330664147E+13	9.586757089E+13	1.542839000E+14	8.325341682E+14
98526895E+13	8.077638769E+13	9.295580684E+13	1.4959787E+14	8.072477962E+14
91171392E+12	1.666132829E+13	1.917351418E+13	3.085678000E+13	1.665068336E+14
71237307E+08	5.399568035E+08	6.213711922E+08	**1E+09**	5.396118248E+09
98526895E+07	8.077638769E+07	9.295580684E+07	1.495979E+08	8.072477962E+08
7123730746	5.39956803456	6.21371192237	**10**	53.96118248377
3718761599	4.00733855292	4.61156284797	7.42159100000	40.04778262709
5151515152	3.00191792657	**38/11**	5.55955200000	**30**
5077944802	3	3.45233834407	**5.556**	29.98083298798
	2.60692872570	3	4.82803200000	26.05263157895
97862483E-01	2.60637149028	2.99935874493	**4.827**	26.04706278492
84949230E-01	2.15982721382	2.48548476895	4	21.58447299351
73683024E-01	2.10475161987	2.42210490734	3.898	21.03406893217
42803030E-01	1.00179136069	1.15284090909	1.85531760000	10.01151315789
55183426E-04	1.00063930886	**38/33**	**1.853184**	**10**
55183426E-04	1.00063930886	**38/33**	**1.853184**	**10**
50611976E-04	**1**	1.15077944802	**1.852**	9.99361099599
13724350E-04	8.68977980E-01	1.00000200001	1.60934721870	8.68422789480
33339547E-01	8.68977862E-01	**1**	**1.609347**	8.68422671467
13711922E-04	8.68976242E-01	**1**	**1.609344**	8.68421052632
61021585E-04	5.39956803E-01	6.21371192E-01	**1**	5.39611824838
56722076E-02	1.18790497E-01	1.36701662E-01	**2.20E-01**	1.18714601464
67139903E-05	1.08622030E-01	**1/8**	2.011680000E-01	1.08552631579
3838383E-02	1.00063931E-01	**3/26**	1.853184000E-01	1
50611976E-05	**1/10**	1.150779448E-01	**1/5**	9.93610996E-01
4	5.924838013E-02	6.818181818E-02	1.097280000E-01	5.921052632E-01
11079587E-02	3.157198164E-02	3.633238761E-02	5.847131000E-02	3.155181029E-01
12207255E-05	1.974946004E-02	**1/44**	**3/82**	1.973684211E-01
76839379E-05	1.645788337E-02	1.893939394E-02	**3.048E-02**	1.644736842E-01
81818182E-03	1.481209503E-02	1.704545455E-02	**2/73**	1.480263158E-01
67139903E-06	1.086220302E-02	**1/80**	**2.01168E-02**	1.085526316E-01
40	1.086220302E-02	**1/80**	**2.01168E-02**	1.085526316E-01
53678758E-06	3.291576674E-03	3.787878788E-03	**6.096E-03**	3.289473684E-02
11052454E-03	3.157127430E-03	3.633157361E-03	5.847E-03	3.155110340E-02
41784976E-06	2.715550756E-03	**1/320**	5.029E-03	2.713815789E-02
42474615E-04	1.079913607E-03	1.242742384E-03	**2E-03**	1.079223650E-02
36917320E-04	1.052395572E-03	1.242742384E-03	**2E-03**	1.079223650E-02
95060884E-04	1.015414611E-03	1.168518265E-03	1.880547859E-03	1.014765862E-02
61036275E-07	9.874730022E-04	1.136363636E-03	1.828800000E-03	9.868421053E-03
13147672E-07	6.171706263E-04	7.102272727E-04	1.143000000E-03	6.167763158E-03
71237307E-04	5.399568035E-04	6.213711922E-04	**1E-03**	5.396118248E-03
93939394E-04	4.937365011E-04	5.681818182E-04	**9.144E-04**	4.934210526E-03
3508311E-05	1.781857451E-04	2.050524934E-04	**3.30E-04**	1.780719022E-03
13131313E-05	1.645788337E-04	1.893939394E-04	**3.048E-04**	1.644736842E-03
13131313E-05	1.645788337E-04	1.893939394E-04	**3.048E-04**	1.644736842E-03
34848485E-05	1.234341253E-04	1.420454545E-04	**2.286E-04**	1.233552632E-03
66666667E-05	1.086220302E-04	**1/8000**	**2.01168E-04**	1.085526316E-03
95902596E-06	1.484881210E-05	1.708770779E-05	**2.75E-05**	1.483932518E-04
60942761E-06	1.371490281E-05	1.578282828E-05	**2.54E-05**	1.370614035E-04

(Continued overleaf)

Table 4.2.2.2 *(continued)* Conversion table: units of length (>1 m)

Unit	Symbol	cable length (UK)	cable length (int.)	chain (engineer'
hubble	hubble	5.105013857E+22	5.108277538E+22	3.103848425E+23
siriusweit	–	8.325341682E+14	8.330664147E+14	5.061807743E+15
siriometre	srm	8.072477962E+14	8.077638769E+14	4.908066601E+15
parsec	pc	1.665068336E+14	1.666132829E+14	1.012361549E+15
spat	spat	5.396118248E+09	5.399568035E+09	3.280839895E+10
astronomic unit	AU	8.072477962E+08	8.077638769E+08	4.908066601E+09
myriametre	mym	53.96118248377	53.99568034557	328.08398950131
mile (geographical)	mi (geogr.)	40.04778262709	40.07338552916	243.49051837270
league (UK, naut.)	lg (UK, naut.)	**30**	30.01917926566	**182.4**
league (int., naut.)	lg (int., naut.)	29.98083298798	**30**	182.28346456693
league (US, stat.)	lg (US, stat.)	26.05263157895	26.06928725702	**158.4**
league (Canadian)	lg (Can.)	26.04706278492	26.06371490281	158.36614173228
lieue (metric)	lieue (metric)	21.58447299351	21.59827213823	131.23359580053
lieue (de Poste)	lieue (de Poste)	21.03406893217	21.04751619870	127.88713910761
mile (telegr., naut.)	mi (tel., naut.)	10.01151315789	10.01791360691	**60.87**
mile (UK, naut.)	mi (UK, naut.)	**10**	10.00639308855	**60.8**
mile (US, naut.)	mi (US, naut.)	**10**	10.00639308855	**60.8**
mile (int., naut.)	mi (int., naut.)	9.99361099599	**10**	60.76115485564
mile (US, survey)	mi (US, surv.)	8.68422789480	8.68977979860	52.80010560039
mile (int., statute)	mi (int., stat.)	8.68422671467	8.68977861771	52.80009842520
mile (US, statute)	mi (US, stat.)	8.68421052632	8.68976241901	**52.8**
kilometre	km	5.39611824838	5.39956803456	32.80839895013
encablure	encablure	1.18714601464	1.18790496760	7.21784776903
furlong (UK)	fur	1.08552631579	1.08622030238	**6.6**
cable length (UK)	cbl (UK, naut.)	**1**	1.00063930886	**6.08**
cable length (int.)	cbl (int.)	9.993610996E-01	**1**	6.07611548556
skein (UK)	–	5.921052632E-01	5.924838013E-01	**3.6**
arpent (Québec)	–	3.155181029E-01	3.157198164E-01	1.91835006562
bolt (US, cloth)	–	1.973684211E-01	1.974946004E-01	**1.20**
chain (engineer's) or (Ramsden's)	ch (eng.), ch (Ramsden's)	1.644736842E-01	1.645788337E-01	**1**
shacles (UK)	–	1.480263158E-01	1.481209503E-01	**9E-01**
chain (surveyor's)	ch (surv.)	1.085526316E-01	1.086220302E-01	**2/3**
chain (Gunter's)	ch (Gunt.)	1.085526316E-01	1.086220302E-01	**2/3**
rope (UK)	rope	3.289473684E-02	3.291576674E-02	**2E-01**
perche (de Paris)	perche	3.155110340E-02	3.157127430E-02	1.918307087E-01
perch (pole, rod)	rod	2.713815789E-02	2.715550756E-02	**1/6**
toise (metric)	toise (metric)	1.079223650E-02	1.079913607E-02	6.561679790E-02
toise (de Paris)	toise	1.051723196E-02	1.052395572E-02	6.394477034E-02
aune (de Paris)	aune	1.014765862E-02	1.015414611E-02	6.169776440E-02
fathom	fath	9.868421053E-03	9.874730022E-03	**6E-02**
ell	ell	6.167763158E-03	6.171706263E-03	**3/80**
metre	m	5.396118248E-03	5.399568035E-03	3.280839895E-02

hain (surveyor's)	metre	yard	foot	inch
702800644E+23	9.4605297E+24	1.034616142E+25	3.103848E+25	3.724618E+26
669405671E+15	1.54283900E+17	1.687269248E+17	5.061808E+17	6.074169E+18
4364645E+15	1.4959787E+17	1.636022200E+17	4.908067E+17	5.889680E+18
533881134E+15	3.0856780E+16	3.374538495E+16	1.012362E+17	1.214834E+18
970969538E+10	1E+12	1.093613298E+12	3.280840E+12	3.937008E+13
436464547E+09	1.4959787E+11	1.636022200E+11	4.908067E+11	5.889680E+12
7.09695378987	1E+04	1.093613298E+04	3.280840E+04	3.937008E+05
8.92502783743	7.421591E+03	8116.350612423	2.434905E+04	2.921886E+05
6.36363636364	5559.552	6080	18240	2.18880E+05
6.18706752565	5556	6076.115485564	1.822835E+04	2.187402E+05
40	4828.032	5280	15840	1.90080E+05
39.94869959437	4827	5278.871391076	1.583661E+04	1.900394E+05
8.83878151595	4000	4374.453193351	1.312336E+04	1.574803E+05
3.76839258729	3898	4262.904636920	1.278871E+04	1.534646E+05
29/22	1855.3176	2029	6087	73044
40/33	1853.184	6080/3	6080	72960
40/33	1853.184	6080/3	6080	72960
.06235584188	1852	2025.371828521	6076.115486	7.291339E+04
.00016000060	1609.3472187	1760.003520013	5280.010560	6.336013E+04
.00014912909	1609.347	1760.003280840	5280.009843	6.336012E+04
)	1609.344	1760	5280	63360
9.70969537899	1000	1093.613298338	3280.839895	3.937008E+04
.93613298338	220	240.594925634	721.78477690	8661.41732283
0	201.168	220	660	7920
04/33	185.3184	608/3	608	7296
20623558419	185.2	202.53718285214	607.611549	7291.33858268
45454545455	109.728	120	360	4320
90659100851	58.47131	63.94500218723	191.83500656	2302.02007874
/11	36.576	40	120	1440
/33	30.48	100/3	100	1200
5/11	27.432	30	90	1080
	20.1168	22	66	792
	20.1168	22	66	792
/33	6.096	20/3	20	240
906525889E-01	5.847	6.394	19.183	230.197
5E-01	5.0292	5.5	16.5	198
941939076E-02	2	2.18722659668	6.56167979	78.74015748
688601567E-02	1.949036600	2.13149234471	6.394477034	76.73372441
348146121E-02	1.880547859	2.05659214653	6.16977644	74.03731728
11	1.8288	2	6	72
681818182E-02	1.143	1.25	3.75	45
970969538E-02	1	1.093613298	3.280839895	39.37007874

Table 4.2.2.3 Conversion table: units of length (<1 m)

Unit	Symbol	metre	yard	foot
metre	m	1	1.0936132983	3.2808398950
yard (US)	yd (US)	0.91440183	1.0000020013	3.0000060039
yard (1963)	yd	**0.9144**	**1**	**3**
yard (old)	yd (old)	0.9143992000	9.99999125E-01	2.9999973753
pace (UK)	pace (UK)	**0.762**	**5/6**	**5/2**
pas (de Paris)	pas	0.6240000000	6.82414698E-01	2.0472440945
coudée (de Paris)	coudée	0.5	5.46806649E-01	1.6404199475
cubit (UK)	cu (UK)	**0.4572**	**1/2**	**3/2**
pied (de Paris)	pied	0.3248394000	3.55248688E-01	1.0657460630
pied (metric)	pied (metric)	**0.333**	3.64173228E-01	1.0925196850
foot (old)	ft (old)	0.3048799700	3.33420790E-01	1.0002623688
foot (US, survey)	ft (US, survey)	0.3048006096	0.3333340000	1.0000020000
foot (WMA, 1963)	ft	**0.3048**	**1/3**	**1**
link (engineer's)	lk (engineer's)	**0.3048**	**1/3**	**1**
cup (US)	cup (US)	0.2365882	2.58736002E-01	7.76208005E-01
nail (UK)	nail (UK)	**2.29E-01**	**1/4**	**3/4**
span (UK)	sp (UK)	**2.2860E-01**	**2.5E-01**	**7.5E-01**
link (surveyor's)	lk (surveyor's)	**2.011680E-01**	**2.20E-01**	**6.6E-01**
cup (metric)	cup (metric)	**1/5**	2.18722660E-01	6.56167979E-01
hand (UK)	hd (UK)	**1.016E-01**	**1/9**	**1/3**
palm (UK)	plm (UK)	**7.62E-02**	**1/12**	**1/4**
pouce (metric)	pouce (metric)	**2.75E-02**	3.00743657E-02	9.02230971E-02
pouce (de Paris)	pouce	2.706995E-02	2.96040573E-02	8.88121719E-02
inch (US, survey)	in (US, survey)	2.54000508E-02	2.77778333E-02	8.33335000E-02
inch	in	**2.54E-02**	**1/36**	**1/12**
inch (old)	in (old)	2.53999800E-02	2.77777559E-02	8.33332677E-02
centimetre	cm	**1E-02**	1.09361330E-02	3.28083990E-02
barleycorn (UK)	-	8.46666667E-03	**1/108**	**1/36**
ligne (de Paris)	ligne	2.55830000E-03	2.79779090E-03	8.39337270E-03
ligne (metric)	ligne (metric)	**2.30E-03**	2.51531059E-03	7.54593176E-03
line (UK), button	li (UK)	2.11666667E-03	**1/432**	**1/144**
millimetre	mm	**1E-03**	1.09361330E-03	3.28083990E-03
baromil	baromil	7.500626793E-04	8.20278521E-04	2.46083556E-03
line (US)	li (US)	**6.35E-04**	**1/1440**	**1/480**
iron	iron	5.3E-04	5.79615048E-04	1.73884514E-03
arcmin	arcmin	2.90888209E-04	3.18119213E-04	9.54357640E-04
calibre (centinch)	cin	**2.54E-04**	**1/3600**	**1/1200**
mil	mil	**2.54E-05**	2.77777778E-05	8.33333333E-05
thou (millinch)	thou	**2.54E-05**	2.77777778E-05	8.33333333E-05
micrometre	µm	**1E-06**	1.09361330E-06	3.28083990E-06
micron	µ	**1E-06**	1.09361330E-06	3.28083990E-06
microinch	µin	**2.54E-08**	2.77777778E-08	8.33333333E-08
millimicron	mµ	**1E-09**	1.09361330E-09	3.28083990E-09
nanometre (nanon)	nm	**1E-09**	1.09361330E-09	3.28083990E-09
ångström* (star)	Å*	1.00001482E-10	1.09362950E-10	3.28088851E-10
X unit	XU	1.00020779E-10	1.09384054E-10	3.28152162E-10
ångström	Å	**1E-10**	1.09361330E-10	3.28083990E-10
1st Bohr radius	a_0	5.29177249E-11	5.78715277E-11	1.73614583E-10
a.u. of length	a.u.	5.29177249E-11	5.78715277E-11	1.73614583E-10
bicron (sigma)	µµ	**1E-12**	1.09361330E-12	3.28083990E-12
picometre	pm	**1E-12**	1.09361330E-12	3.28083990E-12
electron radius	r_0	2.81794092E-15	3.08173766E-15	9.24521299E-15
fermi	F	**1E-15**	1.09361330E-15	3.28083990E-15
femtometre	fm	**1E-15**	1.09361330E-15	3.28083990E-15
attometre	am	**1E-18**	1.09361330E-18	3.28083990E-18

inch	cm	mm	calibre (cin)	mil (thou)
.3700787402	1E+02	1E+03	3937.007874016	3.93700787E+04
.0000720472	91.4401830000	914.4018300000	3600.007204724	3.60000720E+04
5	91.44	914.4	3600	3.6E+04
.9999685039	91.439920	914.399200	3599.996850394	3.59999685E+04
0	76.2	762	3000	3E+04
.5669291339	62.4	624	2456.692913386	2.45669291E+04
.6850393701	50	500	1968.503937008	1.96850394E+04
8	45.72	457.2	1800	1.80E+04
2.7889527559	32.483940	324.839400	1278.895275591	1.27889528E+04
3.1102362205	33.3	333	1311.023622047	1.31102362E+04
2.0031484252	30.487997	304.879970	1200.314842520	1.20031484E+04
2.0000240000	30.4800609601	304.8006096012	1.20000240E+03	1.20000240E+04
2	30.48	304.8	1200	1.2E+04
2	30.48	304.8	1200	1.2E+04
.3144960630	23.658820	236.588200	931.4496062992	9.31449606E+03
	22.86	228.6	900	9000
	22.86	228.6	900	9000
.92	20.1168	201.168	792	7920
.8740157480	2E+01	2E+02	787.4015748032	7.87401575E+03
	10.16	101.6	400	4000
	7.62	76.2	300	3000
.0826771654	2.75	27.5	108.2677165354	1082.677165354
.06574606E+00	2.70699500E+00	2.70699500E+01	1.06574606E+02	1.06574606E+03
.00000200E+00	2.54000508E+00	2.54000508E+01	1.00000200E+02	1.00000200E+03
	2.54	25.4	1E+02	1E+03
.99999213E-01	2.5399980000	25.3999800000	99.9999212598	999.9992125984
.93700787E-01	1	10	39.3700787402	393.7007874016
/3	8.46666667E-01	8.46666667E+00	100/3	1000/3
.00720472E-01	2.55830000E-01	2.55830000E+00	1.00720472E+01	1.00720472E+02
.05511811E-02	20/87	23/10	9.0551181102	90.5511811024
/12	2.11666667E-01	2.1166666667	25/3	250/3
.93700787E-02	1E-01	1	3.9370078740	39.3700787402
.95300267E-02	7.50062679E-02	7.50062679E-01	2.9530026743	29.5300267432
/40	6.35E-02	6.35E-01	2.5	25
.08661417E-02	5.30E-02	5.3E-01	2.0866141732	20.8661417323
.14522917E-02	2.90888209E-02	2.90888209E-01	1.14522917E+00	1.14522917E+01
E-02	2.54E-02	2.54E-01	1	10
E-03	2.54E-03	2.54E-02	1E-01	1
E-03	2.54E-03	2.54E-02	1E-01	1
3.93700787E-05	1E-04	1E-03	3.93700787E-03	3.93700787E-02
3.93700787E-05	1E-04	1E-03	3.93700787E-03	3.93700787E-02
1E-06	2.54E-06	2.54E-05	1E-04	1E-03
3.93700787E-08	1E-07	1E-06	3.93700787E-06	3.93700787E-05
3.93700787E-08	1E-07	1E-06	3.93700787E-06	3.93700787E-05
3.93706622E-09	1.00001482E-08	1.00001482E-07	3.93706622E-07	3.93706622E-06
3.93782594E-09	1.00020779E-08	1.00020779E-07	3.93782594E-07	3.93782594E-06
3.93700787E-09	1E-08	1E-07	3.93700787E-07	3.93700787E-06
2.08337500E-09	5.29177249E-09	5.29177249E-08	2.08337500E-07	2.08337500E-06
2.08337500E-09	5.29177249E-09	5.29177249E-08	2.08337500E-07	2.08337500E-06
3.93700787E-11	1E-10	1E-09	3.93700787E-09	3.93700787E-08
3.93700787E-11	1E-10	1E-09	3.93700787E-09	3.93700787E-08
1.10942556E-13	2.81794092E-13	2.81794092E-12	1.10942556E-11	1.10942556E-10
3.93700787E-14	1E-13	1E-12	3.93700787E-12	3.93700787E-11
3.93700787E-14	1E-13	1E-12	3.93700787E-12	3.93700787E-11
3.93700787E-17	1E-16	1E-15	3.93700787E-15	3.93700787E-14

(Continued overleaf)

Table 4.2.2.3 *(continued)* Conversion table: units of length (<1 m)

Unit	Symbol	mil (thou)	μm (μ)	μinch
metre	m	3.93700787E+04	1000000	3.93700787E+07
yard (US)	yd (US)	3.60000720E+04	9.14401830E+05	3.60000720E+07
yard (1963)	yd	**3.6E+04**	9.144E+05	**3.6E+07**
yard (old)	yd (old)	3.59999685E+04	9.14399200E+05	3.59999685E+07
pace (UK)	pace (UK)	**3E+04**	**7.62E+05**	**3E+07**
pas (de Paris)	pas	2.45669291E+04	**6.24E+05**	2.45669291E+07
coudée (de Paris)	coudée	1.96850394E+04	**5E+05**	1.96850394E+07
cubit (UK)	cu (UK)	**1.80E+04**	**4.572E+05**	**1.8E+07**
pied (de Paris)	pied	1.27889528E+04	3.24839400E+05	1.27889528E+07
pied (metric)	pied (metric)	1.31102362E+04	**3.33E+05**	1.31102362E+07
foot (old)	ft (old)	1.20031484E+04	3.04879970E+05	1.20031484E+07
foot (US, survey)	ft (US, survey)	1.20000240E+04	3.04800610E+05	1.20000240E+07
foot (WMA, 1963)	ft	**1.2E-04**	**3.048E+05**	**1.2E+07**
link (engineer's)	lk (Ramsden's)	**1.2E+04**	**3.048E+05**	**1.2E+07**
cup (US)	cup (US)	9.31449606E+03	2.36588200E+05	9.31449606E+06
nail (UK)	nail (UK)	**9000**	**2.286E+05**	**9E+06**
span (UK)	sp (UK)	**9000**	**2.286E+05**	**9E+06**
link (surveyor's)	lk (surveyor's)	**7920**	2.01168E+05	7.92000E+06
cup (metric)	cup (metric)	7.87401575E+03	**2E+05**	7.87401575E+06
hand (UK)	hd (UK)	**4000**	1.016E+05	**4E+06**
palm (UK)	plm (UK)	**3000**	7.620E+04	**3E+06**
pouce (metric)	pouce (metric)	1082.677165354	2.750E+04	1.08267717E+06
pouce (de Paris)	pouce	1.06574606E+03	2.70699500E+04	1.06574606E+06
inch (US, survey)	in (US, survey)	1.00000200E+03	2.54000508E+04	1.00000200E+06
inch	in	**1E+03**	**2.54E+04**	**1E+06**
inch (old)	in (old)	999.9992125984	2.53999800E+04	9.99999213E+05
centimetre	cm	393.7007874016	**1E+04**	3.93700787E+05
barleycorn (UK)	–	**1000/3**	8466.666667	3.33333333E+05
ligne (de Paris)	ligne	1.00720472E+02	2.55830000E+03	1.00720472E+05
ligne (metric)	ligne (metric)	90.5511811024	**2.3E+03**	9.05511811E+04
line (UK), button	li (UK)	**250/3**	2116.666667	8.33333333E+04
millimetre	mm	39.3700787402	**1000**	3.93700787E+04
baromil	baromil	29.5300267432	750.062679	2.95300267E+04
line (US)	li (US)	**25**	**635**	**2.5E+04**
iron	iron	20.8661417323	530	2.08661417E+04
arcmin	arcmin	1.14522917E+01	2.90888209E+02	1.14522917E+04
calibre (centinch)	cin	**10**	254	**1E+04**
mil	mil	**1**	25	**1000**
thou (millinch)	thou	**1**	25	**1000**
micrometre	μm	3.93700787E-02	1	3.93700787E+01
micron	μ	3.93700787E-02	1	3.93700787E+01
microinch	μin	**1E-03**	**2.54E-02**	1
millimicron	mμ	3.93700787E-05	1E-03	3.93700787E-02
nanometre (nanon)	nm	3.93700787E-05	1E-03	3.93700787E-02
ångström* (star)	Å*	3.93706622E-06	1.00001482E-04	3.93706622E-03
X unit	XU	3.93782594E-06	1.00020779E-04	3.93782594E-03
ångström	Å	3.93700787E-06	**1E-04**	3.93700787E-03
1st Bohr radius	a_0	2.08337500E-06	5.29177249E-05	2.08337500E-03
a.u. of length	a.u.	2.08337500E-06	5.29177249E-05	2.08337500E-03
bicron (sigma)	μμ	3.93700787E-08	**1E-06**	3.93700787E-05
picometre	pm	3.93700787E-08	**1E-06**	3.93700787E-05
electron radius	r_0	1.10942556E-10	2.81794092E-09	1.10942556E-07
fermi	F	3.93700787E-11	**1E-09**	3.93700787E-08
femtometre	fm	3.93700787E-11	**1E-09**	3.93700787E-08
attometre	am	3.93700787E-14	**1E-12**	3.93700787E-11

nm	Å	pm (μμ)	fm (F)	am
E+09	1E+10	1E+12	1E+15	1E+18
14401830E+08	9.14401830E+09	9.14401830E+11	9.14401830E+14	9.14401830E+17
144E+08	9.144E+09	9.144E+11	9.144E+14	9.144E+17
14399200E+08	9.14399200E+09	9.14399200E+11	9.14399200E+14	9.14399200E+17
.62E+08	7.62E+09	7.62E+11	7.62E+14	7.62E+17
24E+08	6.24E+09	6.24E+11	6.24E+14	6.24E+17
E+08	5E+09	5E+11	5E+14	5E+17
572E+08	4.572E+09	4.572E+11	4.572E+14	4.572E+17
.24839400E+08	3.24839400E+09	3.24839400E+11	3.24839400E+14	3.24839400E+17
.330E+08	3.330E+09	3.33E+11	3.33E+14	3.33E+17
.04879970E+08	3.04879970E+09	3.04879970E+11	3.04879970E+14	3.04879970E+17
.04800610E+08	3.04800610E+09	3.04800610E+11	3.04800610E+14	3.04800610E+17
.048E+08	3.048E+09	3.048E+11	3.048E+14	3.048E+17
.048E+08	3.048E+09	3.048E+11	3.048E+14	3.048E+17
.36588200E+08	2.36588200E+09	2.36588200E+11	2.36588200E+14	2.36588200E+17
.286E+08	2.286E+09	2.286E+11	2.286E+14	2.286E+17
.286E+08	2.286E+09	2.286E+11	2.286E+14	2.286E+17
.01168E+08	2.01168E+09	2.01168E+11	2.01168E+14	2.01168E+17
E+08	2E+09	2E+11	2E+14	2E+17
.016E+08	1.016E+09	1.016E+11	1.016E+14	1.016E+17
.62E+07	7.62E+08	7.62E+10	7.62E+13	7.62E+16
.75E+07	2.75E+08	2.75E+10	2.75E+13	2.75E+16
.70699500E+07	2.70699500E+08	2.70699500E+10	2.70699500E+13	2.70699500E+16
.54000508E+07	2.54000508E+08	2.54000508E+10	2.54000508E+13	2.54000508E+16
.54E+07	2.54E+08	2.54E+10	2.54E+13	2.54E+16
.53999800E+07	2.53999800E+08	2.53999800E+10	2.53999800E+13	2.53999800E+16
E+07	1E+08	1E+10	1E+13	1E+16
.46666667E+06	8.46666667E+07	8.46666667E+09	8.46666667E+12	8.46666667E+15
.55830000E+06	2.55830000E+07	2.55830000E+09	2.55830000E+12	2.55830000E+15
.30E+06	2.30E+07	2.3E+09	2.3E+12	2.3E+15
2.11666667E+06	2.11666667E+07	2.11666667E+09	2.11666667E+12	2.11666667E+15
E+06	1E+07	1E+09	1E+12	1E+15
.50062679E+05	7.50062679E+06	7.50062679E+08	7.50062679E+11	7.50062679E+14
.35E+05	6.35E+06	6.35E+08	6.35E+11	6.35E+14
5.30E+05	5.30E+06	5.30E+08	5.30E+11	5.3E+14
2.90888209E+05	2.90888209E+06	2.90888209E+08	2.90888209E+11	2.90888209E+14
2.54E+05	2.54E+06	2.54E+08	2.54E+11	2.54E+14
2.54E+04	2.54E+05	2.54E+07	2.54E+10	2.54E+13
2.54E+04	2.54E+05	2.54E+07	2.54E+10	2.54E+13
000	1E+04	1E+06	1E+09	1E+12
000	1E+04	1E+06	1E+09	1E+12
25	254	2.54E+04	2.54E+07	2.54E+10
	10	1000	1E+06	1E+09
	10	1000	1E+06	1E+09
1.00001482E-01	1.00001482E+00	1.00001482E+02	1.00001482E+05	1.00001482E+08
1.00020779E-01	1.00020779E+00	1.00020779E+02	1.00020779E+05	1.00020779E+08
1E-01	1	100	1E+05	1E+08
5.29177249E-02	5.29177249E-01	5.29177249E+01	5.29177249E+04	5.29177249E+07
5.29177249E-02	5.29177249E-01	5.29177249E+01	5.29177249E+04	5.29177249E+07
1E-03	1E-02	1	1E+03	1E+06
1E-03	1E-02	1	1E+03	1E+06
2.81794092E-06	2.81794092E-05	2.81794092E-03	2.81794092E+00	2.81794092E+03
1E-06	1E-05	1E-03	1	1000
1E-06	1E-05	1E-03	1	1000
1E-09	1E-08	1E-06	1E-03	1

Table 4.2.2.4 Conversion table: units of length (Typography)

Unit	Symbol	m	inch	cm	pica	mm	point Didot	point (I.N.)	point (US)	mil (thou)	µm
metre	**m**	1	39.37007874	**100**	237.1062791	1000	2659.775877	2845.275619004	2845.275619004	3.93700787E+04	1E+06
inch	**in**	**2.54E-02**	1	**2.54**	6.022499489	**25.4**	67.558307270	72.270000723	72.270000723	**1000**	25400
cicéro	-	4.51278000E-03	1.77668504E-01	4.51278000E-01	1.070008474	2825/626	**12**	12.840102908	12.840102908	177.668503937	4.51278000E+03
pica	**pica**	4.21751800E-03	1.66044016E-01	4.21751800E-01	**1**	**2889/685**	11.217652636	**12**	**12**	166.044015748	4.21751800E+03
philoso-phia	–	4.13568681E-03	1.62822315E-01	4.13568681E-01	9.80597310E-01	4.135686806	**11**	11.767168836	11.767168836	162.822315189	4.13568681E+03
petit romain	–	3.38374375E-03	1.33218258E-01	3.38374375E-01	8.02306890E-01	3.383743750	**9**	9.627683593	9.627683593	133.218257882	3.38374375E+03
gaillarde	–	3.00777222E-03	1.18416229E-01	3.00777222E-01	7.13161680E-01	3.007772222	**8**	8.557940972	8.557940972	118.416229228	3.00777222E+03
petit texte	–	2.81978646E-03	1.11015215E-01	2.81978646E-01	6.68589075E-01	2.819786459	**7.5**	8.023069661	8.023069661	111.015214902	2.81978646E+03
mignonne	–	2.63180069E-03	1.03614201E-01	2.63180069E-01	6.24016470E-01	2.631800695	**7**	7.488198350	7.488198350	103.614200575	2.63180069E+03
point (Didot)	**pt (Didot)**	3.75971528E-04	1.48020287E-02	3.75971528E-02	8.91452100E-02	3.75971528E-01	**1**	1.069742621	1.069742621	14.802028654	375.971527800
point (I.N.)*	**pt (I.N.)**	3.51459800E-04	1.38370000E-02	3.51459800E-02	8.33333254E-02	3.51459800E-01	9.34804298E-01	**1**	**1**	13.837	351.459800000
point (US printers)	**pt (US)**	3.51459800E-04	1.38370000E-02	3.51459800E-02	8.33333254E-02	3.51459800E-01	9.34804298E-01	**1**	**1**	13.837	351.459800000

*I.N. = Imprimerie Nationale

Table 4.2.3.1 Conversion table: units of area: multiples and submultiples (SI)

Unit	Symbol	Ym^2	Zm^2	Em^2	Pm^2	Tm^2	Gm^2	Mm^2	km^2	hm^2	dam^2	m^2	dm^2	cm^2	mm^2	μm^2	nm^2	pm^2	fm^2	am^2	zm^2	ym^2
square yottametre	Ym^2	1	1E+06	1E+12	1E+18	1E+24	1E+30	1E+36	1E+42	1E+44	1E+46	1E+48	1E+50	1E+52	1E+54	1E+60	1E+66	1E+72	1E+78	1E+84	1E+90	1E+96
square zetametre	Zm^2	1E-06	1	1E+06	1E+12	1E+18	1E+24	1E+30	1E+36	1E+38	1E+40	1E+42	1E+44	1E+46	1E+48	1E+54	1E+60	1E+66	1E+72	1E+78	1E+84	1E+90
square exametre	Em^2	1E-12	1E-06	1	1E+06	1E+12	1E+18	1E+24	1E+30	1E+32	1E+34	1E+36	1E+38	1E+40	1E+42	1E+48	1E+54	1E+60	1E+66	1E+72	1E+78	1E+84
square petametre	Pm^2	1E-18	1E-12	1E-06	1	1E+06	1E+12	1E+18	1E+24	1E+26	1E+28	1E+30	1E+32	1E+34	1E+36	1E+42	1E+48	1E+54	1E+60	1E+66	1E+72	1E+78
square terametre	Tm^2	1E-24	1E-18	1E-12	1E-06	1	1E+06	1E+12	1E+18	1E+20	1E+22	1E+24	1E+26	1E+28	1E+30	1E+36	1E+42	1E+48	1E+54	1E+60	1E+66	1E+72
square gigametre	Gm^2	1E-30	1E-24	1E-18	1E-12	1E-06	1	1E+06	1E+12	1E+14	1E+16	1E+18	1E+20	1E+22	1E+24	1E+30	1E+36	1E+42	1E+48	1E+54	1E+60	1E+66
square megametre	Mm^2	1E-36	1E-30	1E-24	1E-18	1E-12	1E-06	1	1E+06	1E+08	1E+10	1E+12	1E+14	1E+16	1E+18	1E+24	1E+30	1E+36	1E+42	1E+48	1E+54	1E+60
square kilometre	km^2	1E-42	1E-36	1E-30	1E-24	1E-18	1E-12	1E-06	1	1E+02	1E+04	1E+06	1E+08	1E+10	1E+12	1E+18	1E+24	1E+30	1E+36	1E+42	1E+48	1E+54
square hectometre	hm^2	1E-44	1E-38	1E-32	1E-26	1E-20	1E-14	1E-08	1E-02	1	1E+02	1E+04	1E+06	1E+08	1E+10	1E+16	1E+22	1E+28	1E+34	1E+40	1E+46	1E+52
square decametre	dam^2	1E-46	1E-40	1E-34	1E-28	1E-22	1E-16	1E-10	1E-04	1E-02	1	1E+02	1E+04	1E+06	1E+08	1E+14	1E+20	1E+26	1E+32	1E+38	1E+44	1E+50
square metre	m^2	1E-48	1E-42	1E-36	1E-30	1E-24	1E-18	1E-12	1E-06	1E-04	1E-02	1	1E+02	1E+04	1E+06	1E+12	1E+18	1E+24	1E+30	1E+36	1E+42	1E+48
square decimetre	dm^2	1E-50	1E-44	1E-38	1E-32	1E-26	1E-20	1E-14	1E-08	1E-06	1E-04	1E-02	1	1E+02	1E+04	1E+10	1E+16	1E+22	1E+28	1E+34	1E+40	1E+46
square centimetre	cm^2	1E-52	1E-46	1E-40	1E-34	1E-28	1E-22	1E-16	1E-10	1E-08	1E-06	1E-04	1E-02	1	1E+02	1E+08	1E+14	1E+20	1E+26	1E+32	1E+38	1E+44
square millimetre	mm^2	1E-54	1E-48	1E-42	1E-36	1E-30	1E-24	1E-18	1E-12	1E-10	1E-08	1E-06	1E-04	1E-02	1	1E+06	1E+12	1E+18	1E+24	1E+30	1E+36	1E+42
square micrometre	μm^2	1E-60	1E-54	1E-48	1E-42	1E-36	1E-30	1E-24	1E-18	1E-16	1E-14	1E-12	1E-10	1E-08	1E-06	1	1E+06	1E+12	1E+18	1E+24	1E+30	1E+36
square nanometre	nm^2	1E-66	1E-60	1E-54	1E-48	1E-42	1E-36	1E-30	1E-24	1E-22	1E-20	1E-18	1E-16	1E-14	1E-12	1E-06	1	1E+06	1E+12	1E+18	1E+24	1E+30
square picometre	pm^2	1E-72	1E-66	1E-60	1E-54	1E-48	1E-42	1E-36	1E-30	1E-28	1E-26	1E-24	1E-22	1E-20	1E-18	1E-12	1E-06	1	1E+06	1E+12	1E+18	1E+24
square femtometre	fm^2	1E-78	1E-72	1E-66	1E-60	1E-54	1E-48	1E-42	1E-36	1E-34	1E-32	1E-30	1E-28	1E-26	1E-24	1E-18	1E-12	1E-06	1	1E+06	1E+12	1E+18
square attometre	am^2	1E-84	1E-78	1E-72	1E-66	1E-60	1E-54	1E-48	1E-42	1E-40	1E-38	1E-36	1E-34	1E-32	1E-30	1E-24	1E-18	1E-12	1E-06	1	1E+06	1E+12
square zeptometre	zm^2	1E-90	1E-84	1E-78	1E-72	1E-66	1E-60	1E-54	1E-48	1E-46	1E-44	1E-42	1E-40	1E-38	1E-36	1E-30	1E-24	1E-18	1E-12	1E-06	1	1E+06
square yoctometre	ym^2	1E-96	1E-90	1E-84	1E-78	1E-72	1E-66	1E-60	1E-54	1E-52	1E-50	1E-48	1E-46	1E-44	1E-42	1E-36	1E-30	1E-24	1E-18	1E-12	1E-06	1

Table 4.2.3.2 Conversion table: units of area (>1 m^2)

Unit	Symbol	township (US)	sq. mi (int., naut.)
township (US)	township (US)	1	27.1843095236
square mile (geographical)	sq. mi (geograph.)	5.907364417E-01	16.0587622777
square league (UK, naut.)	sq. lg (UK, naut.)	3.314967860E-01	9.0115112378
square league (US, naut.)	sq. lg (US, naut.)	3.314967860E-01	9.0115112378
square league (int., naut.)	sq. lg (int., naut.)	3.310733345E-01	9
square league (statute)	sq. lg (stat.)	1/4	6.7960773809
square league (Canadian)	sq. lg (Canadian)	2.498931356E-01	6.7931723453
square league (metric)	sq. lg (metric)	1.716009594E-01	4.6648535936
square mile (telegr., naut.)	sq. mi (telegr., naut.)	3.691783782E-02	1.0035859304
square mile (UK, naut.)	sq. mi (UK, naut.)	3.683297623E-02	1.0012790264
square mile (US, naut.)	sq. mi (US, naut.)	3.683297623E-02	1.0012790264
square mile (int., naut.)	sq. mi (int., naut.)	3.678592606E-02	1
circular mile (US, naut.)	c. mi (US, naut.)	2.892855188E-02	7.864027084E-01
circular mile (int., naut.)	c. mi (int., naut.)	2.889159876E-02	7.853981634E-01
square mile (US, survey)	sq. mi (US, survey)	2.777788889E-02	7.551227295E-01
square mile (int.)	sq. mi (int.)	2.777788134E-02	7.551225242E-01
square mile (US, stat.)	sq. mi (US, stat.)	1/36	7.551197090E-01
circular mile (int.)	c. mi (int.)	2.181669699E-02	5.930718437E-01
circular mile (statute, land)	c. mi (stat.)	2.181661565E-02	5.930696326E-01
kilometre	km	1.072505996E-02	2.915533496E-01
quarter section	quart sect	1/144	1.887799272E-01
hectare	ha	1.072505996E-04	2.915533496E-03
acre (Ireland)	ac (Ireland)	7.000448029E-05	1.903023460E-03
acre (Cunningham)	ac (Cunningham)	5.564458689E-05	1.512659673E-03
acre (plantation)	ac (plantation)	5.564458689E-05	1.512659673E-03
acre (Scotland)	ac (Scotland)	5.293021680E-05	1.438871397E-03
acre (US, survey)	ac (US, survey)	4.340295139E-05	1.179879265E-03
acre	ac	4.340277778E-05	1.179874545E-03
arpent (de Paris)	arpent	3.666758574E-05	9.967830003E-04
arpent (Québec)	arp (Québec)	3.666784315E-05	9.967899976E-04
vergees (Jersey)	–	1.929012346E-05	5.243886868E-04
vergees (Guernsey)	–	1.653439153E-05	4.494760173E-04
rood (UK)	rood (UK)	1.085069444E-05	2.949686363E-04
square chain (engineer's)	sq. ch (eng.)	9.963906744E-06	2.708619250E-04
square chain (Ramsden's)	sq. ch (Ramsden's)	9.963906744E-06	2.708619250E-04
square chain (surveyor's)	sq. ch (surv.)	4.340277778E-06	1.179874545E-04
square chain (Gunter's)	sq. ch (Gunter's)	4.340277778E-06	1.179874545E-04
skein (UK)	skein (UK)	1.176839379E-06	3.199156594E-05
are	a	1.072505996E-06	2.915533496E-05
perche (de Paris)	perche	3.666758574E-07	9.967830003E-06
square rod (perch, pole)	rod^2	2.712673611E-07	7.374215908E-06
barrel (US, petrol.) per inch	bbl (US, petrol.).in^{-1}	6.713182169E-08	1.824932220E-06
barrel (US, liq.) per inch	bbl (US, liq.).in^{-1}	5.034886627E-08	1.368699165E-06
square metre	m^2	1.072505996E-08	2.915533496E-07

sq. mi (int.)	sq. mi (US, stat.)	km^2	hectare
5.9998657842	36	93.2395719721	9323.9571972096
1.2664326146	21.2665119008	55.0800129713	5508.0012971281
1.9338398054	11.9338842975	30.9086184407	3090.8618440704
1.9338398054	11.9338842975	30.9086184407	3090.8618440704
1.9185956067	11.9186400419	30.8691360000	3086.9136000000
.9999664460	9	23.3098929930	2330.9892993024
.9961193412	8.9961528808	23.2999290000	2329.9929000000
.1776115051	6.1776345367	**16**	1600
.3290372067	1.3290421617	3.4422033969	344.2203396870
.3259822006	1.3259871442	3.4342909379	343.4290937856
.3259822006	1.3259871442	3.4342909379	343.4290937856
.3242884007	1.3242933380	3.4299040000	342.99040
.0414239851	1.0414278677	2.6972857952	269.7285795165
.0400936778	1.0400975555	2.6938403022	269.3840302230
.0000002718	1.0000040000	2.5899984703	258.9998470337
	1.0000037282	2.5899977664	258.9997766409
.999962718E-01	1	2.5899881103	258.9988110336
.853981634E-01	7.854010915E-01	2.0341794889	203.4179488941
.853952353E-01	7.853981634E-01	2.0341719051	203.4171905079
.861007191E-01	3.861021585E-01	1	**100**
.499990679E-01	**2.50E-01**	6.474970276E-01	64.7497027584
.861007191E-03	3.861021585E-03	**1/100**	1
.520151895E-03	2.520161290E-03	6.527187778E-03	6.527187778E-01
.003197660E-03	2.003205128E-03	5.188277465E-03	5.188277465E-01
.003197660E-03	2.003205128E-03	5.188277465E-03	5.188277465E-01
.905480701E-03	1.905487805E-03	4.935190759E-03	4.935190759E-01
.562500425E-03	1.562506250E-03	4.046872610E-03	4.046872610E-01
.562494175E-03	**1/640**	4.046856422E-03	4.046856422E-01
.320028165E-03	1.320033087E-03	3.418870000E-03	3.418870000E-01
.320037432E-03	1.320042353E-03	3.418894000E-03	3.418894000E-01
.944418554E-04	6.944444444E-04	1.798602854E-03	1.798602854E-01
.952358761E-04	5.952380952E-04	1.541659589E-03	1.541659589E-01
.906235437E-04	3.906250000E-04	1.011714106E-03	1.011714106E-01
.586993055E-04	3.587006428E-04	9.290304000E-04	9.290304000E-02
.586993055E-04	3.587006428E-04	9.290304000E-04	9.290304000E-02
.562494175E-04	1.5625E-04	4.046856422E-04	4.046856422E-02
.562494175E-04	1.5625E-04	4.046856422E-04	4.046856422E-02
.236605970E-05	4.236621765E-05	1.097280000E-04	1.097280000E-02
.861007191E-05	3.861021585E-05	**1E-04**	**1/100**
.320028165E-05	1.320033087E-05	3.418870000E-05	3.418870000E-03
.765588592E-06	9.765625000E-06	2.529285264E-05	2.529285264E-03
.416736571E-06	2.416745581E-06	6.259342320E-06	6.259342320E-04
.812552428E-06	1.812559186E-06	4.694506740E-06	4.694506740E-04
.861007191E-07	3.861021585E-07	**1E-06**	**1E-04**

(Continued overleaf)

Table 4.2.3.2 *(continued)* Conversion table: units of area (>1 m^2)

Unit	Symbol	hectare	acre
township (US)	township (US)	9323.9571972096	**2.304E+04**
square mile (geographical)	sq. mi (geograph.)	5508.0012971281	13610.5676164848
square league (UK, naut.)	sq. lg (UK, naut.)	3090.8618440704	7637.6859504132
square league (US, naut.)	sq. lg (US, naut.)	3090.8618440704	7637.6859504132
square league (int., naut.)	sq. lg (int., naut.)	3086.9136000000	7627.9296268418
square league (statute)	sq. lg (stat.)	2330.9892993024	**5760**
square league (Canadian)	sq. lg (Canadian)	2329.9929000000	5757.5378437029
square league (metric)	sq. lg (metric)	1600	3953.6861034747
square mile (telegr., naut.)	sq. mi (telegr., naut.)	344.2203396870	850.5869834711
square mile (UK, naut.)	sq. mi (UK, naut.)	343.4290937856	848.6317722681
square mile (US, naut.)	sq. mi (US, naut.)	343.4290937856	848.6317722681
square mile (int., naut.)	sq. mi (int., naut.)	342.99040	847.5477363158
circular mile (US, naut.)	c. mi (US, naut.)	269.7285795165	666.5138353401
circular mile (int., naut.)	c. mi (int., naut.)	269.3840302230	665.6624354941
square mile (US, survey)	sq. mi (US, survey)	258.9998470337	640.0025600121
square mile (int.)	sq. mi (int.)	258.9997766409	640.0023860676
square mile (US, stat.)	sq. mi (US, stat.)	258.9988110336	**640**
circular mile (int.)	c. mi (int.)	203.4179488941	502.6566985875
circular mile (statute, land)	c. mi (stat.)	203.4171905079	502.6548245744
kilometre	km	**100**	247.1053814672
quarter section	quart sect	64.7497027584	**160**
hectare	ha	**1**	2.4710538147
acre (Ireland)	ac (Ireland)	6.527187778E-01	1.6129032258
acre (Cunningham)	ac (Cunningham)	5.188277465E-01	1.2820512821
acre (plantation)	ac (plantation)	5.188277465E-01	1.2820512821
acre (Scotland)	ac (Scotland)	4.935190759E-01	**50/41**
acre (US, survey)	ac (US, survey)	4.935190759E-01	1.0000040000
acre	ac	4.046872610E-01	**1**
arpent (de Paris)	arpent	3.418870000E-01	8.448211755E-01
arpent (Québec)	arp (Québec)	3.418894000E-01	8.448271061E-01
vergees (Jersey)	–	1.798602854E-01	**4/9**
vergees (Guernsey)	–	1.541659589E-01	3.809523810E-01
rood (UK)	rood (UK)	1.011714106E-01	**1/4**
square chain (engineer's)	sq. ch (eng.)	9.290304000E-02	2.295684114E-01
square chain (Ramsden's)	sq. ch (Ramsden's)	9.290304000E-02	2.295684114E-01
square chain (surveyor's)	sq. ch (surv.)	4.046856422E-02	**1/10**
square chain (Gunter's)	sq. ch (Gunter's)	4.046856422E-02	**1/10**
skein (UK)	skein (UK)	1.097280000E-02	2.711437930E-02
are	a	**1/100**	2.471053815E-02
perche (de Paris)	perche	3.418870000E-03	8.448211755E-03
square rod (perch, pole)	rod^2	2.529285264E-03	**1/160**
barrel (US, petrol.) per inch	bbl (US, petrol.).in^{-1}	6.259342320E-04	1.546717172E-03
barrel (US, liq.) per inch	bbl (US, liq.).in^{-1}	4.694506740E-04	1.160037879E-03
square metre	m^2	1E-04	2.471053815E-04

acre (US, surv.)	are	square rod	m^2
303990784E+04	9.323957197E+05	3.6864E+06	9.323957197E+07
361051317E+04	5.508001297E+05	2.177690819E+06	5.508001297E+07
37.6553997000	3.090861844E+05	1.222029752E+06	3.090861844E+07
537.6553997000	3.090861844E+05	1.222029752E+06	3.090861844E+07
527.8991151538	3.086913600E+05	1.220468740E+06	3.086913600E+07
759.9769600230	2.330989299E+05	**921600**	2.330989299E+07
757.5148135745	2.329992900E+05	9.212060550E+05	2.329992900E+07
953.6702887461	**160000**	6.325897766E+05	**1.60E+07**
50.5835811265	3.442203397E+04	1.360939174E+05	3.442203397E+06
48.6283777444	3.434290938E+04	1.357810836E+05	3.434290938E+06
48.6283777444	3.434290938E+04	1.357810836E+05	3.434290938E+06
47.5443461282	3.429904000E+04	1.356076378E+05	3.429904000E+06
66.5111692874	2.697285795E+04	1.066422137E+05	2.697285795E+06
65.6597728470	2.693840302E+04	1.065059897E+05	2.693840302E+06
40.0000000044	2.589998470E+04	1.024004096E+05	2.589998470E+06
39.9998260606	2.589997766E+04	1.024003818E+05	2.589997766E+06
39.9974400026	2.589988110E+04	**102400**	2.589988110E+06
02.6546879627	2.034179489E+04	8.042507177E+04	2.034179489E+06
02.6528139571	2.034171905E+04	8.042477193E+04	2.034171905E+06
47.1043930466	**1E+04**	3.953686103E+04	**1E+06**
59.9993600006	6.474970276E+03	**25600**	6.474970276E+05
.4710439305	**100**	395.3686103475	**1E+04**
.6128967742	65.2718777806	258.0645161290	6527.18777806
.2820461539	51.8827746462	205.1282051282	5188.27746462
.2820461539	51.8827746462	205.1282051282	5188.27746462
.2195073171	49.3519075902	195.1219512195	4935.19075902
	40.4687260987	160.0006400019	4046.87260987
.999960000E-01	40.4685642240	**160**	4046.85642240
.448177963E-01	34.1887000000	135.1713880859	3418.87000000
.448237268E-01	34.1889400000	135.1723369705	3418.89400000
.444426667E-01	17.9860285440	**640/9**	1798.60285440
.809508571E-01	15.4165958949	60.9523809524	1541.65958949
2.499990000E-01	10.1171410560	**40**	1011.71410560
2.295674931E-01	9.2903040000	36.7309458219	929.0304000000
2.295674931E-01	9.2903040000	36.7309458219	929.0304000000
.999960000E-02	4.0468564224	**16**	404.6856422400
9.999960000E-02	4.0468564224	**16**	404.6856422400
2.711427084E-02	1.0972800000	4.3383006876	109.7280000000
2.471043930E-02	**1**	3.9536861035	**100**
8.448177963E-03	3.418870000E-01	1.3517138809	34.1887
6.249975000E-03	2.529285264E-01	**1**	25.2928526400
1.546710985E-03	6.259342320E-02	**1/4**	6.2593423200
1.160033239E-03	4.694506740E-02	1.856060606E-01	4.6945067400
2.471043930E-04	**1E-02**	3.953686103E-02	**1**

Table 4.2.3.3 Conversion table: units of area (<1 m^2)

Unit	Symbol	m^2	gal (UK).in^{-1}	gal (US).in^{-1}	ft^2	gal (UK).ft^{-1}	gal (US).ft^{-1}	in^2
square metre	m^2	1	5.587216449	6.709970130	10.763910417	67.046597385	80.519641559	1550.003100006
square yard (US)	sq. yd (US)	8.361307045E-01	4.671643226	5.610412052	9.000036000	56.059718707	67.324944624	1296.005184016
square yard (WMA, 1963)	sq. yd, yd^2	8.361273600E-01	4.671624539	5.610389610	9	56.059494469	67.324675325	1296.000000000
square yard (old)	sq. yd (old)	8.361257141E-01	4.671615343	5.610378566	8.999982283	56.059384115	67.324542796	1295.997448820
barrel (US, petr.) per foot	bbl (US, petr.).ft^{-1}	5.216118600E-01	2.914358364	7/2	5.614583333	34.972300369	42	808.5
barrel (US, liq.) per foot	bbl (US, liq.).ft^{-1}	3.912088950E-01	2.185768773	21/8	4.210937500	26.229225277	31.5	606.375
gallon (UK) per inch	gal (UK).in^{-1}	1.789800000E-01	1	1.200950454	1.926524686	12	14.411405446	277.419554839
gallon (US) per inch	gal (US).in^{-1}	1.490319600E-01	8.326738183E-01	1	1.604166667	9.992085820	12	231
square foot (US, survey)	ft^2(US, survey)	9.290341161E-02	5.190714695E-01	6.233791169E-01	1	6.228857634	7.480549403	144.000576002
square foot	sq. ft, ft^2	9.2903040E-02	5.190693932E-01	6.233766234E-01	1	6.228832719	7.480519481	144
gallon (UK) per foot	gal (UK).ft^{-1}	1.491500000E-02	1/12	1.000792045E-01	1.605437239E-01	1	1.200950454	23.11829624
gallon (US) per foot	gal (US).ft^{-1}	1.241933000E-02	6.938948486E-02	1/12	1.336805556E-01	8.326738183E-01	1	19.25000000
square decimetre	dm^2	1E-02	5.587216449E-02	6.709970130E-02	1.076391042E-01	6.704659739E-01	8.051964156E-01	15.50003100
square inch	sq. in, in^2	6.451600E-04	3.604648564E-03	4.329004329E-03	7E-03	4.325578277E-02	5.194805195E-02	1
circular inch	cin	5.067074791E-04	2.831084362E-03	3.399992049E-03	5.454153912E-03	3.397301234E-02	4.079990459E-02	7.853981634E-01
square centimetre	cm^2	1E-04	5.587216449E-04	6.709970130E-04	1.076391042E-03	6.704659739E-03	8.051964156E-03	1.550003100E-01
gallon (UK) per mile (int.)	gal (UK).mi^{-1}	2.824805340E-06	1.578279886E-05	1.895435946E-05	3.040595163E-05	1.893935863E-04	2.274523135E-04	4.378457034E-03
gallon (US) per mile (int.)	gal (US).mi^{-1}	2.352141449E-06	1.314192339E-05	1.578279886E-05	2.531823984E-05	1.577030807E-04	1.893935863E-04	3.645826537E-03
square millimetre	mm^2	1E-06	5.587216449E-06	6.709970130E-06	1.076391042E-05	6.704659739E-05	8.051964156E-05	1.550003100E-03
circular millimetre	cmm	7.853981634E-07	4.388189537E-06	5.269998216E-06	8.453955472E-06	5.265827445E-05	6.323997860E-05	1.217369588E-03
circular mil	cmil	5.067074791E-10	2.831084362E-09	3.399992049E-09	5.454153912E-09	3.397301234E-08	4.079990459E-08	7.853981634E-07
square micron	μm^2, μ^2	1E-12	5.587216449E-12	6.709970130E-12	1.076391042E-11	6.704659739E-11	8.051964156E-11	1.550003100E-09
square nanometre	nm^2	1E-18	5.587216449E-18	6.709970130E-18	1.076391042E-17	6.704659739E-17	8.051964156E-17	1.550003100E-15
square ångström	$Å^2$	1E-20	5.587216449E-20	6.709970130E-20	1.076391042E-19	6.704659739E-19	8.051964156E-19	1.550003100E-17
square bohr (a.u.)	a.u., a_0^2	2.800285609E-21	1.564580181E-20	1.878983279E-20	3.014202343E-20	1.877496218E-19	2.254779935E-19	4.340451374E-18
square picometre	pm^2	1E-24	5.587216449E-24	6.709970130E-24	1.076391042E-23	6.704659739E-23	8.051964156E-23	1.550003100E-21
barn	b	1E-28	5.587216449E-28	6.709970130E-28	1.076391042E-27	6.704659739E-27	8.051964156E-27	1.550003100E-25
square fermi	F^2	1E-30	5.587216449E-30	6.709970130E-30	1.076391042E-29	6.704659739E-29	8.051964156E-29	1.550003100E-27
square femtometre	fm^2	1E-30	5.587216449E-30	6.709970130E-30	1.076391042E-29	6.704659739E-29	8.051964156E-29	1.550003100E-27
square attometre	am^2	1E-36	5.587216449E-36	6.709970130E-36	1.076391042E-35	6.704659739E-35	8.051964156E-35	1.550003100E-33
shed	shed	1E-52	5.587216449E-52	6.709970130E-52	1.076391042E-51	[illegible]	[illegible]	[illegible]

Table 4.2.3.3 *(continued)*

Unit	Symbol	cm^2	gal (UK) .mi^{-1}	gal (US).mi^{-1}	mm^2	circular mm	circular mil	μm^2
square metre	m^2	**10000**	3.540066941E+05	4.251444999E+05	**1E+06**	1.273239545E+06	1.973525241E+09	**1E+12**
square yard (US)	sq. yd (US)	8361.307045195	2.959958665E+05	3.554763703E+05	8.361307045E+05	1.064594678E+06	1.650125050E+09	8.361307045E+11
square yard (WMA, 1963)	sq. yd, yd^2	8361.273600000	2.959946826E+05	3.554749484E+05	8.361273600E+05	1.064590419E+06	1.650118450E+09	8.361273600E+11
square yard (old)	sq. yd (old)	8361.257140808	2.959940999E+05	3.554742486E+05	8.361257141E+05	1.064588324E+06	1.650115202E+09	8.361257141E+11
barrel (US, petr.) per foot	bbl (US, petr.).ft^{-1}	5216.118600000	1.846540902E+05	2.217604134E+05	5.216118600E+05	6.641368472E+05	1.029414172E+09	5.216118600E+11
barrel (US, liq.) per foot	bbl (US, liq.).ft^{-1}	3912.088950000	1.384905676E+05	1.663203100E+05	3.912088950E+05	4.981026354E+05	7.720606289E+08	3.912088950E+11
gallon (UK) per inch	gal (UK).in^{-1}	1789.800000000	6.336011811E+04	7.609236260E+04	1.789800000E+05	2.278844137E+05	3.532215477E+08	1.789800000E+11
gallon (US) per inch	gal (US).in^{-1}	1490.319600000	5.275831148E+04	6.336011811E+04	1.490319600E+05	1.897533849E+05	2.941183348E+08	1.490319600E+11
square foot (US, survey)	ft^2 (US,Survey)	929.0341	3.288842962E+04	3.949737447E+04	9.290341161E+04	1.182882975E+05	1.833472278E+08	9.290341161E+10
square foot	sq. ft, ft^2	929.0304	3.288829806E+04	3.949721648E+04	9.290304000E+04	1.182878244E+05	1.833464944E+08	9.290304000E+10
gallon (UK) per foot	gal (UK).ft^{-1}	149.150000000	5280.009842520	6341.030216701	1.491500000E+04	1.899036781E+04	2.943512898E+07	1.491500000E+10
gallon (US) per foot	gal (US).ft^{-1}	124.193300000	4396.525956252	5280.009842520	1.241933000E+04	1.581278208E+04	2.450986124E+07	1.241933000E+10
square decimetre	dm^2	**100**	3540.066941012	4251.444999464	**1E+04**	1.273239545E+04	1.973525241E+07	**1E+10**
square inch	sq. in, in^2	**6.4516**	228.390958766	274.286225585	**645.16**	821.443224681	1.273239545E+06	6.4516E+08
circular inch	cin	5.067074791	179.377839552	215.423897820	506.70747910	**645.16**	**1E+06**	5.067074791E+08
square centimetre	cm^2	1	35.400669410	42.514449995	**100**	127.323954474	1.973525241E+05	**1E+08**
gallon (UK) per mile(int.)	gal (UK).mi^{-1}	2.824805340E-02	1	1.200950454	2.824805340	3.596653865	5574.824641100	2.824805340E+06
gallon(US) per mile(int.)	gal (US).mi^{-1}	2.352141449E-02	8.326738183E-01	1	2.352141449	2.994839507	4642.010520279	2.352141449E+06
square millimetre	mm^2	**1E-02**	3.540066941E-01	4.251444999E-01	1	1.273239545	1973.525241390	**1E+06**
circular millimetre	cmm	7.853981634E-03	2.780362074E-01	3.339077094E-01	7.853981634E-01	1	1550.003100006	7.853981634E+05
circular mil	cmil	5.067074791E-06	1.793778396E-04	2.154238978E-04	5.067074791E-04	6.451600000E-04	1	506.707479097
square micron	μm^2, μ^2	**1E-08**	3.540066941E-07	4.251444999E-07	**1E-06**	1.273239545E-06	1.973525241E-03	1
square nanometre	nm^2	**1E-14**	3.540066941E-13	4.251444999E-13	**1E-12**	1.273239545E-12	1.973525241E-09	**1E-06**
square ångström	$Å^2$	**1E-16**	3.540066941E-15	4.251444999E-15	**1E-14**	1.273239545E-14	1.973525241E-11	**1E-08**
square bohr (a.u.)	a.u., a_0^2	2.800285609E-17	9.913198508E-16	1.190526025E-15	2.800285609E-15	3.565434373E-15	5.526434332E-12	2.800285609E-09
square picometre	pm^2	**1E-20**	3.540066941E-19	4.251444999E-19	**1E-18**	1.273239545E-18	1.973525241E-15	**1E-12**
barn	b	**1E-24**	3.540066941E-23	4.251444999E-23	**1E-22**	1.273239545E-22	1.973525241E-19	**1E-16**
square fermi	F^2	**1E-26**	3.540066941E-25	4.251444999E-25	**1E-24**	1.273239545E-24	1.973525241E-21	**1E-18**
square femtometre	fm^2	**1E-26**	3.540066941E-25	4.251444999E-25	**1E-24**	1.273239545E-24	1.973525241E-21	**1E-18**
square attometre	am^2	**1E-32**	3.540066941E-31	4.251444999E-31	**1E-30**	1.273239545E-30	1.973525241E-27	**1E-24**
shed	shed	**1E-48**	3.540066941E-47	4.251444999E-47	**1E-46**	1.273239545E-46	1.973525241E-43	**1E-40**

Table 4.2.4.1 Conversion table: units of volume: multiples and submultiples (SI)

Unit	Symbol	Ym^3	Zm^3	Em^3	Pm^3	Tm^3	Gm^3	Mm^3	km^3	hm^3
cubic yottametre	Ym^3	1	1E+09	1E+18	1E+27	1E+36	1E+51	1E+54	1E+63	1E+66
cubic zetametre	Zm^3	1E-09	1	1E+09	1E+18	1E+27	1E+42	1E+45	1E+54	1E+57
cubic exametre	Em^3	1E-18	1E-09	1	1E+09	1E+18	1E+33	1E+36	1E+45	1E+48
cubic petametre	Pm^3	1E-27	1E-18	1E-09	1	1E+09	1E+24	1E+27	1E+36	1E+39
cubic terametre	Tm^3	1E-36	1E-27	1E-18	1E-09	1	1E+15	1E+18	1E+27	1E+30
cubic gigametre	Gm^3	1E-51	1E-42	1E-33	1E-24	1E-15	1	1E+03	1E+12	1E+15
cubic megametre	Mm^3	1E-54	1E-45	1E-36	1E-27	1E-18	1E-03	1	1E+09	1E+12
cubic kilometre	km^3	1E-63	1E-54	1E-45	1E-36	1E-27	1E-12	1E-09	1	1E+03
cubic hectometre	hm^3	1E-66	1E-57	1E-48	1E-39	1E-30	1E-15	1E-12	1E-03	1
cubic decametre	dam^3	1E-69	1E-60	1E-51	1E-42	1E-33	1E-18	1E-15	1E-06	1E-03
cubic metre	m^3	1E-72	1E-63	1E-54	1E-45	1E-36	1E-21	1E-18	1E-09	1E-06
cubic decimetre	dm^3	1E-75	1E-66	1E-57	1E-48	1E-39	1E-24	1E-21	1E-12	1E-09
cubic centimetre	cm^3	1E-78	1E-69	1E-60	1E-51	1E-42	1E-27	1E-24	1E-15	1E-12
cubic millimetre	mm^3	1E-81	1E-72	1E-63	1E-54	1E-45	1E-30	1E-27	1E-18	1E-15
cubic micrometre	μm^3	1E-90	1E-81	1E-72	1E-63	1E-54	1E-39	1E-36	1E-27	1E-24
cubic nanometre	nm^3	1E-93	1E-84	1E-75	1E-66	1E-57	1E-42	1E-39	1E-30	1E-27
cubic picometre	pm^3	1E-108	1E-99	1E-90	1E-81	1E-72	1E-57	1E-54	1E-45	1E-42
cubic femtometre	fm^3	1E-117	1E-108	1E-99	1E-90	1E-81	1E-66	1E-63	1E-54	1E-51
cubic attometre	am^3	1E-126	1E-117	1E-108	1E-99	1E-90	1E-75	1E-72	1E-63	1E-60
cubic zeptometre	zm^3	1E-135	1E-126	1E-117	1E-108	1E-99	1E-84	1E-81	1E-72	1E-69
cubic yoctometre	ym^3	1E-144	1E-135	1E-126	1E-117	1E-108	1E-93	1E-90	1E-81	1E-78

am³	m³	dm³	cm³	mm³	µm³	nm³	pm³	fm³	am³	zm³	ym³
+69	1E+72	1E+75	1E+78	1E+81	1E+90	1E+93	1E+108	1E+117	1E+126	1E+135	1E+144
+60	1E+63	1E+66	1E+691	1E+72	1E+81	1E+84	1E+99	1E+108	1E+117	1E+126	1E+135
+51	1E+54	1E+57	1E+60	1E+63	1E+72	1E+75	1E+90	1E+99	1E+108	1E+117	1E+126
+42	1E+45	1E+48	1E+51	1E+54	1E+63	1E+66	1E+81	1E+90	1E+99	1E+108	1E+117
+33	1E+36	1E+39	1E+42	1E+45	1E+54	1E+57	1E+72	1E+81	1E+90	1E+99	1E+108
+18	1E+21	1E+24	1E+27	1E+30	1E+39	1E+42	1E+57	1E+66	1E+75	1E+84	1E+93
+15	1E+18	1E+21	1E+24	1E+27	1E+36	1E+39	1E+54	1E+63	1E+72	1E+81	1E+90
+06	1E+09	1E+12	1E+15	1E+18	1E+27	1E+30	1E+45	1E+54	1E+63	1E+72	1E+81
+03	1E+06	1E+09	1E+12	1E+15	1E+24	1E+27	1E+42	1E+51	1E+60	1E+69	1E+78
	1E+03	1E+06	1E+09	1E+12	1E+21	1E+24	1E+39	1E+48	1E+57	1E+66	1E+75
-03	1	11E+03	1E+06	1E+09	1E+18	1E+21	1E+36	1E+45	1E+54	1E+63	1E+72
-06	1E-03	1	1E+03	1E+06	1E+15	1E+18	1E+33	1E+42	1E+51	1E+60	1E+69
-09	1E-06	1E-03	1	1E+03	1E+12	1E+15	1E+30	1E+39	1E+48	1E+57	1E+66
-12	1E-09	1E-06	1E-03	1	1E+09	1E+12	1E+27	1E+36	1E+45	1E+54	1E+63
-21	1E-18	1E-15	1E-12	1E-09	1	1E+03	1E+18	1E+27	1E+36	1E+45	1E+54
-24	1E-21	1E-18	1E-15	1E-12	1E-03	1	1E+15	1E+24	1E+33	1E+42	1E+51
-39	1E-36	1E-33	1E-30	1E-27	1E-18	1E-15	1	1E+09	1E+18	1E+27	1E+36
-48	1E-45	1E-42	1E-39	1E-36	1E-27	1E-24	1E-09	1	1E+09	1E+18	1E+27
-57	1E-54	1E-51	1E-48	1E-45	1E-36	1E-33	1E-18	1E-09	1	1E+09	1E+18
-66	1E-63	1E-60	1E-57	1E-54	1E-45	1E-42	1E-27	1E-18	1E-09	1	1E+09
-75	1E-72	1E-69	1E-66	1E-63	1E-54	1E-51	1E-36	1E-27	1E-18	1E-09	1

Table 4.2.4.2 Conversion table: units of volume (>1 L)

Unit	Symbol	m^3	ft^3	gallon (UK)
cubic int. mile (cubem)	cu mi (int. stat.)	4.168205135E+09	1.471987752E+11	9.168765470E+11
cubic US int. mile	cu mi (int. US)	4.168181825E+09	1.471979520E+11	9.168714195E+11
acre-foot (US, survey)	ac-ft (US)	1233.4892384682	4.356026136E+04	2.713295812E+05
acre-foot	ac-ft	1233.4818375475	**43560**	2.713279532E+05
acre-inch	ac-in	102.7901531290	**3630**	2.261066277E+04
forty feet equiv. unit	FEU, FEQ	72.4911272755	**2560**	1.594581176E+04
twenty feet equiv. unit	TEU, TEQ	36.2455636378	**1280**	7.972905880E+03
decastère	dast	**10**	353.1466672149	2199.69151526190
standart (Petrograd)	std	4.67227968768	**165**	1027.7573985920
cord	cord	3.624556364	**128**	797.2905879987
last	last	2.909498	102.7479521968	639.9998064271
register ton (UK)	–	2.83168466	**100**	622.8832720499
tonneau de jauge	–	2.83168466	**100**	622.8832720499
wey (UK, capacity)	wey (UK)	1.45474944	51.37399163687	**320**
chaldron (UK)	chal (UK)	1.309274496	46.2365924732	**288**
stère	st	**1**	35.3146667215	219.9691515262
cubic metre	m^3	**1**	35.3146667215	219.9691515262
tun (US)	tu (US)	9.539237696E-01	**101/3**	209.8338022125
cubic yard	cu yd, yd^3	7.645548580E-01	**27**	168.1784834060
perche (masonry)	–	7.008419532E-01	**24.75**	154.1636097994
butt (UK)	but (UK)	0.490977936	17.33872218	**108**
cord-foot	cord-ft	4.530695455E-01	**16**	99.6613234998
puncheon (UK)	pun (UK)	3.182264400E-01	11.2380606706	**70**
quarter (UK)	qtr (UK)	2.909498880E-01	10.2747983274	**64**
seam (UK)	seam (UK)	2.909498880E-01	10.2747983274	**64**
muid (de Paris)	muid	**0.274**	9.6762186817	60.2715475182
hogshead (US, liquid)	hhd, hg (US)	2.384809424E-01	8.4218750000	52.45845055313
barrel (UK, alcohol)	bbl (UK, alc.)	2.273046000E-01	8.0271861933	**50**
cran (mease)	cran	1.704784500E-01	6.0203896449	**37.5**
barrel (UK, beer)	bbl (UK, beer)	1.636593120E-01	5.7795740591	**36**
barrel (US, petrol.)	bbl (Petrol.)	1.589872949E-01	5.6145833333	34.9723003688
coomb (UK)	cmb (UK)	1.454749440E-01	5.13739916369	**32**
barrel (UK, wine)	bbl (UK, wine)	1.432018980E-01	5.0571273018	**31.5**
barrel (US, liquid)	bbl (US, liq.)	1.192404712E-01	4.2109375000	26.2292252766
barrel (US, dry)	bbl (US, dry)	1.154550594E-01	4.0772569444	25.3965514583
sack (UK)	sk (UK)	1.090833300E-01	3.85304937277	**24**
bag (UK)	bg (UK)	1.091061600E-01	3.85304937277	**24**
decistère	dst	**1E-01**	3.5314666721	21.9969151526
barrel (US, cranb.)	bbl (US, cranb.)	9.547100000E-02	3.3715265466	21.0006748654
kilderkin (UK)	kil (UK)	8.182965600E-02	2.8897870296	**18**
firkin (UK)	fir (UK)	4.091482800E-02	1.4448935148	**9**
bushel (UK)	bu (UK)	3.636873600E-02	1.2843497909	**8**
bushel (US, dry)	bu (US, dry)	3.523907200E-02	1.2444560833	7.7515087684
Winchester bushel	bu (Winchester)	3.523907200E-02	1.2444560833	7.7515087684
firkin (US, liquid)	fir (US, liq.)	3.406870606E-02	1.2031250000	7.4940643647
cubic foot	cu. ft, ft^3	2.831684659E-02	1	6.2288327187
bucket (UK)	bk (UK)	1.818436800E-02	6.421748955E-01	**4**
peck (UK)	pk (UK)	9.092184000E-03	3.210874477E-01	**2**
peck (US, dry)	pk (US, dry)	8.809768000E-03	3.111140208E-01	1.9378771921
gallon (imperial, UK)	gal (UK)	4.546092000E-03	1.605437239E-01	1
gallon (Canadian, liquid)	gal (Can.)	4.546092000E-03	1.605437239E-01	1
gallon (US, dry)	gal (US, dry)	4.404884000E-03	1.555570104E-01	9.689385961E-01
gallon (US, liquid)	gal (US, liq.)	3.785411784E-03	1.336805556E-01	8.326738183E-01
board foot measure	B.M., fbm	2.359737216E-03	**1/12**	5.190693932E-01
pottle (UK)	pot (UK)	2.273046000E-03	8.027186193E-02	**1/2**
quart (UK)	qt (UK)	1.136523000E-03	4.013593097E-02	**1/4**
quart (US, dry)	qt (US, dry)	1.101221000E-03	3.888925260E-02	2.422346490E-01
litre (1964)	L, l	**1E-03**	3.531466672E-02	2.199691515E-01
cubic decimetre	dm^3	**1E-03**	3.531466672E-02	2.199691515E-01

gallon (US, dry)	gallon (US, liq.)	dm^3, l	in^3	cm^3
.462689904E+11	1.101123305E+12	4.168205135E+12	2.543594835E+14	4.168205135E+15
.462636985E+11	1.101117147E+12	4.168181825E+12	2.543580611E+14	4.168181825E+15
.800276326E+05	3.258533837E+05	1.233489238E+06	7.527213163E+07	1.233489238E+09
.800259525E+05	3.258514286E+05	1.233481838E+06	7.527168000E+07	1.233481838E+09
.333549604E+04	2.715428571E+04	1.027901531E+05	6.272640000E+06	1.027901531E+08
.645698894E+04	1.915012987E+04	7.249112728E+04	**4423680**	7.249112728E+07
.228494471E+03	9.575064935E+03	3.624556364E+04	**2211840**	3.624556364E+07
270.20734257701	2641.72052358148	**10000**	6.102374409E+05	**1E+07**
060.70436535446	1234.2857142857	4672.279688	**285120**	4.672279688E+06
22.84944706285	957.5064935065	3624.556364	**221184**	3.624556364E+06
60.51637228131	768.6080579919	2909.498	177548.4613961350	2.909498000E+06
42.85113069947	748.0519482633	2831.68466	**172800**	2.831684660E+06
42.85113069947	748.0519482633	2831.68466	**172800**	2.831684660E+06
30.25828602978	384.3041452317	1454.74944	8.877425755E+04	1.454749440E+06
97.23245742680	345.8737307085	1309.274496	7.989683179E+04	1.309274496E+06
27.02073425770	264.1720523581	**1000**	6.102374409E+04	**1E+06**
27.0207342577	264.1720523581	**1000**	6.102374409E+04	**1E+06**
16.56047459320	**252**	953.9237696	**58212**	9.539237696E+05
73.56980523982	201.9740259740	764.554858	**46656**	7.645548580E+05
59.10565481407	185.1428571555	700.8419532	**42768**	7.008419532E+05
11.46217153505	129.702649	490.977936	29961.31192	4.909779360E+05
02.85618088286	119.6883116883	453.0695455	**27648**	4.530695455E+05
2.24400006901	84.0665317694	318.22644	19419.3688387377	3.182264400E+05
6.05165720596	76.8608290463	290.949888	17754.8515097030	2.909498880E+05
6.05165720596	76.8608290463	290.949888	17754.8515097030	2.909498880E+05
2.20368118661	72.3831423461	**274**	16720.5058819566	**274000**
4.14011864830	**63**	238.4809424	**14553**	2.384809424E+05
1.60285719215	60.04752269245	227.3046	13870.97774195550	2.273046000E+05
8.70214289411	45.0356420193	170.47845	10403.2333064666	1.704784500E+05
7.15405717835	43.2342163386	163.659312	9987.1039742080	1.636593120E+05
6.09341243220	**42**	158.9872949	**9702**	1.589872949E+05
3.02582860298	38.4304145232	145.47494	8.877425755E+03	1.454749440E+05
2.50980003106	37.8299392962	143.201898	8738.7159774320	1.432018980E+05
7.07005932415	**31.5**	119.2404712	**7276.5**	1.192404712E+05
6.21069236148	**30.5**	115.4550594	**7045.5**	1.154550594E+05
4.76417767188	28.8228108924	109.106208	6.658069316E+03	1.091062080E+05
4.76936055524	28.8228108924	109.106208	6658.0693161386	1.091062080E+05
2.70207342577	26.4172052358	**100**	6102.3744094732	**1E+05**
1.67389652032	25.2207700107	95.471	5825.9978724682	9.547100000E+04
8.57702858918	21.6171081693	81.829656	4993.5519871040	8.182965600E+04
.28851429459	10.8085540846	40.914828	2496.7759935520	4.091482800E+04
.25645715074	9.6076036308	36.368736	2219.3564387129	3.636873600E+04
	9.3091779734	35.239072	2150.4201118639	3.523907200E+04
	9.3091779734	35.239072	2150.4201118639	3.523907200E+04
.73430266404	**9**	34.06870606	2079.0000000000	3.406870606E+04
.42851130518	7.4805194805	28.31684659	**1728**	2.831684659E+04
.12822857537	4.8038018154	18.184368	1109.6782193564	1.818436800E+04
.06411428769	2.4019009077	9.092184	554.8391096782	9092.184
	2.3272944934	8.809768	537.6050279660	8809.768
.03205714384	1.2009504538	4.546092	277.4195548391	4546.092
.03205714384	1.2009504538	4.546092	277.4195548391	4546.092
	1.1636472467	4.404884	268.8025139830	4404.884
.59366963E-01	1	3.785411784	**231**	3785.411784
.35709275E-01	6.233766234E-01	2.359737216	144	2359.737216
.16028572E-01	6.004752269E-01	2.273046	138.7097774196	2273.046
.58014286E-01	3.002376135E-01	1.136523	69.3548887098	1136.523
/4	2.909118117E-01	1.101221	67.2006284957	1101.221
.27020734E-01	2.641720524E-01	**1**	61.0237440947	**1000**
.27020734E-01	2.641720524E-01	**1**	61.0237440947	**1000**

Table 4.2.4.3 Conversion table: units of volume (<1 L)

Unit	Symbol	m^3	ft^3	dm^3, L	pint (UK)
litre (old)	**L (old)**	1.000002800E-03	3.531476561E-02	1.0000028	1.75975814
litre (1964)	**L, l**	**1E-03**	3.531466672E-02	**1**	1.759753212
cubic decimetre	**dm^3**	**1E-03**	3.531466672E-02	**1**	1.759753212
quart (US liquid)	**qt (US, liq.)**	9.463529460E-04	3.342013889E-02	9.463529460E-01	1.665347637
chopine (UK)	**ch (UK)**	5.682615000E-04	2.006796548E-02	5.682615000E-01	**1**
pint (UK)	**pt (UK)**	5.682615000E-04	2.006796548E-02	5.682615000E-01	**1**
pint (US, dry)	**pt (US, dry)**	5.506105000E-04	1.944462630E-02	5.506105000E-01	9.689385961E-01
chopine (US, dry)	**ch (US, dry)**	5.506105000E-04	1.944462630E-02	5.506105000E-01	9.689385961E-01
chopine (US, liquid)	**ch (US, liq.)**	4.731764730E-04	1.671006944E-02	4.731764730E-01	8.326738183E-01
pint (US, liquid)	**pt (US, liq.)**	4.731764730E-04	1.671006944E-02	4.731764730E-01	8.326738183E-01
démiard	–	2.841300000E-04	1.003395626E-02	2.841300000E-01	4.999986802E-01
noggin (UK)	**ng (UK)**	1.420653750E-04	5.016991371E-03	1.420653750E-01	**1/4**
gill, roquille (UK)	**gi (UK)**	1.420653750E-04	5.016991371E-03	1.420653750E-01	**1/4**
gill, roquille (US)	**gi (US, liq.)**	1.18294125E-04	4.177513185E-03	1.182941183E-01	2.081684546E-01
decilitre	**dl, dL**	**1E-04**	3.531466672E-03	**1E-01**	1.759753212E-01
fluid ounce (US)	**fl. oz (US)**	2.957352956E-05	1.044379340E-03	2.957352956E-02	5.204211364E-02
fluid ounce (UK)	**fl. oz (UK)**	2.841307500E-05	1.003398274E-03	2.841307500E-02	**1/20**
cubic inch	**cu. in, in^3**	1.63870640E-05	5.787037037E-04	1.638706400E-02	2.883718851E-02
tablespoon (metric)	–	1.50E-05	5.297200008E-04	1.50E-02	2.639629818E-02
tablespoon (US)	–	1.478677000E-05	5.221898544E-04	1.478677000E-02	2.602106601E-02
centilitre	**cl, cL**	**1E-05**	3.531466672E-04	**1E-02**	1.759753212E-02
teaspoon (metric)	–	**5E-06**	1.765733336E-04	**5E-03**	8.798766061E-03
teaspoon (US)	–	4.928922000E-06	1.740632377E-04	4.928922000E-03	8.673686322E-03
fluid dram (US)	**fl. dr (US, liq.)**	3.696691195E-06	1.305474175E-04	3.696691195E-03	6.505264205E-03
fluid drachm (UK)	**fl. dr (UK)**	3.551634375E-06	1.254247843E-04	3.551634375E-03	**1/160**
fluid scruple (UK)	**fl. sc (UK)**	1.183878125E-06	4.180826142E-05	1.183878125E-03	**1/480**
Mohr centicube	**cc**	1.002380000E-06	3.539871563E-05	1.002380000E-03	1.763941425E-03
cubic centimetre	**cm^3**	**1E-06**	3.531466672E-05	**1E-03**	1.759753212E-03
millilitre	**ml, mL**	**1E-06**	3.531466672E-05	**1E-03**	1.759753212E-03
minim (US, liquid)	**min (US, liq.)**	6.161151992E-08	2.175790292E-06	6.161151992E-05	1.084210701E-04
minim (UK)	**min (UK)**	5.919390625E-08	2.090413071E-06	5.919390625E-05	**1/9600**
microlitre	**μl, μL**	**1E-09**	3.531466672E-08	**1E-06**	1.759753212E-06
lambda	**λ**	**1E-09**	3.531466672E-08	**1E-06**	1.759753212E-06

pint (US dry)	fl. oz (US)	fl. oz (UK)	in³	cm³	mm³
.81617096	33.814117384	35.195162794	61.023914967	1000.0028001	1.00000280E+06
.816165874	33.814022702	35.195064244	61.023744095	**1000**	**1E+06**
.816165874	33.814022702	35.195064244	61.023744095	**1000**	**1E+06**
	32	33.30695273	**57.75**	946.3529460	9.463529460E+05
.032057144	19.215207262	**20**	34.677444355	568.2615	5.682615E+05
.032057144	19.215207262	**20**	34.677444355	568.2615	5.682615E+05
	18.618355947	19.378771921	33.600314248	550.610500000	5.506105000E+05
	18.618355947	19.378771921	33.600314248	550.610500000	5.506105000E+05
.593669627E-01	**16**	16.653476366	**28.875**	473.176473000	4.731764730E+05
.593669627E-01	**16**	16.653476366	**28.875**	473.176473000	4.731764730E+05
.160272098E-01	9.607578270	9.999973604	17.338676410	284.130000000	2.841300000E+05
2.580142860E-01	4.803801815	5	8.669361089	142.065375000	1.420653750E+05
2.580142860E-01	4.803801815	5	8.669361089	142.065375000	1.420653750E+05
2.148417407E-01	**4**	4.163369092	**7.21875**	118.294118250	1.182941183E+05
1.816165874E-01	3.381402270	3.519506424	6.102374409	**100**	1.000000000E+05
5/93	**1**	1.040842273	1.804687500	29.573529563	2.957352956E+04
5.160285719E-02	9.607603631E-01	**1**	1.733872218	28.413075000	2.841307500E+04
2.976162641E-02	5.541125541E-01	5.767437703E-01	**1**	16.387064000	1.638706400E+04
2.724248811E-02	5.072103405E-01	5.279259637E-01	9.153561614E-01	**15**	1.500000000E+04
2.685522706E-02	5.000001765E-01	5.204213201E-01	9.023440685E-01	14.786770000	1.478677000E+04
1.816165874E-02	3.381402270E-01	3.519506424E-01	6.102374409E-01	**10**	**10000**
9.080829370E-03	1.690701135E-01	1.759753212E-01	3.051187205E-01	**5**	**5000**
8.951739932E-03	1.666666804E-01	1.734737264E-01	3.007812748E-01	4.928922000	4928.922000000
19/2830	**1/8**	1.301052841E-01	2.255859375E-01	3.696691195	3696.691195313
6.450357149E-03	1.200950454E-01	**1/8**	2.167340272E-01	3.551634375	3551.634375000
2.150119050E-03	4.003168179E-02	**1/24**	7.224467574E-02	1.183878125	1183.878125000
1.820488349E-03	3.389450008E-02	3.527882850E-02	6.116898061E-02	1.002380000	1002.380000000
1.816165874E-03	3.381402270E-02	3.519506424E-02	6.102374409E-02	**1**	**1000**
1.816165874E-03	3.381402270E-02	3.519506424E-02	6.102374409E-02	**1**	**1000**
1.118967399E-04	**1/480**	2.168421402E-03	3.759765625E-03	6.161152E-02	61.611519922
1.075059525E-04	2.001584090E-03	**1/480**	3.612233787E-03	5.919391E-02	59.193906250
1.816165874E-06	3.381402270E-05	3.519506424E-05	6.102374409E-05	**1E-03**	**1**
1.816165874E-06	3.381402270E-05	3.519506424E-05	6.102374409E-05	**1E-03**	**1**

Table 4.2.5 Conversion table: units of linear mass density

Unit	Symbol	$kg.m^{-1}$	denier	drex	poumar	tex	$gr.(100\ yd)^{-1}$	$gr.(120\ yd)^{-1}$
kilogram per metre	$kg.m^{-1}$	1	9.0009001E+06	1E+07	2.0159069E+06	1E+06	1.4111348E+11	1.6933618E+11
tex	tex	1E-06	9.000900090	10	2.015906925	1	1.4111348E+05	1.6933618E+05
poumar	–	4.9605465E-07	4.464938325	4.960546479	1	0.496054648	70000	84000
denier	denier	1.1110E-07	1	1.111	2.2396726E-01	0.111100000	1.5677708E+04	1.8813250E+04
drex	drex	1E-07	9.0009E-01	1	2.0159069E-01	0.1	1.4111348E+04	1.6933618E+04
grain (av.) per 100 yards	$gr.(100\ yd)^{-1}$	7.0864950E-12	6.3784833E-05	7.0864950E-05	1.4285714E-05	7.0864950E-06	1	1.2
grain (av.) per 120 yards	$gr.(120\ yd)^{-1}$	5.9054125E-12	5.3154028E-05	5.9054125E-05	1.1904762E-05	5.9054125E-06	8.3333333E-01	1

Unit	Symbol	$kg.m^{-1}$	$kg.dm^{-1}$	$kg.cm^{-1}$	$t.km^{-1}$	$lb.in^{-1}$	$lb.ft^{-1}$	$UK\ ton.mi^{-1}$
ton (metric) per metre	$t.m^{-1}$	1000	100	10	1000	55.997414595	671.968975140	1583.926869660
kilogram per centimetre	$kg.cm^{-1}$	100	10	1	100	5.599741459	67.196897514	158.392686966
pound per inch	$lb.in^{-1}$	17.85796732	1.785796732	1.7857967E-01	17.857967323	1	12	28.285714280
kilogram per decimetre	$kg.dm^{-1}$	10	1	0.1	10	5.5997415E-01	6.719689751	15.839268697
pound per foot	$lb.ft^{-1}$	1.488163944	1.4881639E-01	1.4881639E-02	1.488163944	8.3333333E-02	1	2.357142857
UK ton per 1000 yards	(UK) ton. $(1000\ yd)^{-1}$	1.111162411	1.1111624E-01	1.1111624E-02	1.111162411	6.2222222E-02	7.4666667E-01	1.760000000
kilogram per metre	$kg.m^{-1}$	1	0.10	0.01	1	5.5997415E-02	6.7196898E-01	1.583926870
ton (metric) per kilometre	$t.km^{-1}$	1	0.10	0.01	1	5.5997415E-02	6.7196898E-01	1.583926870
UK ton per mile	$UK\ ton.mi^{-1}$	6.3134228E-01	6.3134228E-02	6.3134228E-03	6.3134228E-01	3.5353535E-02	4.2424242E-01	1
pound per yard	$lb.yd^{-1}$	4.9605465E-01	4.9605465E-02	4.9605465E-03	4.9605465E-01	2.7777778E-02	3.3333333E-01	7.8571429E-01

Table 4.2.6 Conversion table: units of surface mass density

Unit	Symbol	lb.in^{-2}	oz.in^{-2}	g.cm^{-2}	lb.ft^{-2}	kg.m^{-2}	lb.yd^{-2}
pound per square inch	lb.in^{-2}	1	16.000000071	70.306957964	144	703.069579639	1296
(UK) ounce per square inch	oz.in^{-2}	6.249999972E-02	1	4.394184853	8.999999960	43.941848534	80.999999643
gram per square centimetre	kg.cm^{-2}	1.422334331E-02	2.275734939E-01	1	2.048161436	10	18.433452926
pound per square foot	lb.ft^{-2}	6.944444444E-03	1.111111116E-01	4.882427636E-01	1	4.882427636	9
kilogram per square metre	kg.m^{-2}	1.422334331E-03	2.275734939E-02	0.1	2.048161436E-01	1	1.843345293
pound per square yard	lb.yd^{-2}	7.716049383E-04	1.234567907E-02	5.424919596E-02	1.111111111E-01	5.424919596E-01	1
(UK) ounce per square foot	oz.ft^{-2}	4.340277759E-04	6.944444444E-03	3.051517259E-02	6.249999972E-02	3.051517259E-01	5.624999975E-01
(UK) ounce per square yard	oz.yd^{-2}	4.822530843E-05	7.716049383E-04	3.390574733E-03	6.944444414E-03	3.390574733E-02	6.249999972E-02
milligram per square centimetre	mg.cm^{-2}	1.422334331E-05	2.275734939E-04	1E-03	2.048161436E-03	1E-02	1.843345293E-02
gram per square metre	g.m^{-2}	1.422334331E-06	2.275734939E-05	1E-04	2.048161436E-04	1E-03	1.843345293E-03
(UK) ton per square mile (int.)	UK ton.(mi (int.))$^{-2}$	5.579787778E-07	8.927660484E-06	3.922979047E-05	8.034894400E-05	3.922979047E-04	7.231404960E-04
pound per acre	lb.ac^{-1}	1.594225079E-07	2.550760138E-06	1.120851156E-05	2.295684114E-05	1.120851156E-04	2.066115702E-04
kilogram per hectare	kg.ha^{-1}	1.422334331E-07	2.275734939E-06	1E-05	2.048161436E-05	1E-04	1.843345293E-04

Unit	Symbol	lb.yd^{-2}	oz.ft^{-2}	mg.cm^{-2}	UK ton.(mi (int.))$^{-2}$	lb.acre^{-1}	kg.ha^{-1}
pound per square inch	lb.in^{-2}	1296	2304.000010159	70306.957963916	1.792182857E+06	6.27264E+06	7.030695796E+06
(UK) ounce per square inch	oz.in^{-2}	80.999999643	144	4394.184853370	1.120114281E+05	3.920399983E+05	4.394184853E+05
gram per square centimetre	g.cm^{-2}	18.433452926	32.770583124	1000	2.549083204E+04	8.921791216E+04	100000
pound per square foot	lb.ft^{-2}	9	16.000000071	488.242763638	1.244571428E+04	43560	48824.276363831
kilogram per square metre	kg.m^{-2}	1.843345293	3.277058312	100	2549.083204126	8921.791216197	10000
pound per square yard	lb.yd^{-2}	1	1.777777786	54.249195960	1382.857142585	4840	5424.919595981
(UK) ounce per square foot	oz.ft^{-2}	5.624999975E-01	1	30.515172593	777.857139274	2722.499987996	3051.517259285
(UK) ounce per square yard	oz.yd^{-2}	6.249999972E-02	1.111111111E-01	3.390574733	86.428571030	302.499998666	339.057473254
milligram per square centimetre	mg.cm^{-2}	1.843345293E-02	3.277058312E-02	1	25.490832041	89.217912162	100
gram per square metre	g.m^{-2}	1.843345293E-03	3.277058312E-03	1E-01	2.549083204	8.921791216	10
(UK) ton per square mile (int.)	UK ton.(mi (int.))$^{-2}$	7.231404960E-04	1.285583110E-03	3.92297905E-02	1	3.500000001	3.922979047
pound per acre	lb.ac^{-1}	2.066115702E-04	3.673094598E-04	1.12085116E-02	2.857142857E-01	1	1.120851156
kilogram per hectare	kg.ha^{-1}	1.843345293E-04	3.277058312E-04	1E-02	2.549083204E-01	8.921791216E-01	1

Table 4.2.7 Conversion table: units of mass density

Unit	Symbol	$lb.in^{-3}$	$kg.dm^{-3}$
gram per cubic millimetre	$g.mm^{-3}$	36.127292000	1000
pound per cubic inch	$lb.in^{-3}$	1	27.679904710
UK ton per cubic yard	UK ton. yd^{-3}	4.801097394E-02	1.328939184
ton (metric) per cubic metre	$t.m^{-3}$	3.612729200E-02	1
kilogram per cubic decimetre	$kg.dm^{-3}$	3.612729200E-02	1
gram per cubic centimetre	$g.cm^{-3}$	3.612729200E-02	1
pound (avdp) per US gallon	$lb.(US\ gal)^{-1}$	4.329004311E-03	1.198264268E-01
pound (avdp) per UK gallon	$lb.(UK\ gal\}^{-1}$	3.604648564E-03	9.977632877E-02
pound (avdp) per cubic foot	$lb.ft^{-3}$	5.787037037E-04	1.601846337E-02
ounce (avdp) per US gallon	$oz.(US\ gal)^{-1}$	2.705627682E-04	7.489151643E-03
ounce (avdp) per UK gallon	$oz.(UK\ gal)^{-1}$	2.252905343E-04	6.236020520E-03
kilogram per cubic metre	$kg.m^{-3}$	3.612729200E-05	1E-03
gram per cubic decimetre	$g.dm^{-3}$	3.612729200E-05	1E-03
grain (avdp) per US gallon	$gr.(US\ gal\}^{-1}$	6.184291872E-07	1.711806097E-05
grain (avdp) per UK gallon	$gr.(UK\ gal)^{-1}$	5.149497949E-07	1.425376125E-05
grain (avdp) per cubic foot	$gr.ft^{-3}$	8.267195767F-08	2.288351911E-06
gram per cubic metre	$g.m^{-3}$	3.612729200E-08	1E-06

Unit	Symbol	$oz.(US\ gal)^{-1}$	$oz.(UK\ gal)^{-1}$
gram per cubic millimetre	$g.mm^{-3}$	1.335264724E+05	1.603586769E+05
pound per cubic inch	$lb.in^{-3}$	3.696000032E+03	4.438712897E+03
UK ton per cubic yard	UK $ton.yd^{-3}$	177.448561203	213.106929211
ton (metric) per cubic metre	$t.m^{-3}$	133.526472385	160.358676934
kilogram per cubic decimetre	$kg.dm^{-3}$	133.526472385	160.358676934
gram per cubic centimetre	$g.cm^{-3}$	133.526472385	160.358676934
pound (avdp) per US gallon	$lb.(US\ gal)^{-1}$	16	19.215207265
pound (avdp) per UK gallon	$lb.(UK\ gal)^{-1}$	13.322781208	16
pound (avdp) per cubic foot	$lb.ft^{-3}$	2.138888907	2.568699593
ounce (avdp) per US gallon	$oz.(US\ gal)^{-1}$	1	1.200950449
ounce (avdp) per UK gallon	$oz.(UK\ gal)^{-1}$	8.326738218E-01	1
kilogram per cubic metre	$kg.m^{-3}$	1.335264724E-01	1.603586769E-01
gram per cubic decimetre	$g.dm^{-3}$	1.335264724E-01	1.603586769E-01
grain (avdp) per US gallon	$gr.(US\ gal)^{-1}$	2.285714296E-03	2.745029609E-03
grain (avdp) per UK gallon	$gr.(UK\ gal)^{-1}$	1.903254458E-03	2.285714296E-03
grain (avdp) per cubic foot	$gr.ft^{-3}$	3.055555582E-04	3.669570847E-04
gram per cubic metre	$g.m^{-3}$	1.335264724E-04	1.603586769E-04

g.cm^{-3}	lb.(US gal)$^{-1}$	lb.(UK gal)$^{-1}$	lb.ft^{-3}	oz.(US gal)$^{-1}$
000	8.345404487E+03	1.002241726E+04	6.242796058E+04	1.335264724E+05
.679904710	**231**	277.419554839	**1728**	3.696000032E+03
.328939184	11.090535026	13.319183017	82.962962963	177.448561203
	8.345404487	10.022417264	62.427960576	133.526472385
	8.345404487	10.022417264	62.427960576	133.526472385
	8.345404487	10.022417264	62.427960576	133.526472385
.198264268E-01	**1**	1.200950449	7.480519449	**16**
977632877E-02	8.326738218E-01	**1**	6.228832719	13.322781208
.601846337E-02	1.336805561E-01	1.605437239E-01	**1**	2.138888907
489151643E-03	6.249999972E-02	7.505940272E-02	4.675324635E-01	**1**
236020520E-03	5.204211363E-02	6.249999972E-02	3.893020432E-01	8.326738218E-01
E-03	8.345404487E-03	1.002241726E-02	6.242796058E-02	1.335264724E-01
E-03	8.345404487E-03	1.002241726E-02	6.242796058E-02	1.335264724E-01
.711806097E-05	1.428571429E-04	1.715643498E-04	1.068645636E-03	2.285714296E-03
.425376125E-05	1.189534031E-04	1.428571429E-04	8.898332455E-04	1.903254458E-03
.288351911E-06	1.909722230E-05	2.293481770E-05	1.428571429E-04	3.055555582E-04
E-06	8.345404487E-06	1.002241726E-05	6.242796058E-05	1.335264724E-04

kg.m^{-3}	grain.(US gal)$^{-1}$	grain.(UK gal)$^{-1}$	grain.ft^{-3}	g.m^{-3}
E+06	5.841783141E+07	7.015692085E+07	4.369957240E+08	**1E+09**
.767990471E+04	1.617000007E+06	1.941936884E+06	1.2096E+07	2.767990471E+07
.328939184E+03	7.763374518E+04	9.323428112E+04	5.807407407E+05	1.328939184E+06
000	5.841783141E+04	7.015692085E+04	4.369957240E+05	**1E+06**
000	5.841783141E+04	7.015692085E+04	4.369957240E+05	**1E+06**
000	5.841783141E+04	7.015692085E+04	4.369957240E+05	**1E+06**
19.826426810	**7000**	8.406653141E+03	5.236363614E+04	1.198264268E+05
9.776328768	5.828716753E+03	**7000**	4.360182903E+04	9.977632877E+04
6.018463374	935.763892844	1.123806067E+03	**7000**	1.601846337E+04
.489151643	437.499998071	525.415819021	3.272727244E+03	7.489151643E+03
.236020520	364.294795441	437.499998071	2.725114302E+03	6.236020520E+03
	58.417831411	70.156920849	436.995724033	**1E+03**
	58.417831411	70.156920849	436.995724033	**1E+03**
.711806097E-02	**1**	1.200950449	7.480519449	17.118060973
.425376125E-02	8.326738218E-01	**1**	6.228832719	14.253761253
.288351911E-03	1.336805561E-01	1.605437239E-01	**1**	2.288351911
E-03	5.841783141E-02	7.015692085E-02	4.369957240E-01	**1**

Table 4.2.8 Conversion table: hydrometer degrees

Liquids heavier than pure water ($d > 1$) $d = \frac{a}{(b - {}^\circ\text{Bé})}$ or $^\circ\text{Bé} = b - \frac{a}{d}$

Hydrometer degree	Symbol	a	b	Specific gravi
Balling	°Balling	200	200	17.5°C/17.5°C
Baumé (US)	°Bé (US)	145	145	60°F/60°F
Baumé (old)	°Bé (old)	146.78	146.78	17.5°C/17.5°C
Baumé (Gerlach)	°Bé (Gerlach)	146.3	146.3	15°C/15°C
Baumé (rational)	°Bé (rat.)	144.3	144.3	15°C/15°C
Baumé (12.5°C)	°Bé (12.5°C)	145.88	145.88	12.5°C/12.5°C
Baumé-Lunge	°Bé (Lunge)	144.32	144.32	12.5°C/12.5°C
Baumé (French)	°Bé (Fr.)	144.32	144.32	15°C/15°C
Baumé (Holland)	°Bé (Nederl.)	144	144	12.5°C/12.5°C
Baumé (NIST)	°Bé (NIST)	145	145	60°F/60°F
Brix	°Brix	400	400	60°F/60°F
Cartier	°Cart	136.8	126.1	12.5°C/12.5°C
Fisher	°Fisher	400	400	60°F/60°F
Gay-Lussac	°GL	**100**	**100**	12.5°C/12.5°C
Beck	°Beck	170	170	12.5°C/12.5°C
Stoppani	°Stop	166	166	60°F/60°F

(Continued oppo

Table 4.2.8 *(continued)*

Liquids lighter than pure water ($d < 1$) $d = \frac{a}{(b + °\text{Bé})}$ or $°\text{Bé} = \frac{a}{d} - b$

Hydrometer degree	Symbol	*a*	*b*	Specific gravity
API	°API	141.5	131.5	60°F/60°F
Balling	°Balling	200	200	17.5°C/17.5°C
Baumé (US)	°Bé (US)	140	130	60°F/60°F
Baumé (anc.)	°Bé (anc.)	146.78	146.78	17.5°C/17.5°C
Baumé (Gerlach)	°Bé (Gerlach)	146.3	146.3	15°C/15°C
Baumé (rational)	°Bé (rat.)	144.3	144.3	15°C/15°C
Baumé (Holland)	°Bé (Nederl.)	144	144	12.5°C/12.5°C
Baumé (NIST)	°Bé (NIST)	145	145	60°F/60°F
Brix	°Brix	400	400	60°F/60°F
Beck	°Beck	170	170	12.5°C/12.5°C
Fisher	°Fisher	400	400	60°F/60°F
Gay-Lussac	°GL	100	100	12.5°C/12.5°C

Table 4.2.9 Conversion table: units of time

Unit	Symbol	second (s)	minute (min)
gigan, eon, billion years	Ga, eon	3.1557600000E+16	5.2596E+14
megayear, cron, million years	Ma, cron	3.1557600000E+13	5.2596E+11
millenium (1000 y)	–	3.1557600000E+10	5.2596E+08
century (100 y)	–	3.1557600000E+09	5.2596E+07
anomalistic year (1900)	a. anomal	3.1558432900E+07	5.2597388167E+05
Gaussian year	a. gauss	3.1558368787E+07	5.2597281312E+05
sidereal year (1900)	a. sider	3.1558149500E+07	5.2596915833E+05
astronomical year (Bessel)	a. astr	3.1557875590E+07	5.2596459317E+05
Gregorian year	a. greg	3.1557600000E+07	5.2596000000E+05
year (365.25 days)	a, y	3.1557600000E+07	5.2596000000E+05
Julian year	a. jul	3.1556952000E+07	5.2594920000E+05
tropical year	a. trop	3.1556925975E+07	5.2594876625E+05
civil year (calandar)	a. civil	3.1536000000E+07	5.2560000000E+05
month (30 days)	m	2.5920000000E+06	4.3200000000E+04
mean solar day	–	86400	1440
day (24 h)	j, d	86400	1440
stellar day	–	86164.098900	1436.068315
sidereal day	–	86164.090500	1436.068175
hour (60 min)	h	3600	60
cé	Cé	86.4	1.44
minute (60 s)	min	60	1
second (SI)	s	1	1.6666666667E-02
blink	blink	0.864	1.44E-02
shake	shake	1E-08	1.6666666667E-10
wink	wink	3.3333333333E-12	5.5555555556E-14
u.a. of time	u.a	2.4188843380E-17	4.0314738967E-19
chronon, tempon	–	1E-23	1.6666666667E-25

hour (h)	day (j, d)	month (m)	year (a, y)
.7660000000E+12	**3.6525E+11**	**1.2175E+10**	**1E+09**
.7660000000E+09	**3.6525E+08**	**1.2175E+07**	**1E+06**
.7660000000E+06	**3.6525E+05**	**12175**	**1000**
.7660000000E+05	**3.6525E+04**	**1217.5**	**100**
.7662313611E+03	3.6525964005E+02	1.2175321335E+01	1.0000263930
.7662135519E+03	3.6525889800E+02	1.2175296600E+01	1.0000243614
.7661526389E+03	3.6525635995E+02	1.2175211998E+01	1.0000174126
.7660765528E+03	3.6525318970E+02	1.2175106323E+01	1.0000087329
766	**365.25**	**12.175**	**1**
766	**365.25**	**12.175**	**1**
765.82	**365.2425**	**12.17475**	9.9997946612E-01
.7658127708E+03	365.2421987813	12.1747399594	9.9997864143E-01
760	365	12.1666666667	9.9931553730E-01
720	30	1	8.2135523614E-02
24	1	3.3333333333E-02	2.7378507871E-03
24	1	3.3333333333E-02	2.7378507871E-03
23.9344719167	9.9726966319E-01	3.3242322106E-02	2.7303755324E-03
23.9344695833	9.9726956597E-01	3.3242318866E-02	2.7303752662E-03
	4.1666666667E-02	1.3888888889E-03	1.1407711613E-04
2.40E-02	**1E-03**	3.3333333333E-05	2.7378507871E-06
1.6666666667E-02	6.9444444444E-04	2.3148148148E-05	1.9012852688E-06
2.77777778E-04	1.15740741E-05	3.8580246914E-07	3.1688087814E-08
2.40E-04	**1E-05**	3.3333333333E-07	2.7378507871E-08
2.7777777778E-12	1.1574074074E-13	3.8580246914E-15	3.1688087814E-16
9.2592592593E-16	3.8580246914E-17	1.2860082305E-18	1.0562695938E-19
6.7191231611E-21	2.7996346505E-22	9.3321155015E-24	7.6649819315E-25
2.7777777778E-27	1.1574074074E-28	3.8580246914E-30	3.1688087814E-31

Table 4.2.10 Conversion table: units of dynamic viscosity

Unit	Symbol	$lbf.h.ft^{-2}$	$kip.s.m^{-2}$	$lbf.s.ft^{-2}$	$pdl.s.ft^{-2}$	pascal second	poise	centipoise
pound force hour per square foot	$lbf.h.ft^{-2}$	1	4.8824276364	3600	115826.5748031500	172368.9323292090	1723689.3232920900	172368932.3292090000
kilopound hour per square metre	$kip.s.m^{-2}$	2.0481614362E-01	1	737.3381170411	23723.1523801866	35303.94	353039.4000000000	35303940.0000000000
pièze second	pz.s	5.8015095092E-03	2.8325450360E-02	20.8854342332	671.9689751395	1000	10000	1000000
sthène second per square metre	$sth.s.m^{-2}$	5.8015095092E-03	2.8325450360E-02	20.8854342332	671.9689751395	1000	10000	1000000
pound force second per square foot	$lbf.s.ft^{-2}$	2.7777777778E-04	1.3562298990E-03	1	32.1740485564	47.8802589803	478.8025898034	47880.2589803358
kilopound second per square metre	$kip.s.m^{-2}$	5.6893373228E-05	2.7777777778E-04	2.0481614362E-01	6.5897645501	9.80665	98.0665	9806.65
poundal second per square foot	$pdl.s.ft^{-2}$	8.6335972699E-06	4.2152913912E-05	3.1080950172E-02	1	1.4881639436	14.8816394357	1488.1639435696
newton second per square metre	$N.s.m^{-2}$	5.8015095092E-06	2.8325450360E-05	2.0885434233E-02	6.7196897514E-01	1	10	1000
pascal-second	Pa.s	5.8015095092E-06	2.8325450360E-05	2.0885434233E-02	6.7196897514E-01	1	10	1000
dyne second per square centimetre	dyn.s	5.8015095092E-07	2.8325450360E-06	2.0885434233E-03	6.7196897514E-02	1E-01	1	100
barye second	barye.s	5.8015095092E-07	2.8325450360E-06	2.0885434233E-03	6.7196897514E-02	1E-01	1	100
poise	P, Po	5.8015095092E-07	2.8325450360E-06	2.0885434233E-03	6.7196897514E-02	1E-01	1	100
millipascal second	mPa.s	5.8015095092E-09	2.8325450360E-08	2.0885434233E-05	6.7196897514E-04	1E-03	0.01	1
centipoise	cP, cPo	5.8015095092E-09	2.8325450360E-08	2.0885434233E-05	6.7196897514E-04	1E-03	0.01	1
millipoise	mP, mPo	5.8015095092E-10	2.8325450360E-09	2.0885434233E-06	6.7196897514E-05	1E-04	0.001	0.1

Table 4.2.11 Conversion table: units of kinematic viscosity

Unit	Symbol	$m^2.s^{-1}$	$ft^2.s^{-1}$	$in^2.s^{-1}$	$m^2.h^{-1}$	St, $cm^2.s^{-1}$
square metre per second	$m^2.s^{-1}$	1	10.763910000	1550.003100000	3600	10000
square foot per second	$ft^2.s^{-1}$	9.290304000E-02	1	144	334.450944000	929.030400000
square inch per second	$in^2.s^{-1}$	6.451600000E-04	6.944444176E-03	1	2.322576000	6.451600000
square metre per hour	$m^2.h^{-1}$	2.777777778E-04	2.989975000E-03	4.305564167E-01	1	2.777777778
square centimetre per second	$cm^2.s^{-1}$	1E-04	1.076391000E-03	1.550003100E-01	0.36	1
stokes	St	1E-04	1.076391000E-03	1.550003100E-01	0.36	1
square foot per hour	$ft^2.h^{-1}$	2.580640000E-05	2.777777670E-04	4E-02	9.290304000E-02	2.580640000E-01
square millimetre per second	$mm^2.s^{-1}$	1E-06	1.076391000E-05	1.550003100E-03	3.60E-03	1E-02
centistokes	cSt	1E-06	1.076391000E-05	1.550003100E-03	3.60E-03	1E-02
square inch per hour	$in^2.h^{-1}$	1.792111111E-07	1.929012271E-06	2.777777778E-04	6.451600000E-04	1.792111111E-03
millistokes	mSt	1E-07	1.076391000E-06	1.550003100E-04	3.60E-04	1E-03

Unit	Symbol	St, $cm^2.s^{-1}$	$ft^2.h^{-1}$	cSt, $mm^2.s^{-1}$	$in^2.h^{-1}$	mSt
square metre per second	$m^2.s^{-1}$	10000	38750.1	1E+06	5.58001E+06	1E+07
square foot per second	$ft^2.s^{-1}$	929.030400000	3600	9.29030E+04	5.18400E+05	9.29030E+05
square inch per second	$in^2.s^{-1}$	6.451600000	25	645.16	3600	6451.600000000
square metre per hour	$m^2.h^{-1}$	2.777777778	10.763916667	277.777777778	1550.003100000	2777.777777778
square centimetre per second	$cm^2.s^{-1}$	1	3.875010000	100	558.001116000	1000
stokes	St	1	3.875010000	100	558.001116000	1000
square foot per hour	$ft^2.h^{-1}$	2.580640000E-01	1	25.806400000	144	258.064000000
square millimetre per second	$mm^2.s^{-1}$	1E-02	3.875010000E-02	1	5.580011160	10
centistokes	cSt	1E-02	3.875010000E-02	1	5.580011160	10
square inch per hour	$in^2.h^{-1}$	1.792111111E-03	6.944448477E-03	1.792111111E-01	1	1.792111111
millistokes	mSt	1E-03	3.875010000E-03	1E-01	0. 55800	1

Table 4.2.12 Conversion table: units of linear velocity

Unit	Symbol	$km.s^{-1}$	$m.s^{-1}$	$yd.s^{-1}$	knot (UK)	knot (int.)	mph	$ft.s^{-1}$	$km.h^{-1}$
kilometre per second	**$km.s^{-1}$**	1	**1000**	1093.613298338	1942.606762430	1943.844492441	2236.932122159	3280.839895013	**3600**
metre per second	**$m.s^{-1}$**	**1E-03**	1	1.09361329834	1.94260676243	1.94384449244	2.23693212216	3.28083989501	**3.6**
yard per second	**$yd.s^{-1}$**	**9.144E-04**	**0.9144**	1	1.77631962357	1.77745140389	2.04545073250	**3**	3.291840000
UK knot	**UK knot**	5.14772222E-04	5.14772222E-01	5.629617478E-01	1	1.00063714903	1.15151051948	1.68888524351	1.853184
international knot	**knot (int.)**	5.14444444E-04	5.14444444E-01	5.626032857E-01	9.993632567E-01	1	1.15077730284	1.68780985710	1.852
mile per hour	**mph**	4.47040833E-04	4.47040833E-01	4.888898002E-01	8.684245459E-01	8.689778618E-01	1	1.46666940070	1.609344
foot per second	**$ft.s^{-1}$**	**3.048E-04**	**0.3048**	3.333333333E-01	5.921065412E-01	5.924838013E-01	6.818169108E-01	1	1.097280
kilometre per hour	**$km.h^{-1}$**	2.77777778E-04	2.77777778E-01	3.037814718E-01	5.396129896E-01	5.399568035E-01	6.213700339E-01	9.113444153E-01	1
decimetre per second	**$dm.s^{-1}$**	**1E-04**	**1E-01**	1.093613298E-01	1.942606762E-01	1.943844492E-01	2.236932122E-01	3.280839895E-01	**3.6E-01**
inch per second	**$in.s^{-1}$**	**2.54E-05**	**2.54E-02**	2.777777778E-02	4.934221177E-02	4.937365011E-02	5.681807590E-02	8.333333333E-02	9.1440E-02
yard per minute	**$yd.min^{-1}$**	1.524E-05	1.52400000E-02	1.666666667E-02	2.960532706E-02	2.962419006E-02	3.409084554E-02	**5E-02**	5.4864E-02
foot per minute	**$ft.min^{-1}$**	5.080E-06	5.08000000E-03	5.555555556E-03	9.868442353E-03	9.874730022E-03	1.136361518E-02	1.666666667E-02	1.8288E-02
centimetre per second	**$cm.s^{-1}$**	**1E-05**	**1E-02**	1.093613298E-02	1.942606762E-02	1.943844492E-02	2.236932122E-02	3.280839895E-02	**3.6E-02**
millimetre per second	**$mm.s^{-1}$**	**1E-06**	**1E-03**	1.093613298E-03	1.942606762E-03	1.943844492E-03	2.236932122E-03	3.280839895E-03	**3.6E-03**
inch per minute	**$in.min^{-1}$, ipm**	4.23333333E-07	4.23333333E-04	4.629629630E-04	8.223701961E-04	8.228941685E-04	9.469679317E-04	1.388888889E-03	1.52400000E-03
metre per hour	**$m.h^{-1}$**	2.77777778E-07	2.77777778E-04	3.037814718E-04	5.396129896E-04	5.399568035E-04	6.213700339E-04	9.113444153E-04	**1E-03**
yard per hour	**$yd.h^{-1}$**	**2.54E-07**	**2.54E-04**	2.777777778E-04	4.934221177E-04	4.937365011E-04	5.681807590E-04	8.333333333E-04	9.144E-04
foot per hour	**$ft.h^{-1}$**	8.46666667E-08	8.46666667E-05	9.259259259E-05	1.644740392E-04	1.645788337E-04	1.893935863E-04	2.777777778E-04	**3.048E-04**
inch per hour	**$in.h^{-1}$, iph**	7.05555556E-09	7.05555556E-06	7.716049383E-06	1.370616993E-05	1.371490281E-05	1.578279886E-05	2.314814815E-05	**2.54E-05**

(Continued opposite)

Table 4.2.12 *(continued)*

Unit	Symbol	$km.h^{-1}$	$dm.s^{-1}$	$in.s^{-1}$	$ft.min^{-1}$	$in.min^{-1}$	$yd.h^{-1}$	$ft.h^{-1}$	$in.h^{-1}$
kilometre per second	$km.s^{-1}$	3600	1E+04	39370.078740158	1.96850394E+05	2.36220472E+06	3.937007874E+06	1.181102362E+07	1.417322835E+08
metre per second	$m.s^{-1}$	3.6	10.0	39.37007874016	196.85039370079	2362.204724409	3.937007874E+03	1.181102362E+04	1.417322835E+05
yard per second	$yd.s^{-1}$	3.291840000	9.144	36	180	2160	3600	10800	129600
UK knot	UK knot	1.853184	5.147722222	20.26662292213	101.33311461067	1215.997375328	2026.662292213	6.079986877E+03	7.295984252E+04
international knot	knot (int.)	1.852	5.144444444	20.25371828521	101.26859142607	1215.223097113	2025.371828521	6.076115486E+03	7.291338583E+04
mile per hour	mph	1.609344	4.470408333	17.60003280840	88.00016404199	1056.001968504	1760.003280840	5.280009843E+03	6.336011811E+04
foot per second	$ft.s^{-1}$	1.097280	3.048	12	60	720	1200	3600	43200
kilometre per hour	$km.h^{-1}$	1	2.777777778	10.93613298338	54.68066491689	656.16797900263	1093.613298338	3.280839895E+03	3.937007874E+04
decimetre per second	$dm.s^{-1}$	3.6E-01	1	3.93700787402	19.68503937008	236.22047244095	393.70078740158	1.181102362E+03	1.417322835E+04
inch per second	$in.s^{-1}$	9.1440E-02	2.540E-1	1	5	60	100	300	3600
yard per minute	$yd.min^{-1}$	5.4864E-02	1.524E-01	6E-01	3	36	60	180	2160
foot per minute	$ft.min^{-1}$	1.8288E-02	5.080E-02	2E-01	1	12	20	60	720
centimetre per second	$cm.s^{-1}$	3.6E-02	1E-01	3.937007874E-01	1.96850393701	23.62204724409	39.37007874016	118.11023622047	1417.322834646
millimetre per second	$mm.s^{-1}$	3.6E-03	1E-02	3.937007874E-01	1.968503937E-01	2.362204724409	3.937007874016	11.811023622047	141.7322834646
inch per minute	$in.min^{-1}$, ipm	1.52400000E-03	4.23333333E-03	1.666666667E-02	8.333333333E-02	1	1.66666666667	5	60
metre per hour	$m.h^{-1}$	1E-03	2.77777778E-03	1.093613298E-02	5.468066492E-02	6.561679790E-01	1.09361329834	3.28083989501	39.37007874016
yard per hour	$yd.h^{-1}$	9.144E-04	2.54E-03	1E-02	5E-02	6E-01	1	3	36
foot per hour	$ft.h^{-1}$	3.048E-04	8.46666667E-04	3.333333333E-03	1.666666667E-02	2E-01	3.333333333E-01	1	12
inch per hour	$in.h^{-1}$, iph	2.54E-05	7.05555556E-05	2.777777778E-04	1.388888889E-03	1.666666667E-02	2.777777778E-02	8.333333333E-02	1

Table 4.2.13 Conversion table: units of angular velocity

Unit	Symbol	tr.s^{-1}	rad.s^{-1}	tr.min^{-1}, rpm
turn per second	tr.s^{-1}	1	6.2831853072	60
revolutions per second	rps	1	6.2831853072	60
radian per second	rad.s^{-1}	1.591549431E-01	1	9.5492965855
turn per minute	tr.min^{-1}	1.666666667E-02	1.047197551E-01	1
revolutions per minute	rpm	1.666666667E-02	1.047197551E-01	1
degree per second	°.s^{-1}	2.777777778E-03	1.745329252E-02	1.666666667E-01
radian per minute	rad.min^{-1}	2.652582385E-03	1.666666667E-02	1.591549431E-01
grade per second	gon.s^{-1}	2.50E-03	1.570796327E-02	1.50E-01
degree per minute	°.min^{-1}	4.629629630E-05	2.908882087E-04	2.777777778E-03
grade per minute	gon.min^{-1}	4.166666667E-05	2.617993878E-04	2.50E-03

$°.s^{-1}$	$rad.min^{-1}$	$°.min^{-1}$	$gon.min^{-1}$
360	376.9911184308	21600	24000
360	376.9911184308	21600	24000
57.2957795131	60	3437.7467707849	3819.7186342055
6	6.2831853072	360	400
6	6.2831853072	360	400
1	1.0471975512	60	66.6666666667
9.549296586E-01	1	57.2957795131	63.6619772368
9E-01	9.424777961E-01	54	60
1.666666667E-02	1.745329252E-02	1	1.1111111111
1.50E-02	1.570796327E-02	9E-01	1

Table 4.2.14 Conversion table: units of force and weights

Unit	Symbol	ton-force (long)	ton-force (metric)	ton-force (short)
ton-force (long)	–	1	1.016046909	**1.12**
ton-force (metric)	–	9.842065276E-01	1	1.102311311
ton-force (short)	–	8.928571429E-01	9.071847400E-01	1
kip-force	kipf	4.464285714E-01	**4.5359237E-01**	**5E-01**
funal	funal	**1E-01**	**1E-01**	1.124044715E-01
sthène	sn	**1E-01**	**1E-01**	1.124044715E-01
slug-force	slugf	1.436341521E-02	1.459390363E-02	1.608702504E-02
kilogram-force	kgf	9.842065276E-04	**1E-03**	1.102311311E-03
pound-force	lbf	4.464285714E-04	4.5359237E-04	**5E-04**
newton	N	**1E-04**	**1E-04**	1.124044715E-04
ounce-force	ozf	2.790178571E-05	2.834952313E-05	**3.125E-05**
poundal	pdl	1.387542418E-05	1.409808185E-05	1.554047509E-05
crinal	crinal	**1E-05**	**1E-05**	1.124044715E-05
pond	p	9.842065276E-07	**1E-06**	1.102311311E-06
dyne	dyn	**1E-09**	**1E-09**	1.124044715E-09
u.a. of force	u.a	8.268482492E-12	8.401166076E-12	9.260700391E-12

Unit	Symbol	pound-force	newton	ounce-force
ton-force (long)	–	**2240**	9964.016418184	3.584E+04
ton-force (metric)	–	2204.622621849	9806.65	3.527396195E+04
ton-force (short)	–	**2000**	8896.443230521	**3.2E+04**
kip-force	kipf	**1000**	4448.221615261	**1.6E+04**
funal	funal	224.808943100	**1000**	3596.943089595
sthène	sn	224.808943100	**1000**	3596.943089595
slug-force	slugf	32.174050076	143.117305000	514.784801221
kilogram-force	kgf	2.204622622	**9.80665**	35.273961950
pound-force	lbf	1	4.448221615	**16**
newton	N	2.248089431E-01	1	**3.59694309**
ounce-force	ozf	**6.25E-02**	2.780138510E-01	1
poundal	pdl	3.108095017E-02	1.382549544E-01	4.972952027E-01
crinal	crinal	2.248089431E-02	**1E-01**	3.596943090E-01
pond	p	2.204622622E-03	**9.80665E-03**	3.527396195E-02
dyne	dyne	2.248089431E-06	**1E-05**	3.596943090E-05
u.a. of force	u.a	1.852140078E-08	8.238729530E-08	2.963424125E-07

kip-force	sthène	slug-force	kg-force
24	9.964016418	69.621325095	1016.046908800
204622622	9.80665	68.521762620	**1000**
	8.896443231	62.161897407	907.184740000
	4.448221615	31.080948703	**453.59237**
E-01	1	6.987275229	101.971621298
E-01	1	6.987275229	101.971621298
.217405008E-02	1.431173050E-01	1	14.593903627
.204622622E-03	**9.80665E-03**	6.852176262E-02	1
E-03	4.448221615E-03	3.108094870E-02	4.5359237E-01
.248089431E-04	**1E-03**	6.987275229E-03	1.019716213E-01
.25E-05	2.780138510E-04	1.942559294E-03	2.834952313E-02
.108095017E-05	1.382549544E-04	9.660254179E-04	1.409808185E-02
.248089431E-05	**1E-04**	6.987275229E-04	1.019716213E-02
.204622622E-06	9.80665E-06	6.852176262E-05	**1E-03**
.248089431E-09	**1E-08**	6.987275229E-08	1.019716213E-06
.852140078E-11	8.238729530E-11	5.756627076E-10	8.401166076E-09

poundal	crinal	pond	dyne
.206986877E+04	9.964016418E+04	1.016046909E+06	9.964016418E+08
.093163528E+04	98066.5	**1E+06**	9.80665E+08
.434809711E+04	8.896443231E+04	9.071847400E+05	8.896443231E+08
.217404856E+04	4.448221615E+04	**4.5359237E+05**	4.448221615E+08
233.013851210	**1E+04**	1.019716213E+05	**1E+08**
233.013851210	**1E+04**	1.019716213E+05	**1E+08**
035.169449413	1431.17305	1.459390363E+04	1.431173050E+07
0.931635284	**98.0665**	**1E+03**	**9.80665E+05**
2.174048556	44.482216153	**453.59237**	4.448221615E+05
.233013851	**10**	101.971621298	**1E+05**
.010878035	2.780138510	28.349523125	2.780138510E+04
1	1.382549544	14.098081850	1.382549544E+04
7.233013851E-01	**1**	10.197162130	**1E+04**
7.093163528E-02	9.80665E-02	**1**	**9.80665E+02**
7.233013851E-05	**1E-04**	1.019716213E-03	**1**
5.959084481E-07	8.238729530E-07	8.401166076E-06	8.238729530E-03

Table 4.2.15 Conversion table: units of pressure and stress

Unit	Symbol	atm	bar
gigapascal	GPa	9.86923267E+03	**1E+04**
kilogram-force per square millimetre	kgf.mm^{-2}	96.784110535	**98.0665**
kip-force per square inch	KSI, ksi, kipf.in^{-2}	68.045963910	68.94757293
megapascal	MPa	9.869232667	**10**
standard atmosphere	atm	1	1.01325
bar	bar	9.86923267E-01	1
kilogram-force per square centimetre	at, kgf.cm^{-2}	9.67841105E-01	**0.980665**
foot of mercury (0°C, 32°F)	ftHg (0°C, 32°F)	4.01049739E-01	4.06363648E-01
foot of mercury (15.56°C, 60°F)	ftHg (15.56°C, 60°F)	3.99962376E-01	4.05261878E-01
metre of water (4°C, 39.2°F)	mH_2O (4°C, 32°F), mCE	9.67812070E-02	9.80635580E-02
metre of water (15.56°C, 60°F)	mH_2O (15.56°C, 60°F), mCE	9.66894557E-02	9.79705910E-02
pound-force per square inch	PSI, psi, lbf.in^{-2}	6.80459639E-02	6.89475729E-02
inch of mercury (0°C, 32°F)	inHg (0°C, 32°F)	3.34208116E-02	3.38636373E-02
inch of mercury (15.56°C, 60°F)	inHg (15.56°C, 60°F)	3.33301980E-02	3.37718231E-02
foot of water (4°C, 39.2°F)	ftH_2O (4°C, 39.2°F)	2.94989119E-02	2.98897725E-02
foot of water (15.56°C, 60°F)	ftH_2O (15.56°C, 60°F)	2.94709461E-02	2.98614361E-02
centimetre of mercury (0°C, 32°F)	cmHg (0°C, 32°F)	1.31578947E-02	1.33322368E-02
centimetre of mercury (15.56°C, 60°F)	cmHg (15.56°C, 60°F)	1.31221252E-02	1.32959934E-02
sthène per square metre	sn.m^{-2}	9.86923267E-03	**0.01**
pièze	pz	9.86923267E-03	**0.01**
kilogram-force per square decimetre	kgf.dm^{-2}	9.67841105E-03	9.80665E-03
ounce-force per square inch	OSI, osi, ozf.in^{-2}	4.25287273E-03	4.30922329E-03
inch of water (4°C, 39.2°F)	inH_2O (4°C, 39.2°F)	2.45824266E-03	2.49081437E-03
inch of water (15.56°C, 60°F)	inH_2O (15.56°C, 60°F)	2.45591217E-03	2.48845301E-03
torr	torr, mmHg (0°C, 32°F)	1.31578947E-03	1.33322368E-03
millimetre of mercury (0°C, 32°F)	torr, mmHg (0°C, 32°F)	1.31578947E-03	1.33322368E-03
millimetre of mercury (4°C, 39.2°F)	mmHg (4°C, 39.2°F)	1.31221252E-03	1.32959934E-03
millibar, hectopascal	mbar, hPa	9.86923267E-04	**1E-03**
centimetre of water (4°C, 39.2°F)	cmH_2O (4°C, 39.2°F)	9.67812070E-04	9.80635580E-04
centimetre of water (15.56°C, 60°F)	cmH_2O (15.56°C, 60°F)	9.66894557E-04	9.79705910E-04
pound-force per square foot	PSF, psf, lbf.ft^{-1}	4.72541416E-04	4.78802590E-04
kilogram-force per square metre	kgf.m^{-2}	9.67841105E-05	**9.80665E-05**
millimetre of water (4°C, 39.2°F)	mmH_2O (4°C, 39.2°F)	9.67814006E-05	9.80637541E-05
millimetre of water (15.56°C, 60°F)	mmH_2O (15.56°C, 60°F)	9.66894557E-05	9.79705910E-05
pascal	Pa	9.86923267E-06	**1E-05**
newton per square metre	N.m^{-2}	9.86923267E-06	**1E-05**
barye	barye	9.86923267E-07	**1E-06**
dyne per square centimetre	dyn.cm^{-2}	9.86923267E-07	**1E-06**
microbar	μbar	9.86923267E-07	**1E-06**

at, kgf.cm^{-2}	psi, PSI	inHg (0°C)	ftH_2O (4°C)	pièze
.01971621E+04	1.45037738E+05	2.95300267E+05	3.34561931E+05	**1E+06**
00	1.42233433E+03	2.89590637E+03	3.28093176E+03	**9806.65**
0.306957964	**1000**	2.03602367E+03	2.30672331E+03	6.89475729E+03
0.197162130	145.037737730	295.300267432	334.561931047	**1000**
.033227453	14.695948776	29.921299598	33.899487663	**101.325**
.019716213	14.503773773	29.530026743	33.456193105	**100**
	14.223343307	28.959063676	32.809317611	**98.0665**
.14375600E-01	5.893806417	12	13.595380671	40.636364777
.13252107E-01	5.877826593	11.967394087	13.558519639	40.526187771
.99970000E-02	1.422291661	2.895819490	3.280833333	9.806355801
.99022000E-02	1.420943288	2.893074171	3.277723010	9.797059096
.03069580E-02	1	2.036023673	2.306723314	6.894757293
.45313000E-02	4.91150535E-01	1	1.132948389	3.386363731
.44376756E-02	4.89818883E-01	9.97282841E-01	1.129876637	3.377182314
.04790856E-02	4.33514498E-01	8.82645781E-01	1	2.988977248
.04501906E-02	4.33103514E-01	8.81809007E-01	9.99049973E-01	2.986143613
.35950981E-02	1.93367747E-01	3.93701310E-01	4.46045890E-01	1.333223684
.35581400E-02	1.92842080E-01	3.92631040E-01	4.44833321E-01	1.329599336
.01971621E-02	1.45037738E-01	2.95300267E-01	3.34561931E-01	1
.01971621E-02	1.45037738E-01	2.95300267E-01	3.34561931E-01	1
E-02	1.42233433E-01	2.89590637E-01	3.28093176E-01	**9.80665E-01**
4.39418485E-03	6.24999997E-02	1.27251479E-01	1.44170206E-01	4.30922329E-01
2.53992380E-03	3.61262082E-02	7.35538151E-02	8.33331667E-02	2.49081437E-01
2.53751588E-03	3.60919595E-02	7.34840839E-02	8.32541644E-02	2.48845301E-01
1.35950981E-03	1.93367747E-02	3.93701310E-02	4.46045890E-02	1.33322368E-01
1.35950981E-03	1.93367747E-02	3.93701310E-02	4.46045890E-02	1.33322368E-01
1.35581400E-03	1.92842080E-02	3.92631040E-02	4.44833321E-02	1.32959934E-01
1.01971621E-03	1.45037738E-02	2.95300267E-02	3.34561931E-02	**1E-01**
9.99970000E-04	1.42229166E-02	2.89581949E-02	3.28083333E-02	9.80635580E-02
9.99022000E-04	1.42094329E-02	2.89307417E-02	3.27772301E-02	9.79705910E-02
4.88242764E-04	6.94444444E-03	1.41390533E-02	1.60189119E-02	4.78802590E-02
1E-04	1.42233433E-03	2.89590637E-03	3.28093176E-03	**9.80665E-03**
9.99972000E-05	1.42229451E-03	2.89582528E-03	3.28083990E-03	9.80637541E-03
9.99022000E-05	1.42094329E-03	2.89307417E-03	3.27772301E-03	9.79705910E-03
1.01971621E-05	1.45037738E-04	2.95300267E-04	3.34561931E-04	**1E-03**
1.01971621E-05	1.45037738E-04	2.95300267E-04	3.34561931E-04	**1E-03**
1.01971621E-06	1.45037738E-05	2.95300267E-05	3.34561931E-05	**1E-04**
1.01971621E-06	1.45037738E-05	2.95300267E-05	3.34561931E-05	**1E-04**
1.01971621E-06	1.45037738E-05	2.95300267E-05	3.34561931E-05	**1E-04**

(Continued overleaf)

Table 4.2.15 *(continued)* Conversion table: units of pressure and stress

Unit	Symbol	pièze	ozf.in^{-2}
gigapascal	GPa	1E+06	2.32060381E+06
kilogram-force per square millimetre	kgf.mm^{-2}	9806.65	2.27573494E+04
kip-force per square inch	KSI, ksi, kipf.in^{-2}	6.89475729E+03	1.60E+04
megapascal	MPa	1000	2.32060381E+03
standard atmosphere	atm	101.325	235.135181445
bar	bar	100	232.060381392
kilogram-force per square centimetre	at, kgf.cm^{-2}	98.0665	227.573493917
foot of mercury (0°C, 32°F)	ftHg (0°C, 32°F)	40.636364777	94.300903086
foot of mercury (15.56°C, 60°F)	ftHg (15.56°C, 60°F)	40.526187771	94.045225904
metre of water (4°C, 39.2°F)	mH_2O (4°C, 32°F), mCE	9.806355801	22.756666671
metre of water (15.56°C, 60°F)	mH_2O (15.56°C, 60°F), mCE	9.797059096	22.735092704
pound-force per square inch	PSI, psi, lbf.in^{-2}	6.894757293	16
inch of mercury (0°C, 32°F)	inHg (0°C, 32°F)	3.386363731	7.858408591
inch of mercury (15.56°C, 60°F)	inHg (15.56°C, 60°F)	3.377182314	7.837102159
foot of water (4°C, 39.2°F)	ftH_2O (4°C, 39.2°F)	2.988977248	6.936232001
foot of water (15.56°C, 60°F)	ftH_2O (15.56°C, 60°F)	2.986143613	6.929656256
centimetre of mercury (0°C, 32°F)	cmHg (0°C, 32°F)	1.333223684	3.093883966
centimetre of mercury (15.56°C, 60°F)	cmHg (15.56°C, 60°F)	1.329599336	3.085473291
sthène per square metre	sn.m^{-2}	1	2.320603814
pièze	pz	1	2.320603814
kilogram-force per square decimetre	kgf.dm^{-2}	9.80665E-01	2.275734939
ounce-force per square inch	OSI, osi, ozf.in^{-2}	4.30922329E-01	1
inch of water (4°C, 39.2°F)	inH_2O (4°C, 39.2°F)	2.49081437E-01	5.78019333E-01
inch of water (15.56°C, 60°F)	inH_2O (15.56°C, 60°F)	2.48845301E-01	5.77471355E-01
torr	torr, mmHg (0°C, 32°F)	1.33322368E-01	3.09388397E-01
millimetre of mercury (0°C, 32°F)	torr, mmHg (0°C, 32°F)	1.33322368E-01	3.09388397E-01
millimetre of mercury (15.56°C, 39.2°F)	mmHg (15.56°C, 39.2°F)	1.32959934E-01	3.08547329E-01
millibar, hectopascal	mbar, hPa	1E-01	2.32060381E-01
centimetre of water (4°C, 39.2°F)	cmH_2O (4°C, 39.2°F)	9.80635580E-02	2.27566667E-01
centimetre of water (15.56°C, 60°F)	cmH_2O (15.56°C, 60°F)	9.79705910E-02	2.27350927E-01
pound-force per square foot	PSF, psf, lbf.ft^{-1}	4.78802590E-02	1.11111112E-01
kilogram-force per square metre	kgf.m^{-2}	9.80665E-03	2.27573494E-02
millimetre of water (4°C, 39.2°F)	mmH_2O (4°C, 39.2°F)	9.80637541E-03	2.27567122E-02
millimetre of water (15.56°C, 60°F)	mmH_2O (15.56°C, 60°F)	9.79705910E-03	2.27350927E-02
pascal	Pa	1E-03	2.32060381E-03
newton per square metre	N.m^{-2}	1E-03	2.32060381E-03
barye	barye	1E-04	2.32060381E-04
dyne per square centimetre	dyn.cm^{-2}	1E-04	2.32060381E-04
microbar	μbar	1E-04	2.32060381E-04

inH$_2$O (4°C)	torr, mmHg(0°C)	cmH$_2$O (4°C)	pascal	barye
01474317E+06	7.50061683E+06	1.01974477E+07	**1E+09**	**1E+10**
93711811E+04	7.35559240E+04	1.00002800E+05	**9.80665E+06**	9.80665E+07
76806798E+04	5.17149326E+04	7.03089266E+04	6.89475729E+06	6.89475729E+07
01474317E+03	7.50061683E+03	1.01974477E+04	**1E+06**	**1E+07**
6.793851960	**760**	1.03325638E+03	**101325**	1.01325E+06
1.474317256	750.061682704	1.01974477E+03	**1E+05**	**1E+06**
3.711811332	735.559240069	1.00002800E+03	**9.80665E+04**	**9.80665E+05**
3.144568048	304.797801439	414.387202842	4.06366039E+04	4.06363648E+05
2.702235663	303.971405929	413.263678583	4.05261878E+04	4.05261878E+05
.369999998	73.553717329	**100**	9.80637541E+03	9.80635580E+04
.332676118	73.483986313	99.904997340	9.79705910E+03	9.79705910E+04
.680679769	51.714932572	70.308926614	6.89475729E+03	6.89475729E+04
.595380671	25.399816787	34.532266903	3.38638366E+03	3.38636373E+04
.558519639	25.330950494	34.438639882	3.37718231E+03	3.37718231E+03
2	22.419173042	**30.48**	2.98898323E+03	2.98897725E+04
1.988599681	22.397919028	30.451043189	2.98614361E+03	2.98614361E+04
352550684	**10**	13.595478737	1.33322368E+03	1.33322368E+04
337999858	9.972815155	13.558519639	1.32959934E+03	1.32959934E+04
014743173	7.500616827	10.197447658	**1000**	10000
014743173	7.500616827	10.197447658	**1000**	10000
.937118113	7.355592401	10.000280008	**980.665**	**9.80665E+03**
.730042478	3.232183271	4.394307894	430.922328923	4.30922329E+03
	1.868264420	**2.54**	249.081935511	2.49081437E+03
.99049973E-01	1.866493252	2.537586932	248.845301046	2.48845301E+03
.35255068E-01	1	1.359547874	133.322368421	1.33322368E+03
.35255068E-01	1	1.359547874	133.322368421	1.33322368E+03
.33799986E-01	9.97281516E-01	1.355851964	132.959933631	1.32959934E+03
.01474317E-01	7.50061683E-01	1.019744766	**100**	**1000**
.93700000E-01	7.35537173E-01	**1**	98.063754138	980.635580050
.93326761E-01	7.34839863E-01	9.99049973E-01	97.970590963	979.705909630
.92226943E-01	3.59131476E-01	4.88256435E-01	47.880258980	478.802589803
.93711811E-02	7.35559240E-02	1.00002800E-01	9.80665	98.0665
.93700787E-02	7.35538644E-02	**0.1**	9.806375414	98.063754138
.93326761E-02	7.34839863E-02	9.99049973E-02	9.797059096	97.970590963
.01474317E-03	7.50061683E-03	1.01974477E-02	**1**	**10**
.01474317E-03	7.50061683E-03	1.01974477E-02	**1**	**10**
.01474317E-04	7.50061683E-04	1.01974477E-03	**0.1**	**1**
.01474317E-04	7.50061683E-04	1.01974477E-03	**0.1**	**1**
.01474317E-04	7.50061683E-04	1.01974477E-03	**0.1**	**1**

Table 4.2.16 Conversion table: units of energy, heat and work

Unit	Symbol	toe (tep)	tce (tec)	kWh
megaton (TNT)	Mt (TNT)	9.983281586E+12	1.427108228E+13	1.161111111E+15
Q unit	Q	2.519703845E+12	3.601911915E+12	2.930555556E+14
kiloton (TNT)	kt (TNT)	9.983281586E+09	1.427108228E+10	1.161111111E+12
quad (quadrillion Btu)	quad	2.519837241E+09	3.602102604E+09	2.930710702E+11
barrel oil equivalent	bboe	14.616670647	20.894503243	1700
ton oil equivalent	toe (tep)	1	1.429498122	116.305555556
ton coal equivalent	tce (tec)	6.995462145E-01	1	81.361111111
therm (EEG)	therm (EEG)	2.519837241E-01	3.602102604E-01	29.307107017
therm (US)	therm (US)	2.519235730E-01	3.601242745E-01	29.300111111
thermie (15°C)	th (15°C)	9.996417483E-03	1.428986002E-02	1.162638889
kilowatt-hour	kWh	8.598041557E-03	1.229088426E-02	1
kilocalorie (therm.)	kcal (therm.)	9.992834965E-06	1.428473882E-05	1.162222222E-03
Celsius heat unit	Chu	4.535896824E-06	6.484055992E-06	5.275500000E-04
pound centigrade unit (15°C)	pcu (15°C)	4.535705756E-06	6.483782861E-06	5.275277778E-04
Celsius heat unit (15°C)	Chu (15°C)	4.535705756E-06	6.483782861E-06	5.275277778E-04
British thermal unit (39.2°F, 4°C)	Btu (39.2°F)	2.530857416E-06	3.617855924E-06	2.943527778E-04
British thermal unit (mean)	Btu (mean)	2.521781705E-06	3.604882212E-06	2.932972222E-04
British thermal unit (ISO/TC 12)	Btu (ISO)	2.519847146E-06	3.602116763E-06	2.930722222E-04
British thermal unit (Intern. Steam Tables)	Btu (IT)	2.519837241E-06	3.602102604E-06	2.930710702E-04
British thermal unit (UK, gas industry)	Btu (Gas)	2.519130642E-06	3.601092523E-06	2.929888889E-04
British thermal unit (thermochemical)	Btu (therm.)	2.518151421E-06	3.599692728E-06	2.928750000E-04
British thermal unit (60°F, 15.56°C)	Btu (60°F)	2.518934798E-06	3.600812564E-06	2.929661111E-04
British thermal unit (Gas Inspec. Act Reg.)	Btu (GIAR)	2.518784332E-06	3.600597474E-06	2.929486111E-04
litre-atmosphere	L.atm	2.419990447E-07	3.459371799E-07	2.814583333E-05
kilogram-force-metre	kgf-m	2.342166229E-08	3.348122226E-08	2.724069444E-06
calorie (4°C)	cal (4°C)	1.004179604E-08	1.435472858E-08	1.167916667E-06
calorie (mean)	cal. mean	1.000721280E-08	1.430529191E-08	1.163894444E-06
calorie (Intern. Steam Tables)	cal (IT)	9.999379030E-09	1.429409355E-08	1.162983333E-06
calorie (15°C)	cal (15°C)	9.996417483E-09	1.428986002E-08	1.162638889E-06
calorie (thermochemical)	cal. therm	9.992834965E-09	1.428473882E-08	1.162222222E-06
calorie (20°C)	cal (20°C)	9.987819441E-09	1.427756914E-08	1.161638889E-06
foot-pound force (duty)	ft-lbf	3.238160851E-09	4.628944856E-09	3.766160968E-07
joule (international)	J (int.)	2.388738954E-09	3.414697849E-09	2.778236111E-07
joule	J	2.388344877E-09	3.414134517E-09	2.777777778E-07
foot-poundal	ft-pdl	1.006451161E-10	1.438720044E-10	1.170558614E-08
erg	erg	2.388344877E-16	3.414134517E-16	2.777777778E-14
billion electronvolt	BeV	3.826552018E-19	5.470048925E-19	4.450492583E-17
prout	prout	7.078576546E-23	1.011881188E-22	8.232777778E-21
hartree (u.a. of energy)	E_H, a.u.	4.359748200E-18	4.359748200E-18	4.359748200E-18
rydberg	Ry	2.179874100E-18	2.179874100E-18	2.179874100E-18
electronvolt	eV	3.826552018E-28	5.470048925E-28	4.450492583E-2
frigorie	fg	-9.996417483E-06	-1.428986002E-05	-1.162638889E-0

kcal (therm)	pcu (15°C)	Btu (ISO)	joule (J)	erg
.990439771E+17	2.201042599E+18	3.961859989E+18	**4.18E+21**	**4.18E+28**
.521510516E+17	5.555263019E+17	9.999431312E+17	**1.055E+21**	**1.055E+28**
.990439771E+14	2.201042599E+15	3.961859989E+15	**4.18E+18**	**4.18E+25**
.521644007E+14	5.555557120E+14	9.999960691E+14	1.055055853E+18	1.055055853E+25
.462715105E+06	3.222579116E+06	5.800617974E+06	**6.12E+09**	**6.12E+16**
.000717017E+05	2.204728556E+05	3.968494683E+05	**4.187E+08**	**4.187E+15**
.000478011E+04	1.542309515E+05	2.776145432E+05	**2.929E+08**	**2.929E+15**
.521644007E+04	5.555557120E+04	9.999960691E+04	1.055055853E+08	1.055055853E+15
.521042065E+04	5.554230952E+04	9.997573598E+04	1.054804000E+08	1.054804000E+15
000.358508604	2.203938708E+03	3.967072963E+03	4.185500000E+06	4.185500000E+13
60.420650096	1.895634774E+03	3412.128220196	**3.60E+06**	**3.60E+13**
	2.203148860	3.965651243	**4184**	4.1840E+10
.539149140E-01	1.000042125	1.800068243	1899.18	1.899180000E+10
.538957935E-01	1	1.799992417	1899.1	1.8991E+10
.538957935E-01	1	1.799992417	1899.10	1.899100000E+10
.532672084E-01	5.579853615E-01	1.004369420	1059.6700	1.059670000E+10
.523589866E-01	5.559844137E-01	1.000767729	1055.8700	1.055870000E+10
.521653920E-01	5.555578958E-01	1	1055.0600	1.055060000E+10
.521644007E-01	5.555557120E-01	9.999960691E-01	1055.055852620	1.055055853E+10
2.520936902E-01	5.553999263E-01	9.997156560E-01	1054.7600	1.054760000E+10
2.519956979E-01	5.551840345E-01	9.993270525E-01	1054.3500	1.054350000E+10
2.520740918E-01	5.553567479E-01	9.996379353E-01	1054.6780	1.054678000E+10
2.520590344E-01	5.553235743E-01	9.995782230E-01	1054.6150	1.054615000E+10
2.421725621E-02	5.335422042E-02	9.603719220E-02	**101.3250**	1.013250000E+09
2.343845602E-03	5.163840767E-03	9.294874225E-03	**9.80665**	**9.80665E+07**
.004899618E-03	2.213943447E-03	3.985081417E-03	4.20450	4.204500000E+07
.001438815E-03	2.206318783E-03	3.971357079E-03	4.19002	4.190020000E+07
1.000654876E-03	2.204591649E-03	3.968248251E-03	4.18674	4.186740000E+07
1.000358509E-03	2.203938708E-03	3.967072963E-03	4.18550	4.185500000E+07
1E-03	2.203148860E-03	3.965651243E-03	4.184	4.184E+07
9.994980880E-04	2.202043073E-03	3.963660834E-03	4.18190	4.181900000E+07
3.240482668E-04	7.139265696E-04	1.285062412E-03	1.355817948	1.355817948E+07
2.390451721E-04	5.266520984E-04	9.479697837E-04	1.000165	1.000165000E+07
2.390057361E-04	5.265652151E-04	9.478133945E-04	**1**	**1E+07**
.007172803E-05	2.218951614E-05	3.994096079E-05	4.214011009E-02	4.214011009E+05
2.390057361E-11	5.265652151E-11	9.478133945E-11	**1E-07**	**1**
3.829295722E-14	8.436508504E-14	1.518565134E-13	1.602177330E-10	1.602177330E-03
7.083652008E-18	1.560633985E-17	2.809129339E-17	2.963800000E-14	2.963800000E-07
4.359748200E-18	4.359748200E-18	4.359748200E-18	4.359748200E-18	4.359748200E-18
2.179874100E-18	2.179874100E-18	2.179874100E-18	2.179874100E-18	2.179874100E-18
3.829295722E-23	8.436508504E-23	1.518565134E-22	1.602177330E-19	1.602177330E-12
-1.000358509	-2.203938708	-3.967072963E+00	-4.185500000E+03	-4.185500000E+10

Table 4.2.17 Conversion table: units of power

Unit	Symbol	HP (boiler)
megawatt	MW	101.941995005
horsepower (boiler)	HP (boiler)	1
commercial ton of refrigeration (UK)	CTR (UK)	3.998875295E-01
commercial ton of refrigeration (US)	CTR (US)	3.585096080E-01
kilowatt	kW	1.019419950E-01
poncelet	poncelet	9.997094653E-02
horsepower (water)	HP (water)	7.605311178E-02
horsepower (electric)	HP (electric)	7.604872827E-02
horsepower (UK)	BHP, hp	7.601814568E-02
horsepower (550 ft.lbf.s^{-1})	HP (550)	7.601813258E-02
horsepower (cheval vapeur)	HP, cv	7.497820990E-02
horsepower (metric)	HP (metric)	7.497820990E-02
donkey	donkey	2.548549875E-02
prony	prony	9.997094653E-03
British thermal unit (39°F) per minute	Btu (39°F).min^{-1}	1.800414564E-03
British thermal unit (mean) per minute	Btu (mean).min^{-1}	1.793958238E-03
British thermal unit (ISO) per minute	Btu (ISO).min^{-1}	1.792582021E-03
British thermal unit (IT) per minute	Btu (IT).min^{-1}	1.792574974E-03
British thermal unit (60°F) per minute	Btu (60°F).min^{-1}	1.791932990E-03
British thermal unit (therm.) per minute	Btu (therm.).min^{-1}	1.791375707E-03
kilogram-force metre per second	kgf.m.s^{-1}	9.997094653E-04
foot pound-force per second	ft.lbf.s^{-1}	1.382147865E-04
watt (int. mean)	W (int. mean)	1.019613640E-04
watt (int. US)	W (int. US)	1.019588154E-04
volt-ampere	VA	1.019419950E-04
watt	W	1.019419950E-04
British thermal unit (39°F) per hour	Btu (39°F).h^{-1}	3.000690940E-05
British thermal unit (mean) per hour	Btu (mean).h^{-1}	2.989930396E-05
British thermal unit (ISO) per hour	Btu (ISO).h^{-1}	2.987636701E-05
British thermal unit (IT) per hour	Btu (IT).h^{-1}	2.987624957E-05
British thermal unit (60°F) per hour	Btu (60°F).h^{-1}	2.986554984E-05
British thermal unit (therm.) per hour	Btu (therm.).h^{-1}	2.985626179E-05
foot poundal per second	ft.pdl.s^{-1}	4.295846893E-06
foot pound-force per minute	ft.lbf.min^{-1}	2.303579775E-06
inch ounce-force per second	in.ozf.s^{-1}	9.598249064E-07
clusec	clusec	1.325245935E-07
calorie (4°C) per hour	cal.(4°C).h^{-1}	1.190597550E-07
calorie (mean) per hour	cal.(mean).h^{-1}	1.186497216E-07
calorie (IT) per hour	cal.(IT).h^{-1}	1.185568412E-07
calorie (15°C) per hour	cal.(15°C).h^{-1}	1.185217278E-07
calorie (therm.) per hour	cal.(therm.).h^{-1}	1.184792520E-07
foot pound-force per hour	ft.lbf.h^{-1}	3.776360287E-08
erg per second	erg.s^{-1}	1.019419950E-11
abwatt (emu of power)	emu	1.019419950E-12

CTR (UK)	CTR (US)	HP (water)	HP (metric)
254.926666825	284.349408553	1340.405311758	1.359621155E+03
2.500703138	2.789325523	13.148705906	13.337203722
1	1.115416492	5.258003521	5.333381447
8.965261019E-01	1	4.713937400	4.781515678
2.549266668E-01	2.843494086E-01	1.340405312	1.359621155
2.499976597E-01	2.788515127E-01	1.314488575	1.333332880
1.901862553E-01	2.121368858E-01	1	1.014335845
1.901752935E-01	2.121246588E-01	9.999423626E-01	1.014277382
1.900988155E-01	2.120393540E-01	9.995402410E-01	1.013869495
1.900987827E-01	2.120393174E-01	9.995400688E-01	1.013869321
1.874982448E-01	2.091386346E-01	9.858664313E-01	9.999996601E-01
1.874982448E-01	2.091386346E-01	9.858664313E-01	9.999996601E-01
6.373166671E-02	7.108735214E-02	3.351013279E-01	3.399052888E-01
2.499976597E-02	2.788515127E-02	1.314488575E-01	1.333332880E-01
4.502302351E-03	5.021942296E-03	2.367312161E-02	2.401249582E-02
4.486156995E-03	5.003933500E-03	2.358822928E-02	2.392638649E-02
4.482715485E-03	5.000094783E-03	2.357013380E-02	2.390803160E-02
4.482697864E-03	5.000075128E-03	2.357004115E-02	2.390793762E-02
4.481092452E-03	4.998284425E-03	2.356159989E-02	2.389937534E-02
4.479698853E-03	4.996729982E-03	2.355427234E-02	2.389194275E-02
2.499976597E-03	2.788515127E-03	1.314488575E-02	1.333332880E-02
3.456341504E-04	3.855260317E-04	1.817345580E-03	1.843398765E-03
2.549751029E-04	2.844034349E-04	1.340659989E-03	1.359879483E-03
2.549687297E-04	2.843963262E-04	1.340626479E-03	1.359845493E-03
2.549266668E-04	2.843494086E-04	1.340405312E-03	1.359621155E-03
2.549266668E-04	2.843494086E-04	1.340405312E-03	1.359621155E-03
7.503837251E-05	8.369903827E-05	3.945520269E-04	4.002082637E-04
7.476928325E-05	8.339889167E-05	3.931371546E-04	3.987731081E-04
7.471192475E-05	8.333491305E-05	3.928355634E-04	3.984671933E-04
7.471163106E-05	8.333458547E-05	3.928340192E-04	3.984656270E-04
7.468487420E-05	8.330474042E-05	3.926933315E-04	3.983229224E-04
7.466164755E-05	8.327883303E-05	3.925712057E-04	3.981990458E-04
1.074263781E-05	1.198251538E-05	5.648482741E-05	5.729458516E-05
5.760569173E-06	6.425433862E-06	3.028909300E-05	3.072331275E-05
2.400237155E-06	2.677264109E-06	1.262045541E-05	1.280138031E-05
3.314046669E-07	3.696542311E-07	1.742526905E-06	1.767507502E-06
2.977331030E-07	3.320964134E-07	1.565481704E-06	1.587924207E-06
2.967077313E-07	3.309526969E-07	1.560090296E-06	1.582455509E-06
2.964754647E-07	3.306936230E-07	1.558869037E-06	1.581216743E-06
2.963876567E-07	3.305956804E-07	1.558407342E-06	1.580748429E-06
2.962814372E-07	3.304772015E-07	1.557848840E-06	1.580181920E-06
9.443556022E-08	1.053349813E-07	4.965425081E-07	5.036608648E-07
2.549266668E-11	2.843494086E-11	1.340405312E-10	1.359621155E-10
2.549266668E-12	2.843494086E-12	1.340405312E-11	1.359621155E-11

(Continued overleaf)

Table 4.2.17 *(continued)* Conversion table: units of power

Unit	Symbol	HP (metric)
megawatt	MW	1.359621155E+03
horsepower (boiler)	HP (boiler)	13.337203722
commercial ton of refrigeration (UK)	CTR (UK)	5.333381447
commercial ton of refrigeration (US)	CTR (US)	4.781515678
kilowatt	kW	1.359621155
poncelet	poncelet	1.333332880
horsepower (water)	HP (water)	1.014335845
horsepower (electric)	HP (electric)	1.014277382
horsepower (UK)	BHP, hp	1.013869495
horsepower (550 $ft.lbf.s^{-1}$)	HP (550)	1.013869321
horsepower (cheval vapeur)	HP, cv	9.999996601E-01
horsepower (metric)	HP (metric)	9.999996601E-01
donkey	donkey	3.399052888E-01
prony	prony	1.333332880E-01
British thermal unit (39°F) per minute	Btu (39°F).min^{-1}	2.401249582E-02
British thermal unit (mean) per minute	Btu (mean).min^{-1}	2.392638649E-02
British thermal unit (ISO) per minute	Btu (ISO).min^{-1}	2.390803160E-02
British thermal unit (IT) per minute	Btu (IT).min^{-1}	2.390793762E-02
British thermal unit (60°F) per minute	Btu (60°F).min^{-1}	2.389937534E-02
British thermal unit (therm.) per minute	Btu (therm.).min^{-1}	2.389194275E-02
kilogram-force metre per second	$kgf.m.s^{-1}$	1.333332880E-02
foot pound-force per second	$ft.lbf.s^{-1}$	1.843398765E-03
watt (int. mean)	W (int. mean)	1.359879483E-03
watt (int. US)	W (int. US)	1.359845493E-03
volt-ampere	VA	1.359621155E-03
watt	W	1.359621155E-03
British thermal unit (39°F) per hour	Btu (39°F).h^{-1}	4.002082637E-04
British thermal unit (mean) per hour	Btu (mean).h^{-1}	3.987731081E-04
British thermal unit (ISO) per hour	Btu (ISO).h^{-1}	3.984671933E-04
British thermal unit (IT) per hour	Btu (IT).h^{-1}	3.984656270E-04
British thermal unit (60°F) per hour	Btu (60°F).h^{-1}	3.983229224E-04
British thermal unit (therm.) per hour	Btu (therm.).h^{-1}	3.981990458E-04
foot poundal per second	$ft.pdl.s^{-1}$	5.729458516E-05
foot pound-force per minute	$ft.lbf.min^{-1}$	3.072331275E-05
inch ounce-force per second	$in.ozf.s^{-1}$	1.280138031E-05
clusec	clusec	1.767507502E-06
calorie (4°C) per hour	cal.(4°C).h^{-1}	1.587924207E-06
calorie (mean) per hour	cal.(mean).h^{-1}	1.582455509E-06
calorie (IT) per hour	cal.(IT).h^{-1}	1.581216743E-06
calorie (15°C) per hour	cal.(15°C).h^{-1}	1.580748429E-06
calorie (therm.) per hour	cal.(therm.).h^{-1}	1.580181920E-06
foot pound-force per hour	$ft.lbf.h^{-1}$	5.036608648E-07
erg per second	$erg.s^{-1}$	1.359621155E-10
abwatt (emu of power)	emu	1.359621155E-11

ft.lbf.s^{-1}	watt	Btu (mean).h^{-1}	ft.lbf.min^{-1}
.375621493E+05	**1E+06**	3.409510641E+06	4.425372896E+07
.235115903E+03	9.809500000E+03	3.344559463E+04	4.341069542E+05
.893232624E+03	3.922696721E+03	1.337447621E+04	1.735939575E+05
.593858567E+03	3.516800000E+03	1.199056702E+04	1.556315140E+05
37.562149277	**1000**	3.409510641E+03	4.425372896E+04
23.301385121	**980.665**	3.343587752E+03	4.339808311E+04
50.253078533	746.043000000	2.543641547E+03	3.301518471E+04
50.221363361	746.000000000	2.543494938E+03	3.301328180E+04
50.000094716	745.700000000	2.542472085E+03	3.300000568E+04
50	745.699871582	2.542471647E+03	**33000**
42.476038841	735.498750000	2.507690814E+03	3.254856233E+04
42.476038841	735.498750000	2.507690814E+03	3.254856233E+04
84.390537319	**250**	852.377660129	1.106343224E+04
2.330138512	**98.0665**	334.358775228	4.339808311E+03
3.026208045	17.661166667	60.215935674	781.572482725
2.979495776	17.597833333	**60**	778.769746557
2.969538687	17.584333333	59.953971606	778.172321216
2.969487704	17.584264210	59.953735931	778.169262266
2.964842875	17.577966667	59.932264389	777.890572475
2.960810868	17.572500000	59.913625730	777.648652090
.233013851	9.806650000	33.435877523	433.980831073
[illegible]	1.355817948	4.622675721	**60**
7.377022861E-01	1.000190000	3.410158448	44.262137165
7.376838470E-01	1.000165000	3.410073210	44.261030822
7.375621493E-01	**1**	3.409510641	44.253728957
7.375621493E-01	**1**	3.409510641	44.253728957
2.171034674E-01	2.943527778E-01	1.003598928	13.026208045
2.163249296E-01	2.932972222E-01	**1**	12.979495776
2.161589781E-01	2.930722222E-01	9.992328601E-01	12.969538687
2.161581284E-01	2.930710702E-01	9.992289322E-01	12.969487704
2.160807146E-01	2.929661111E-01	9.988710731E-01	12.964842875
2.160135145E-01	2.928750000E-01	9.985604288E-01	12.960810868
3.108095017E-02	4.214011009E-02	1.436771538E-01	1.864857010
1/60	2.259696581E-02	7.704459536E-02	**1**
1/144	9.415402419E-03	3.210191473E-02	4.166666667E-01
9.588307941E-04	1.3E-03	4.432363833E-03	5.752984764E-02
8.614111268E-04	1.167916667E-03	3.982024302E-03	5.168466761E-02
8.584444880E-04	1.163894444E-03	3.968310493E-03	5.150666928E-02
8.577724869E-04	1.162983333E-03	3.965204050E-03	5.146634921E-02
8.575184377E-04	1.162638889E-03	3.964029663E-03	5.145110626E-02
8.572111202E-04	1.162222222E-03	3.962609033E-03	5.143266721E-02
2.732240437E-04	3.704420624E-04	1.263026153E-03	1.639344262E-02
7.375621493E-08	**1E-07**	3.409510641E-07	4.425372896E-06
7.375621493E-09	**1E-08**	3.409510641E-08	4.425372896E-07

Table 4.2.18 Conversion table: units of plane and solid angle

Unit	Symbol	radian (rad)
round	**tr, r**	6.283185307180
circumference	**circ**	6.283185307180
plane angle	–	3.141592653590
right angle	⊥	1.570796326795
quadrant	**quadr**	1.570796326795
sign	**sign**	5.235987756E-01
radian	**rad**	1
degree	°	1.7453292520E-02
grade (gon)	**g**	1.5707963268E-02
percent	%	9.9996666867E-03
thousandth (US)	‰ (US)	1.5707963270E-03
thousandth (French)	‰ (Fra.)	9.9999966000E-04
thousandth (USSR)	‰ (USSR)	9.9733101100E-04
thousandth (NATO)	‰ (NATO)	9.8174770400E-04
minute of angle	′	2.9088820867E-04
minute (new)	c	1.5707963268E-04
second of angle	″	4.8481368111E-06
second (new)	cc	1.5707963268E-06

Unit	Symbol	steradian (sr)
spat	**spat**	12.5663
steradian	**sr**	1
square degree	$(^{\circ})^2$	3.046174198E-04
square gon	$(g)^2$	2.467401101E-04

degree (°)	minute (′)	second (″)	grade (g)
60	2.16E+04	1.2960E+06	400
60	2.16E+04	1.2960E+06	400
80	10800	6.4800E+05	200
0	5400	3.2400E+05	100
0	5400	3.2400E+05	100
0	1800	1.0800E+05	33.33333333
7.295779513082	3437.74677078494	206264.806247096	63.661977236758
	60	3600	1.11111
.90	54	3240	1
.729386977E-01	34.376321861009	2062.579311660550	6.365985530E-01
.09	5.40	324	0.10
.729576003E-02	3.437745601951	206.264736117062	6.366195559E-02
.714285771E-02	3.428571462469	205.714287748136	6.349206412E-02
.624999999E-02	3.374999999152	202.499999949092	6.249999998E-02
.666666667E-02	1	60	1.851851852E-02
.00E-03	0.54	32.4	0.01
.777777778E-04	1.666666667E-02	1	3.086419753E-04
.00E-05	5.40E-03	0.324	1E-04

spat	square degree $(°)^2$	square gon $(g)^2$
	41252.72943435	50929.29558517
7.957791872E-02	3282.80635	4052.847344
2.424082027E-05	1	1.234568
1.963506443E-05	8.100000002E-01	1

Table 4.2.19 Conversion table: units of thermal conductivity

Unit	Symbol	cal (IT). $cm^{-1}.°C^{-1}$	cal (therm.). $cm^{-1}.°C^{-1}$	$W.cm^{-1}.K^{-1}$
cal (IT) per centimetre per second per Celsius degree	cal (IT). $cm^{-1}.°C^{-1}$	1	1.000654876	4.18674
cal (therm.) per centimetre per second per Celsius degree	cal (therm.). $cm^{-1}.°C^{-1}$	9.993455529E-01	1	4.184
watt per centimetre per kelvin	$W.cm^{-1}.K^{-1}$	2.388493195E-01	2.390057361E-01	1
Btu (IT) per inch per hour per Fahrenheit degree	Btu (IT). $in^{-1}.h^{-1}.°F$	4.960618261E-02	4.963866849E-02	2.076881890E-01
Btu (therm.) per inch per hour per Fahrenheit degree	Btu (therm.). $in^{-1}.h^{-1}.°F^{-1}$	4.957298820E-02	4.960545234E-02	2.075492126E-01
Btu (IT) per foot per hour per Fahrenheit degree	Btu (IT). $ft^{-1}.h^{-1}.°F^{-1}$	4.133848551E-03	4.136555708E-03	1.730734908E-02
Btu (therm.) per foot per hour per Fahrenheit degree	Btu (therm.). $ft^{-1}.h^{-1}.°F^{-1}$	4.131082350E-03	4.133787695E-03	1.729576772E-02
joule per second per centimetre per kelvin	$J.s^{-1}.cm^{-1}.K^{-1}$	2.388493195E-03	2.390057361E-03	1E-02
watt per metre per kelvin	$W.m^{-1}.K^{-1}$	2.388493195E-03	2.390057361E-03	1E-02

(IT).in^{-1}. h^{-1}.°F	Btu (therm.).in^{-1}. h^{-1}.°F^{-1}	Btu (IT).ft^{-1}. h^{-1}.°F^{-1}	Btu (therm.).ft^{-1}. h^{-1}.°F^{-1}	W.m^{-1}.K^{-1}
.158777544	20.172275999	241.905330523	242.067311993	418.674
.145584689	20.159074311	241.747016272	241.908891734	418.4
814910299	4.818134396	57.778923583	57.817612747	100
	1.000669607	12	12.008035282	20.768818898
993308412E-01	1	11.991970094	12	20.754921260
333333333E-02	8.338913391E-02	1	1.000669607	1.730734908
327757010E-02	8.333333333E-02	9.993308412E-01	1	1.729576772
814910299E-02	4.818134396E-02	5.777892358E-01	5.781761275E-01	1
814910299E-02	4.818134396E-02	5.777892358E-01	5.781761275E-01	1

Table 4.2.20 Conversion table: units of concentration

Physical quantities	Symbol	Equation		
Molarity (molar concentration)	c_i	$c_i = \frac{n_i}{V_{sol}}$	c_i n_i V_{Sol}	molarity in mol.m^{-3}, amount of substance in mo volume of the solution in m
Particle density	C_i	$C_i = \frac{N_i}{V}$	C_i N_i V	density of particles i in mol.m^{-3}, number of particles i, volume of the solution in m
Molarity of solute	b_i	$b_i = \frac{n_i}{M_s}$	b_i n_i M_s	molarity in mol.kg^{-1}, amount of substance in mo mass of solvent in kg.
Normality (number of equivalents)	N_i	$N_i = \frac{Z_i n_i}{V_{Sol}}$	N_i n_i Z_i V_{Sol}	normality of the species i in eq.m^{-3}, amount of substance in mo valency of the ionic species volume of the solution in m
Mass fraction	w_i	$w_i = \frac{m_i}{\sum_{i=1}^{n} m_i}$	w_i m_i	mass fraction of the species mass of the species i in kg.
Volumetric fraction	ϕ_i	$\phi_i = \frac{v_i}{\sum_{i=1}^{n} v_i}$	ϕ_i v_i	volumetric fraction of the species i, volume of the species i in m
Molar fraction	x_i	$x_i = \frac{n_i}{\sum_{i=1}^{n} n_i}$	x_i n_i	molar fraction of the specie amount of substance of the species i.
Atomic fraction	a_i	$a_i = \frac{N_i}{\sum_{i=1}^{n} N_i}$	a_i N_i	atomic fraction of the atom number of atoms.

Table 4.2.21 Conversion table: units of temperature

Temperature scale	T [K]	t_C [°C]	$t_{Ré}$ [°Ré]	t_F [°F]	t_R [°R]
Absolute temperature T (K)	T	$T - 273.15$	$\frac{4}{5}(T - 273.15)$	$\frac{9}{5}(T - 273.15) + 32$	$\frac{9}{5}$
Celsius temperature t_C (°C)	$t_C + 273.15$	t_C	$\frac{4}{5}t_C$	$\frac{9}{5}t_C + 32$	$\frac{9}{5}t_C + 491.67$
Réamur temperature $t_{Ré}$ (°Ré)	$\frac{5}{4}t_{Ré} + 273.15$	$\frac{5}{4}t_{Ré}$	$t_{Ré}$	$\frac{9}{5}t_{Ré} + 32$	$\frac{9}{5}t_{Ré} + 491.67$
Fahrenheit temperature t_F (°F)	$\frac{5}{9}(t_F - 32) + 273.15$	$\frac{5}{9}(t_F - 32)$	$\frac{4}{9}(t_F - 32)$	t_F	$t_F + 459.67$
Rankine temperature t_R (°R)	$\frac{5}{9}t_R$	$\frac{5}{9}(t_R - 491.67)$	$\frac{4}{9}t_R - 491.67$	$t_R - 459.67$	t_R

Unit (symbol)	K	°C	°Ré	°F	°R
1 kelvin (K)	1	1	4/5	9/5	9/5
1 Celsius degree (°C)	1	1	4/5	9/5	9/5
1 Réamur degree (°Ré)	5/4	5/4	1	9/4	9/4
1 Fahrenheit degree (°F)	5/9	5/9	4/9	1	1
1 Rankine degree (°R)	5/9	5/9	4/9	1	1

Table 4.2.22 Conversion table: units used in electricity

Electric potential						
Unit	**Symbol**	**esu cgs**	**V (int. mean)**	**V (int. US)**	**volt**	**emu cgs**
statvolt	esu cgs	1	299.6905632	299.6935591	**299.792458**	2.997925E+
u.a. of potential	u.a.	7.408924E-02	22.20384691	22.20406888	22.21139622	2.221140E+
volt (int. mean)	V (int. mean)	3.336775E-03	1	1.000009997	1.00034	1.000340E+
volt (int. US)	V (int. US)	3.336742E-03	9.999900E-01	1	1.00033	1.000330E+
volt	V	3.335641E-03	9.996601E-01	9.996701E-01	1	**1E+08**
abvolt	emu cgs	3.335641E-11	9.996601E-09	9.996701E-09	**1E-08**	1

Electric charge						
Unit	**Symbol**	**faraday**	**Ah**	**emu cgs**	**coulomb**	**esu cgs**
faraday (C12)	F	1	2.680147E+01	9.648531E+03	9.648531E+04	2.875317E+1
ampere-hour	Ah	3.731138E-02	1	**360**	**3600**	1.072820E+1
abcoulomb	emu cgs	1.036427E-04	2.777778E-03	1	**10**	2.980057E+1
coulomb	C	1.036427E-05	2.777778E-04	1.000000E-01	1	2.980057E+0
statcoulomb	esu cgs, Fr	3.477878E-15	9.321225E-14	3.355641E-11	3.355641E-10	1
u.a. charge	u.a.	1.660540E-24	4.450493E-23	1.602177E-20	1.602177E-19	4.774579E-1

Electric current intensity						
Unit	**Symbol**	**esu cgs**	**ampere**	**A (int. mean)**	**A (int. US)**	**emu cgs**
abampere	emu cgs	1	**10**	10.00150023	10.00165027	2.997925E+1
biot	biot	1	**10**	10.00150023	10.00165027	2.997925E+1
ampere	A	**0.1**	1	1.000150023	1.000165027	2.997925E+0
amp. (int. mean)	A (int. mean)	9.998500E-02	0.99985	1	1.000015002	2.997476E+0
amp. (int. US)	A (int. US)	9.998350E-02	0.999835	9.999850E-01	1	2.997431E+0
u.a. of current	u.a.	6.623621E-04	6.623621E-03	6.624615E-03	6.624714E-03	1.985712E+0
statampere	esu cgs	3.335640E-11	3.335640E-10	3.336140E-10	3.336190E-10	1

Electric capacitance						
Unit	**Symbol**	**emu cgs**	**farad**	**F (int.)**	**esu cgs**	**"cm"**
abfarad	emu cgs	1	**1E+09**	1.000490E+09	8.98473E+20	**9E+20**
farad	F	**1E-09**	1	1.00049024	8.98473E+11	**9E+11**
farad (int.)	F (int.)	9.9951E-10	0.99951	1	8.98032E+11	8.995590E+11
jar	jar	1.111111E-18	1.111111E-09	1.111656E-09	998.3028851	**1000**
statfarad	esu cgs	1.113000E-21	1.113000E-12	1.113546E-12	1	1.0017
"cm"	"cm"	1.111111E-21	1.111111E-12	1.111656E-12	0.998302885	1
puff	puff	**1E-21**	**1E-12**	1.000490E-12	0.898472597	0.9

Electric resistance						
Unit	**Symbol**	**esu cgs**	**Ω (int. US)**	**Ω (int. mean)**	**ohm**	**emu cgs**
statohm	esu cgs	1	8.98255E+11	8.982599E+11	8.98700E+11	8.987000E+20
preece	preece	1.11272E-06	9.995052E+05	9.995102E+05	**1E+06**	**1E+15**
lenz	lenz	8.90175E-08	7.996042E+04	7.996082E+04	8E+04	**8E+13**
ohm (int. US)	Ω (int.US)	1.11327E-12	1	1.000004998	1.000495	1.000495E+09
ohm (Int. mean)	Ω (int.mean)	1.11326E-12	9.999950E-01	1	1.00049	1.000490E+09
ohm	Ω	1.11272E-12	0.999505245	0.99951024	1	**1E+09**
ohm (legal)	Ω (legal)	1.1096E-12	9.967066E-01	0.996711611	0.9972	9.972000E+08
jacobi	–	7.1214E-13	6.396834E-01	0.639686554	0.64	**6.40E+08**
ring. equ. number	REN	2.7818E-16	2.498763E-04	0.000249878	**2.50E-04**	**2.50E+05**
wheathstone	–	2.7818E-15	2.498763E-03	0.002498776	**2.50E-03**	**2.50E+06**
abohm	emu cgs	**1E-21**	**1E-09**	**1E-09**	**1E-09**	1

Table 4.2.23 Conversion table: units used in magnetism

Magnetic field					
Unit	**Symbol**	**praoersted**	**Oe**	**A.m^{-1}**	**lenz**
praoersted	–	1	143.999049	11459.08	11459.08
oersted	Oe	6.944490E-03	1	79.57747	79.57747
ampere per metre	A.m^{-1}	8.726704E-05	1.256637E-02	1	1
lenz	lenz	8.726704E-05	1.256637E-02	1	1

Magnetic induction field strength					
Unit	**Symbol**	**tesla**	**gauss**	**u.a.**	**gamma**
weber per square metre	Wb.m^{-2}	1	10000	42543.77755	1E+09
tesla	T	1	10000	42543.77755	1E+09
gauss	G	1E-04	1	4.254377755	1E+05
maxwell per square centimetre	Mx.cm^{-2}	1E-04	1	4.254377755	1E+05
line per square centimetre	line.cm^{-2}	1E-04	1	4.254377755	1E+05
u.a.	u.a.	2.35052E-05	0.235052	1	23505.2
gamma	γ	1E-09	1E-05	4.25438E-05	1

Magnetic flux					
Unit	**Symbol**	**weber**	**kapp line**	**unit pole**	**maxwell**
weber	Wb	1	1.666667E+04	7.957748E+06	1E+08
kapp line	–	6E-05	1	477.4648526	6000
unit pole	unit pole	1.256637E-07	2.094395E-03	1	12.56637
maxwell	Mx	1E-08	1.666667E-04	7.957748E-02	1
line	line	1E-08	1.666667E-04	7.957748E-02	1

Electric inductance					
Unit	**Symbol**	**esu cgs**	**henry**	**"cm"**	**emu cgs**
stathenry	esu cgs	1	8.98755E+11	8.987552E+20	8.987552E+20
henry (int. US)	H (int. US)	1.113201E-12	1.000495	1.000495E+09	1.000495E+09
henry (int. mean)	H (int. mean)	1.113195E-12	1.00049	1.000490E+09	1.000490E+09
henry	H	1.112650E-12	1	1E+09	1E+09
mic	mic	1.112650E-18	1E-06	1000	1000
cm	cm	1.112650E-21	1E-09	1	1
abhenry	emu cgs	1.112650E-21	1E-09	1	1

Magnetomotive force					
Unit	**Symbol**	**A-turn**	**emu cgs**	**ampere**	**gilbert**
ampere-turn	A-tour	1	286.3959814	2864.77	3599.997487
abampere-turn	emu cgs	3.491669E-03	1	10.0028289	12.57
ampere	A	3.490682E-04	9.997172E-02	1	1.256644508
gilbert	Gb	2.777780E-04	7.955449E-02	7.957700E-01	1

Table 4.2.24 Conversion table: units used in photometry

Luminous intensity					
Unit	**Symbol**	**carcel**	**candle (int.)**	**candela**	**Hefner un**
carcel	carcel	1	9.810000672	10	11.07419712
candle (int.)	Cd (int.)	0.101936792	1	1.019367924	1.12886813
candela	cd	0.1	0.981000067	1	1.107419712
Hefner unit	HK	0.0903	0.885843061	0.903	1

Illuminance					
Unit	**Symbol**	**phot**	**Ft-C**	**lux**	**nox**
phot	ph	1	929.030436	1E+04	1E+07
foot-candle	ft-C	0.001076391	1	10.76391	10763.91
lux	lx	1E-04	9.290304E-02	1	1E+03
skot	skot	1E-07	9.290304E-05	1E-03	1
nox	nox	1E-07	9.290304E-05	1E-03	1

Luminous luminance					
Unit	**Symbol**	**lambert**	**ft-L**	**cd.m^{-2}**	**asb**
stilb	sb	3.141592654	2918.635164	1E+04	31415.92654
luxon (troland)	luxon	3.141592654	2918.635164	1E+04	31415.92654
lambert	L	1	9E+02	3183.098862	1E+04
foot-lambert	ft-L	1.076391E-03	1	3.426259	10.7639101
candela per square metre	cd.m^{-2}	3.141593E-04	2.918635E-01	1	3
nit	nit	3.141593E-04	2.918635E-01	1	3.141592654
apostilb	asb	1E-04	9.290304E-02	3.183099E-01	1
blondel	Bl	1E-04	9.290304E-02	3.183099E-01	1

Table 4.2.25 Conversion table: units used in nuclear sciences

Radioactivity					
Unit	**Symbol**	**curie**	**rutherford**	**Mache unit**	**becquerel**
curie	**Ci**	1	37000	2.803030E+09	**3.70E+10**
rutherford	**Rd**	2.702703E-05	1	7.575758E+04	**1E+06**
Mache unit	**Mache**	3.567568E-10	1.320000E-05	1	13.2
becquerel	**Bq**	2.702703E-11	**1E-06**	7.575758E-02	1
stat	**stat**	3.630000E-27	1.343100E-24	1.017500E-17	1.343100E-16

Absorbed dose					
Unit	**Symbol**	**gray**	**rad**	**g-rad**	**erg.g^{-1}**
gray	**Gy**	**1**	**100**	**100**	**10000**
rad	**rad**	**1E-02**	**1**	**1**	**100**
erg per gram	**erg.g^{-1}**	**1E-04**	**1E-02**	**1E-02**	**1**
gram-rad	**g-rad**	**1E-02**	**1**	**1**	**100**

Dose equivalent			
Unit	**Symbol**	**sievert**	**rem**
sievert	**Sv**	**1**	**100**
rem	**rem**	**0.01**	**1**

Exposure					
Unit	**Symbol**	**C.kg^{-1}**	**B unit**	**D unit**	**röntgen**
coulomb per kg	**C.kg^{-1}**	1	7.751937984	38.59811641	3875.968992
pastille dose (B unit)	**B unit**	1.29E-01	1	4.979157017	**500**
D unit	**D unit**	2.59E-02	2.008372E-01	1	100.4186047
röntgen	**R**	**2.58E-04**	**2E-03**	9.958314E-03	**1**

5 Fundamental Constants

1 Fundamental Mathematical Constants

e knowledge of fundamental mathematical constants is needed to compute urately several conversion factors. In spite of the increased use of compus in most of scientific calculus, it seems to be very important to remind the der of some approximate values of a selection of numerical mathematical nstants. Accurate values are given in several mathematical treatises and ndbooks.[12] Numerical values of the following constants are accurate to the mber of significant digits given below.

able 5-1 Fundamental mathematical constants

i[13]	π	$\approx$ **3.14159265358979323846...**
aperian (natural) logarithm base	e	$\approx$ **2.71828182845904523536...**
quare root of two	$\sqrt{2}$	$\approx$ **1.41421356237309504880...**
quare root of three	$\sqrt{3}$	$\approx$ **1.73205080756887729353...**
aperian (natural) logarithm of two	ln 2	$\approx$ **0.69314718055994530942...**
aperian (natural) logarithm of ten	ln 10	$\approx$ **2.30258509299404568402...**
riggsian (common) logarithm of two	$\log_{10} 2$	$\approx$ **0.30102999566398119521...**
uler's constant	γ	$\approx$ **0.57721566490153286061...**
old number	ϕ	$\approx$ **1.61803398874989484820...**

2 Fundamental Physical Constants

e numerical values of fundamental physical constants were recommended CODATA task group on fundamental constants in 1986.[14] For each connt the standard deviation uncertainty in the least significant digits is given parentheses.

Abramowitz, M., and Stegun, I.A., *Handbook of Mathematical Functions, with Formulas, phs, and Mathematics Tables*, Dover Publications Inc., New York, 1972.

A mnemonic for π, based on the number of letters in English words, is "How I like a nk, alcoholic of course, after the heavy lectures involving quantum mechanics!"

Cohen, E.R., and Taylor, B.N., *Journal of Research of the National Bureau of Standards,* (1987) 85, and Cohen, E. R., and Taylor, B. N., The 1986 Adjustment of the Fundamental ysical Constants, *CODATA Bull.*, **63** (1986)1–49.

Table 5-2 Universal constants

speed of light in vacuum	c_0	$c_0 = 2.99792458 \times 10^8$ m.s^{-1}
permeability of vacuum	μ_0	$\mu_0 = 4\pi \times 10^{-7}$ F.m^{-1}
permittivity of vacuum	$\varepsilon_0 = 1/\mu_0 c^2$	$\varepsilon_0 = 8.854187817 \times 10^{-12}$ F.m^{-1}
Newtonian gravitational constant	G	$G = 6.67259(85) \times 10^{-11}$ N.kg^{-2}.m^2
Planck's constant	h	$h = 6.6260755(40) \times 10^{-34}$ J.s $h = 4.1356692(12) \times 10^{-15}$ eV.s
Planck's constant (rationalized)	$\hbar = h/2\pi$	$\hbar = 1.05457266(63) \times 10^{-34}$ J.s $\hbar = 6.5821220(20) \times 10^{-16}$ eV.s
elementary electric charge	e	$e = 1.60217733(49) \times 10^{-19}$ C
standard acceleration of gravity	g_n	$g_n = 9.80665$ m.s^{-2}
Planck's mass	$m_p = (\hbar c/G)^{1/2}$	$m_p = 2.17671(14) \times 10^{-8}$ kg
Planck's length	$L_p = \hbar/m_p c = (\hbar G/c^3)^{1/2}$	$L_p = 1.61605(10) \times 10^{-35}$ m
Planck's time	$t_p = L_p/c = (\hbar G/c^5)^{1/2}$	$t_p = 5.39056(34) \times 10^{-44}$ s
quantum of magnetic flux	$\Phi_0 = h/2e$	$\Phi_0 = 2.06783461(61) \times 10^{-15}$ Wb

Table 5-3 Physico-chemical constants

Avogadro's number	N_A, L	$N_A = 6.0221367(36) \times 10^{23}$ mol^{-1}
atomic mass unit	u, u.m.a.	$u = 1.6605402(10) \times 10^{-27}$ kg $u = 931.49432(28)$ MeV.c^{-2}
Faraday's constant	$F = N_A \times E$	$F = 96485.309(29)$ C.mol^{-1}
Boltzmann's constant	$k = R/N_A$	$k = 1.380658(12) \times 10^{-23}$ J.K^{-1} $k = 8.617385(73) \times 10^{-5}$ eV.K^{-1}
ideal gas constant	R	$R = 8.314510(70)$ J.K^{-1}.mol^{-1}
molar Planck constant	$N_A h$	$N_A h = 3.99031323(36) \times 10^{-10}$ J.s.mol^{-1}
standard atmosphere	P_0	$P_0 = 101325$ Pa
standard molar volume (STP) (Ideal gas)	$V_0 = RT_0/P_0$	$V_0 = 22.41410(19) \times 10^{-3}$ m^3.mol^{-1} [273.15 K, 1 atm] $V_1 = 22.71108(19) \times 10^{-3}$ m^3.mol^{-1} [273.15 K, 1 bar]
Lochsmidt's constant	$n_0 = N_A/V_m$	$n_0 = 2.686763(23) \times 10^{25}$ m^{-3}
Sackur-Tetrode's constant (absolute entropy)	S_0/R[15]	$S_0/R = -1.151693(21)$ [T = 1 K, P = 10^5 Pa] $S_0/R = -1.164856(21)$ [T = 1 K, P = 101325 Pa]
Stefan-Boltzmann's constant	$\sigma = (\pi^2/60)(k^4/\hbar^3 c^2)$	$\sigma = 5.67051(19) \times 10^{-8}$ W.m^{-2}.K^{-4}
first radiation constant	$c_1 = 2\pi hc^2$	$c_1 = 3.7417749(22) \times 10^{-16}$ W.m^{-2}
second radiation constant	$c_2 = hc/k$	$c_2 = 0.01438769(12)$ m.K
Wien displacement law constant	$b = c_2/4.96511423$	$b = 2.897756(24) \times 10^{-3}$ m.K

[15] $S_0/R = 5/2 + \ln\left[(2\pi m_u kT/b^2)^{3/2} kT/P_0\right]$

Table 5-4 Electromagnetic and atomic constants

electron rest mass	m_e	$m_e = 9.1093897(54) \times 10^{-31}$ kg $m_e = 5.48579903(13) \times 10^{-4}$ u $m_e = 0.51099906(15) \times 10^{-31}$ MeV.c^{-2}
Bohr magneton (B.M., β, μ_β)	$\mu_\beta = e\hbar/2m_e$	$\mu_\beta = 9.2740154(31) \times 10^{-24}$ J.T^{-1} $\mu_\beta = 5.78838263(52) \times 10^{-5}$ eV.T^{-1}
fine structure constant	$\alpha = \mu_0 e^2 c/2h$	$\alpha = 7.29735308(33) \times 10^{-3}$ $\alpha^{-1} = 137.0359895(61)$
Rydberg constant	$R_\infty = E_h/2hc$	$R_\infty = 1.0973731534(13) \times 10^7$ m^{-1}
Rydberg	$R_y = R_\infty hc$	$R_y = 2.1798741(13) \times 10^{-18}$ J $R_y = 13.6056981(40)$ J
first Bohr atomic radius	$a_0 = 4\pi\varepsilon_0\hbar^2/m_0 e^2$	$a_0 = 0.529177249(24) \times 10^{-10}$ m
quantized Hall resistance	$h/e^2 = \mu_0 c/2\alpha$	$h/e^2 = 25812.8056(12)$ W
proton rest mass	M_p	$m_p = 1.6726231(10) \times 10^{-27}$ kg
nuclear magneton (N.M., β_N, μ_β)	$\mu_N = e\hbar/2m_p$	$\mu_p = 5.0507866(17) \times 10^{-27}$ J.T^{-1} $\mu_N = 3.15245166(28) \times 10^{-8}$ eV.T^{-1}
Hartree energy	$E_h = \hbar^2/m_0 a_0{}^2$	$E_h = 4.3597482(26) \times 10^{-18}$ J $E_h = 27.2113961(81)$ eV
Josephson frequency-voltage ratio	$2e/h$	$2e/h = 4.8359767(14) \times 10^{14}$ Hz.V^{-1}
quantum of circulation	$h/2m_e$	$h/2m_e = 3.63694807(33) \times 10^{-4}$ m^2.s^{-1}
quantum of magnetic flux	$h/2e$	$h/2e = 2.06783461(61) \times 10^{-15}$ Wb

Table 5-5 Electron		
electron rest mass	m_e	$m_e = 9.1093897(54) \times 10^{-31}$ kg $m_e = 5.48579903(13) \times 10^{-4}$ u $m_e = 0.51099906(15) \times 10^{-31}$ MeV.c^{-2}
electron molar mass	M_e	$M_e = 5.48579903(13) \times 10^{-7}$ kg.mol^{-1}
electron-proton rest mass ratio	m_e/m_p	$m_e/m_p = 5.44617013(11) \times 10^{-4}$
electron-neutron rest mass ratio	m_e/m_n	$m_e/m_n = 5.4386734 \times 10^{-4}$
electron-muon rest mass ratio	m_e/m_μ	$m_e/m_\mu = 4.83633218(71) \times 10^{-3}$
electron-deuteron rest mass ratio	m_e/m_d	$m_e/m_d = 2.72443707(6) \times 10^{-4}$
electron-helion rest mass ratio	m_e/m_α	$m_e/m_\alpha = 1.37093354(3) \times 10^{-4}$
electron specific charge	e/m_e	$e/m_e = -1.75881962(53) \times 10^{11}$ C.kg^{-1}
electron classical radius	$r_e = \alpha^2/a_0$	$r_e = 2.81794092(38) \times 10^{-15}$ m
electron Thomson cross section	$\sigma_e = (8\pi/3)r_e{}^2$	$\sigma_e = 0.66524616(18) \times 10^{-28}$ m^2
electron Compton wavelength	$\Lambda_C = h/m_0c$	$\Lambda_C = 2.42631058(22) \times 10^{-12}$ m
electron Compton rationalized wavelength	$\bar{\Lambda}_C = \Lambda_C/2\pi$ $= \hbar/m_0c$	$\bar{\Lambda}_C = 3.86159323(35) \times 10^{-13}$ m
magnetic moment of electron	μ_e	$\mu_e = 9.2847701(31) \times 10^{-24}$ J.T^{-1} $\mu_e = 1.001159652193(10)$ B $\mu_e = 1838.282000(37)$ N
electron magnetic moment anomaly	$a_e = (\mu_e/\mu_B) - 1$	$a_e = 1.159652193(10) \times 10^{-3}$
electron Landé factor (g factor)	$g_e = 2(1 + a_e)$	$g_e = 2.002319304386(20)$
electron-proton magnetic moment ratio	μ_e/μ_p	$\mu_e/\mu_p = 658.2106881(66)$
electron-muon magnetic moment ratio	μ_e/μ_μ	$\mu_e/\mu_\mu = 206.766967(30)$

ble 5-6 Proton		
oton rest mass	m_n	$m_n = 1.6726231(10) \times 10^{-27}$ kg $m_n = 1.007276470(12)$ u $m_n = 938.27231(28)$ MeV.c^{-2}
oton molar mass	M_p	$M_p = 1.007246470(12) \times 10^{-3}$ kg.mol^{-1}
ecific charge of oton	e/m_p	$e/m_p = 9.5788309(29) \times 10^{7}$ C.kg^{-1}
oton-electron rest ss ratio	m_p/m_e	$m_p/m_e = 1836.152701(37)$
oton magnetic ment	μ_p	$\mu_p = 1.41060761(47) \times 10^{-26}$ J.T^{-1}
oton Compton velength	$\Lambda_{C,p} = h/m_p c$	$\Lambda_{C,p} = 1.32141002(12) \times 10^{-15}$ m
oton Compton ionalized wavelength	$\bar{\Lambda}_{C,p} = \Lambda_{C,p}/2\pi = \hbar/m_p c$	$\bar{\Lambda}_{C,p} = 2.10308937(19) \times 10^{-16}$ m
oton gyromagnetic io	γ_p	$\gamma_p = 2.67522128(81) \times 10^{8}$ s^{-1}.T^{-1}

ble 5-7 Neutron		
utron rest mass	m_n	$m_n = 1.6749286(10) \times 10^{-27}$ kg $m_n = 1.008664904(14)$ u $m_n = 939.56563(28)$ MeV.c^{-2}
utron molar mass	M_n	$Mn = 1.008664904(14) \times 10^{-3}$ kg.mol^{-1}
utron-proton rest ss ratio	m_n/m_p	$m_n/m_p = 1.001378404(9)$
utron-electron rest ss ratio	m_n/m_e	$m_n/m_e = 1838.683662(40)$
utron mean life time	τ	$\tau = 889.1(21)$ s
utron Compton velength	$\Lambda_{C,n} = h/m_n c$	$\Lambda_{C,n} = 1.31959110(12) \times 10^{-15}$ m
utron Compton tionalized wavelength	$\bar{\Lambda}_{C,n} = \Lambda_{C,n}/2\pi = \hbar/m_n c$	$\bar{\Lambda}_{C,n} = 2.10019445(19) \times 10^{-16}$ m
utron magnetic oment	μ_n	$\mu_n = 0.96623707(40) \times 10^{-26}$ J.T^{-1} $\mu_n = 1.04187563(25) \times 10^{-3}\ \mu_B$ $\mu_n = 1.91304275(45)\ \mu_N$
utron-electron agnetic moment ratio	μ_n/μ_e	$\mu_n/\mu_e = 1.04066882(25) \times 10^{-3}$
utron-proton agnetic moment ratio	μ_n/μ_p	$\mu_n/\mu_p = 0.68497934(16)$

Table 5-8 Muon		
muon rest mass	m_μ	$m_\mu = 1.8835327(11) \times 10^{-28}$ kg $\mu_\mu = 0.113428913(17)$ u $\mu_\mu = 105.658389(34)$ MeV.c^{-2}
muon molar mass	M_μ	$M_\mu = 0.113428913(17) \times 10^{-3}$ kg.mol
muon mean life time	τ	$\tau = 2.19703(4) \times 10^{-6}$ s
muon-electron rest mass ratio	m_μ/m_e	$m_\mu/m_e = 206.768262(30)$
muon magnetic moment ratio	μ_μ	$\mu_\mu = 4.4904514(15) \times 10^{-26}$ J.T^{-1} $\mu_\mu = 4.84197097(71) \times 10^{-3}\ \mu_B$ $\mu_\mu = 8.8905981(13)\ \mu_N$
muon magnetic moment anomaly	$a_\mu = \left[\frac{\mu_\mu}{(e\hbar/2m_\mu)}\right] - 1$	$a_\mu = 1.1659230(84) \times 10^{-3}$
muon Landé factor (g-factor)	$g_\mu = 2(1 + a_\mu)$	$g_\mu = 2.002331846(17)$
muon-proton magnetic moment ratio	μ_μ/μ_p	$\mu_\mu/\mu_p = 3.18334547(47)$

Table 5-9 Deuteron		
deuteron rest mass	m_d	$m_d = 3.3435860(20) \times 10^{-27}$ kg $m_d = 2.013553214(24)$ u $m_d = 1875.61339(57)$ MeV.c^{-2}
deuteron molar mass	M_d	$M_d = 2.013553214(24)$ u
deuteron-electron rest mass ratio	m_d/m_e	$m_d/m_e = 3670.483014(75)$
deuteron-proton rest mass ratio	m_d/m_p	$m_d/m_p = 1.999007496(6)$
deuteron magnetic moment	μ_d	$\mu_d = 0.43307375(15) \times 10^{-26}$ J.T^{-1} $\mu_d = 0.4669754479(91) \times 10^{-3}\ \mu_B$ $\mu_d = 0.857438230(24)\ \mu_N$
deuteron-electron magnetic moment ratio	μ_d/μ_e	$\mu_d/\mu_e = 0.4664345460(91) \times 10^{-3}$
deuteron-proton magnetic moment ratio	μ_d/μ_p	$\mu_n/\mu_p = 0.3070122035(51)$

6 Appendices

6.1 Greek Alphabet

Numerous symbols of units or prefixes and physical quantities are described by a letter derived from the greek alphabet. So, here is the complete greek alphabet which is listed in ***Table 6-1***.

Table 6-1 The Greek alphabet

Capital letter	Small letter	Appellation	Capital letter	Small letter	Appellation
A	α	alpha	O	o	omicron
B	β	beta	Π	π, ϖ	pi (dorian pi)
Γ	γ	gamma	P	ρ	rho
Δ	δ	delta	Σ	σ	sigma
E	ε	epsilon	T	τ	tau
Φ	ϕ, φ	phi	Υ	υ	upsilon
H	η	eta	X	ξ	ksi, xi
I	ι	iota	Y	ψ	psi
K	κ	kappa	Z	ζ	zeta
Λ	λ	lambda	Θ	θ, ϑ	theta
M	μ	mu	X	χ	chi
N	ν	nu	Ω	ω	omega

Greek small letters listed by Latin alphabetical order:

$$\alpha - \beta - \gamma - \delta - \varepsilon - \zeta - \eta - \theta - \iota - \kappa - \chi - \lambda - \mu$$
$$- \nu - \xi - o - \pi - \varpi - \rho - \sigma - \tau - \upsilon - \phi - \psi - \omega$$

6.2 Roman Numerals

The ancient Roman notation for numerals uses seven letters and a bar. A letter with a bar placed over it represents a thousand times as much as it does without the bar.

e.g.: $\bar{\text{V}}$ is equal to 5000

The old signs used for roman numerals are listed below.

Table 6-2 Roman numerals

I	1	LXX	70
II	2	LXXX	80
III	3	XC	90
IV	4	IC	99
V	5	C	100
VI	6	CC	200
VII	7	CCC	300
VIII	8	CD	400
IX	9	D	500
X	10	DC	600
XX	20	DCC	700
XXX	30	DCCC	800
XL	40	CM	900
L	50	XM	990
LX	60	M	1000

Rule 1: If no letter precedes a letter of greater value, add the number represented by all the letters.

e.g.: XX is equal to 20, VII is equal to 7

Rule 2: If a letter precedes a letter of greater value, subtract the smaller from the greater; add the remainder or remainders thus obtained to the number represented by the other letters.

e.g.: IX is equal to 9, MCMXCVI is equal to 1996

6.3 Rules about Large Numbers (>1 000 000)

In France (since May 3rd, 1961), but also in Germany, the UK and other countries, large numbers expressed in powers of ten are given by the ***n* rule**. The powers of ten are written as: $\mathbf{10^{6n}}$. For example, we get according to this rule 10^6: one million, 10^{12}: one billion, 10^{18}: one trillion, 10^{24}: one quadrillion, 10^{36}: one sextillion, etc.
In the USA, but also in Italy, the ***Latin rule*** or **($n-1$) rule** remains in use for large numbers expressed in powers of ten. The powers of ten are written as: $\mathbf{10^{6(n-1)}}$. For example, we get: 10^6: one million, 10^9: one billion, 10^{12}: one trillion, 10^{15}: one quadrillion, 10^{18}: one quintillion, 10^{21} one sextillion, etc.

Note for French readers: the French word *milliard* ($=10^9$) has no legal validity, but it came into use in France as a translation of the American word billion.

Power of ten	UK, France (*n* rule)	USA, Italy (Latin rule)
10^6	million	million
10^9	(*milliard*)	billion
10^{12}	billion	trillion
10^{15}	–	quadrillion
10^{18}	trillion	quintillion

6.4 Date and Time Numerical Representation

The numerical representation of dates and times is very useful for data exchange in computers. Legal rules and procedures are given in the following ISO Standard: **ISO 8601: 1988 [Data elements and interchange formats, information interchange, representation of dates and times].**

For example, the numerical representation of the date **September 19th, 1995** should be written as:

1995-09-19

When exchanging data in electronic files, the separation between digits could be replaced by a blank space or omitted. This does not cause any confusion:

1995 09 19 or **95 09 19** or **19950919**

The numerical representation of time for **5 h 45 min 30 s p.m.** should be written as:

17:45:30

- the first number is written in two digits ranging from 00 to 23,
- the second number is composed of two digits ranging from 00 to 59,
- the third number is composed of two digits ranging from 00 to 59.

Time can also be written as a decimal number (according to second, minute or hour):

17:45:30.0 s; 17:45.5 min; 17.75833 h

Time and date can also appear together, for example September 19th, 1995 at 5 h 45 min and 30 s p.m should be written as follows:

1995 09 19 17:45:30 or in compressed mode 19950919174530.

If the time refers to the Universal Time Coordinate (UTC) the capital letter Z which describes the Greenwich meridian should be written at the end of the numerical time representation. If the previous described time corresponds to the time in New York City (USA), according to the time difference between the American city and the Greenwich meridian time (5 h) the time in UTC should be written as follows:

22:45:30 Z

6.5 National, Regional, and International Standardization Bodies

This field covers standards and standardization bodies, both national and international.

6.5.1 International Standards Bodies

Table 6-3 International standards bodies

Acronym	Name	Address (headquarters)
BIH	Bureau International de l'Heure	Pavillon de Breuteuil, F-92310 Sèvre FRANCE
BIPM	Bureau International des Poids et Mesures	Pavillon de Breuteuil, F-92310 Sèvre FRANCE Telephone: + 33 1 45 07 70 70 Fax: + 33 1 45 34 20 21
CIE	Commission Internationale de l'Éclairage	Internet: http://www.cie.ch
CODATA	Committee on Data for Science and Technology	51, Boulevard de Montmorency F-75016 Paris, FRANCE Telephone: + 33 1 45 88 77 97 Fax: + 33 1 45 25 04 96
COPANT	Pan American Standards Commission	Av. Pte., roque Soenz Pena 501 7 Pis OF716, Buenos Aires, ARGENTINA
IAEA	International Atomic Energy Agency	P.O. Box 100, Wagramerstrasse 5, A-1400 Vienna, AUSTRIA Telephone: +43 1 2060 Fax: +43 1 20 607 E-mail: IAEO@iaeal.iaea.or.at. Internet: http://www.iaea.or.at/
IEC	International Electrotechnical Commission	P.O. Box 131, CH-1211 Geneva 20, SWITZERLAND E-mail: customer service@iec.ch Internet: http://www.iec.ch
ISO	International Standardization Organization	1, rue de Varembe, CH-1211 Geneva, SWITZERLAND Internet: http://www.iso.ch
ITU	International Telecommunication Union	Place des Nations, CH-1211 Geneva 20, SWITZERLAND Telephone: +41 22 730 6666 +41 22 730 5554 Fax: +41 22 730 5337 E-mail: helpdesk@itu.ch Internet: http://www.itu.ch/
IUCr	International Union of Crystallography	2 Abbey Square, Chester CH1 2HU, UK Telephone: +44 1 244 345 431 Fax: +44 1 244 344 843 E-mail: execsec@iucr.ac.uk Internet: http://www.iucr.ac.uk/
IUPAC	International Union of Pure and Applied Chemistry	Bank Court Way, Cowley Centre, Oxford OX4 3YF, UK
IUPAP	International Union of Pure and Applied Physics	Department of Physics, University of Manitoba, 301 Allen Building, Winnipeg MB R3T 2N2, Canada. Telephone: +204 474 9817 Fax: + 204 269 8489
NATO	North Atlantic Treaty Organization	Military Commitee, Conference of National Armament Directors, B-1110 Brussels, BELGIUM

.5.2 National Standards Bodies Worldwide

oday, most countries have a national standards organization that both leads d standardizes activities and acts within its own country as sales agents and information centre for other national standardization bodies. The ad- esses of standardization bodies for countries around the world are as llows:

.5.2.1 National Standards Bodies

he following table lists the ISO committee members, correspondent members] and subscribing members [#]. Each national standards body is completely escribed as follows: abbreviation, name, address, telephone and facsimile umbers, and where available, e-mail address and internet server.

Table 6-4 National standards bodies worldwide

Acronym	Name	Address (headquarters)
–	Construction Planning and Research Unit [*]	Ministry of Development, Brunei Darussalam BRUNEI DARUSSALAM Telephone: + 673 2 38 10 33 Fax: + 673 2 38 15 41
–	Department of Standards, Measurements and Consumer Protection [*]	Ministry of Finance, Economy and Commerce, P.O. Box 1968 Doha QATAR Telephone: + 974 40 85 55 Fax: + 974 42 54 49
–	Directorate General for Specifications and Measurements [*]	Ministry of Commerce and Industry, P.O. Box 550 - Postal code No. 113, Muscat, OMAN Telephone: + 968 70 32 38 Fax: + 968 79 59 92
–	Directorate of Standards and Metrology [*]	Ministry of Commerce, P.O. Box 5479, BAHRAIN Telephone: + 973 53 01 00 Fax: + 973 53 07 30
–	Fiji Trade Standards and Quality Control Office [#]	Ministry of Commerce, Industry and Tourism, Nabati House, Government Buildings, P.O. Box 2118 Suva, FIJI Telephone: + 679 30 54 11 Fax: + 679 30 26 17
–	Industry Department[*]	36/F., Immigration Tower, 7 Gloucester Road, Wan Chai Hong Kong, HONG KONG Telephone: + 852 28 29 48 24 Fax: + 852 28 24 13 02
–	Ministry of Commerce and Industry Standards and Metrology Affairs [*]	Post Box 2944 Safat, 13030 Kuwait City, KUWAIT Telephone: + 965 246 51 03 Fax: + 965 243 66 38
–	Ministry of Industry, Mines and Energy [#]	Technical Department, 5, Blvd Norodom, Phnom Penh, CAMBODIA Telephone: + 855 234 278 40 Fax: + 855 234 278 40

Table 6-4 *(continued)* National standards bodies worldwide

Acronym	Name	Address (headquarters)
ABBS	Antigua and Barbuda Bureau of Standards [#]	P.O. Box 1550, Redcliffe Street, St. John's ANTIGUA AND BARBUDA Telephone: + 1 809 462 15 32 Fax: + 1 809 462 16 25
ABNT	Associaçao Brasileira de Normas Téchnicas	Av. 13 de Maio, No.13, 27o andar, Caxa Postal 1680, 200003-900, Rio de Janeiro-RJ, BRAZIL Telephone: +55 21 210 31 22 Fax: +55 21 532 21 43
AENOR	Asoiación Espagñola de Normalización y Certificación	Fernandez de la Hoz, 52 E-28010 Madrid, SPAIN Telephone: + 34 1 432 60 00 Fax: + 34 1 310 49 76
AFNOR	Association Française de Normalisation	Tour Europe F-92049 Paris La Defense Cedex, FRANCE Telephone: +33 1 42 91 55 55 Fax: +33 1 42 91 56 56 Internet: http://www.afnor.fr/
ANSI (ASA)	American National Standards Institute (previously American Standards Association and US of America Standards Institute)	11 West 42nd Street, 13th floor, New York, New York, 10036, USA Telephone: +1 212 642 49 00 Fax: +1 212 398 00 23 Internet: http://www.ansi.org E-mail: info@ansi.org
BASMP	Institute for Standardization, Metrology and Patents c/o permanent Mission of Bosnia and Herzegovina	22, rue Lamartine, CH-1203 Genève, SWITZERLAND Telephone: +387 71 67 06 55 Fax: +387 71 67 06 56
BDS	Committee for Standardization and Metrology at the Council of Ministers	21, 6th September Str., 1000 Sofia, BULGARIA Telephone: +359 2 85 91 Fax: +359 2 80 14 02
BELST	Committee for Standardization, Metrology and Certification	Starovilensky Trakt 93, Minsk 220053, BÉLARUS Telephone: +375 172 37 52 13 Fax: +375 172 37 25 88
BIS	Bureau of Indian Standards	Manak Bhavan, 9 Bahadur Sha Zafar Marg, New Delhi 110002 INDIA Telephone: +91 11 323 79 91 Fax: +91 11 323 40 62
BNSI	Barbados National Standards Institution [*]	Flodden Culloden Road, St. Michael BARBADOS Telephone: + 1 809 426 38 70 Fax: + 1 809 436 14 95
BPS	Bureau of Products Standards	Department of Trade and Industry, 361 Sen. Gil J. Puyat Avenue, Makati, Metro Manila 1200, THE PHILIPPINES Telephone: +63 2 890 51 29 Fax: +63 2 890 49 26
BSI	British Standards Institution	389 Chiswick High Road, London W4 4AL, UK Telephone: +44 181 996 90 00 Fax: +44 181 996 74 01
BSTI	Bangladesh Standards and Testing Institution	116/A, Tejgaon Industrial Area, Dhaka 1208, BANGLADESH Telephone: +880 2 88 14 62

Table 6-4 *(continued)* National standards bodies worldwide

Acronym	Name	Address (headquarters)
COPANIT	Comisión Panameña de Normas Industriales y Técnicas	Ministerio de Comercio e Industrias Apartado Postal 9658, Panama, Zona 4, PANAMA Telephone: +507 2 27 47 49 Fax: +507 2 25 78 53
COSMT	Czech Office for Standards, Metrology and Testing	Biskupsky dvur 5, 113 47 Praha 1, CZECH REPUBLIC Telephone: +42 2 232 44 30 Fax: +42 2 232 43 73
COVENIN	Comisión Venezolana de Normas Industriales	Avenida Andrès Bello-Edf. Torre Fondo, Comùn, Piso 12, Caracas 1050, VENEZUELA Telephone: +58 2 575 22 98 Fax: +58 2 574 13 12 E-mail: covenin@dino.conicit.ve
CSBTS	China State Bureau of Technical Supervision	4, Zhichun Road, Haidan District, P.O. Box 8010, Beijing 100088, PEOPLE'S REPUBLIC OF CHINA Telephone: +86 10 20324 24 Fax: +86 10 203 10 10
CSK	Committee for Standardization of the Democratic People's Republic of Korea	Zung Gu Yok Seungli-Street, Pyongyang, DEMOCRATIC PEOPLE'S REPUBLIC of KOREA Telephone: +85 02 57 15 76
CYS	Cyprus Organization for Standards and Control of Quality	Ministry of Commerce, Industry and tourism, Nicosia 1421, CYPRUS Telephone: +357 2 37 50 53 Fax: +357 2 37 51 20
DGN	Dirección General de Normas	Calle Puente de Tecamachalco No.6 Lomas de Tecamachalco, Seccion Fuentes, Naucalpan de Juárez, 53 950, MEXICO Telephone: +52 5 729 93 00 Fax: +52 5 729 94 94
DIN	Deutsches Institut für Normung	Burggrafenstrasse 6, D-112623 Berlin, GERMANY Telephone: +49 30 26 01 22 60 Fax: +49 30 26 01 12 31 Internet: http://www.din.de/
DS	Dansk Standardiseringsraad	Baunegaardsvej 73, DK-29000 Hellerup, DENMARK Telephone: +45 39 77 01 01 Fax: +45 39 77 02 02
DSC	Drejtoria e Standardizimit dhe Cilesise	Rruga Mine Peza, Tirana, ALBANIA Telephone: +355 42 2 62 55 Fax: +355 42 2 62 55
DSN	Dewan Standardisasi Nasional	C/o Pusat Standardisasi-LIPI, Jalan Jend. gatot Subroto 10, Jakarta 12710, INDONESIA Telephone: +62 21 522 16 86 Fax: +62 21 520 65 74
DSTU	State Committee of Ukraine for Standardization, Metrology and Certification	174 Gorky Street, GSP, Kiev-6, 252650, UKRAINE Telephone: +380 44 226 29 71 Fax: +380 44 226 29 70

Table 6-4 *(continued)* National standards bodies worldwide

Acronym	Name	Address (headquarters)
DZNM	State Office for Standardization and Metrology	Ulica grada Vukovara 78, 10000 Zagreb, CROATIA Telephone: +385 1 53 99 34 Fax: +385 1 53 65 98
ELOT	Hellenic Organization for Standardization	313. Acharnon Street, GR-111 45, Athens, GREECE Telephone: +30 1 228 00 01 Fax: +30 1 202 07 76
EOS	Egyptian Organization for Standardization and Quality Control	2 Latin America Street, Garden City, Cairo, EGYPT Telephone: +20 2 354 97 20 Fax: +20 2 355 78 41
ESA	Ethiopian Authority for Standardization	P.O. Box 2310, Addis Ababa, ETHIOPIA Telephone: +251 1 61 01 11 Fax: +251 1 61 31 77
EVS	National Standards Board of Estonia [*]	Aru 10, EE-0003 Tallinn, ESTONIA Telephone: + 372 2 49 35 72 Fax: + 372 654 13 30 E-mail: @evs.ee
GDBS	Grenada Bureau of Standards [#]	H.A. Blaize Street, St. George's, GRENADA Telephone: + 1 809 440 58 86 Fax: + 1 809 440 41 15
GNBS	Guyana National Bureau of Standards [#]	Sophia Exhibition Centre, Sophia Complex, P.O. Box 10926, Georgetown, GUYANA Telephone: + 592 2 590 41 Fax: + 592 2 574 55
GOST R	Committee of the Russian Federation for Standardization, Metrology and Certification	Leninsky Prospekt 9, Moskva, 117049 RUSSIA Telephone: +7 095 236 40 44 Fax: +7 095 237 60 32
GSB	Ghana Standards Board	P.O. Box M 245, Accra, GHANA Telephone: +233 21 50 00 65 Fax: +233 21 50 00 92
IBN/BIN	Institut Belge de Normalisation/Belgisch Instituut voor Normalisatie	29, avenue de la Brabanconne, B-1040 Bruxelles 4, BELGIUM Telephone: +32 2 738 01 11 Fax: +32 2 733 42 64
IBNORCA	Instituto Boliviano de Normalización y Calidad [#]	Av. Camacho Esq. Bueno No 1488, Casilla 5034, La Paz, BOLIVIA Telephone: + 591 2 31 72 62 Fax: + 591 2 31 72 62
ICONTEC	Instituto Colombiano de Normas Téchnicas	Carrera 37 52-95, Edificio ICONTEC, P.O. Box 14237, Sntafé de Bogotá, COLOMBIA Telephone: +57 1 215 03 77 Fax: +57 1 222 14 35
INAPI	Institut Algérien de Normalisation et de Propriété Industrielle	5, rue Abou Hamou Moussa, B.P. 403-Centre de tri, Alger, ALGERIA Telephone: +213 2 63 96 42 Fax: +213 2 61 09 71

Table 6-4 *(continued)* National standards bodies worldwide

Acronym	Name	Address (headquarters)
INDECOPI	Instituto Nacional de Defensa de la Competencia y de la Protección de la Propiedad Intelectual [*]	Calle La Prosa 138, San Borja, Lima 41 PERU Telephone: + 51 1 224 78 00 Fax: + 51 1 224 03 48 E-mail: postmast@indecopi.gob.pe
INEN	Instituto Ecuatoriano de Normalización	Baquerizo Moreno 454 y, Av. 6 de Diciembre, Casilla 17-01-3999, Quito, ECUADOR Telephone: +593 2 56 56 26 Fax: +593 2 56 78 15
INN	Instituto Nacional de Normalización	Matías Cousino 64 - 6o piso, Castilla 995-Correo Central, Santiago, CHILE Telephone: +56 2 696 81 44 Fax: +56 2 696 02 47
INNOQ	National Institute of Standardization and Quality [*]	C.P. 2983, Maputo MOZAMBIQUE Telephone: + 258 1 42 14 09 Fax: + 258 1 42 45 85
INNORPI	Institut National de la Normalisation et de la Propriété Industrielle	B.P. 23, 1012 Tunis-Belvédère, TUNISIA Telephone: +216 1 78 59 22 Fax: +216 1 78 15 63
INTENCO	Instituto de Normas Téchnicas de Costa Rica	Barrio González Flores Ciudad Cientifica, San Pedro de Montes de Oca, San José, COSTA RICA Telephone: +506 283 45 22 Fax: +506 283 48 31
INTN	Instituto Nacional de Tecnología y Normalización [#]	Casilla de Correo 967, Asunción, PARAGUAY Telephone: + 595 21 29 01 60 Fax: + 595 21 29 08 73
IPQ	Instituto Português da Qualidade	Rua C à Avenida dos Três Vales, P-2825 Monte de Caparica, PORTUGAL Telephone: +351 1 294 81 00 Fax: +351 1 294 81 01
IRAM	Instituto Argentino de Racionalización de Materiales	Chile 1192, 1098 Buenos Aires, ARGENTINA Telephone: +54 1 393 37 51 Fax: +54 1 383 84 63 E-mail: postmaster@iram.org.ar
IRS	Institutul Român de Standardizare	Str. Jean-Louis Calderon Nr.13, Cod 70201, Bucuresti 2, ROMANIA Telephone: +40 1 211 32 96 Fax: +40 1 210 08 33
ISIRI	Intitute of Standards and Industrial Research of Iran	P.O. Box 31585-163, Karaj, IRAN Telephone: +98 261 22 60 31 Fax: +98 261 22 50 15
JBS	Jamaica Bureau of Standards	6 Winchester Road, P.O. Box 113, Kingston 10, JAMAICA Telephone: +1 809 926 31 40-6 Fax: +1 809 929 47 36
JISC (JIS)	Japanese Industrial Standards Committee	Agency of Industrial Science and Technology, Ministry of International Trade and Industry, 1-3-1 Kusumigaseki Chiyoda-Ku, Tokyo 100, JAPAN Telephone: +81 3 35 01 92 95 Fax: +81 3 35 80 14 18

Table 6-4 *(continued)* National standards bodies worldwide

Acronym	Name	Address (headquarters)
JISM	Jordanian Institution for Standards and Metrology [*]	P.O. Box 941287, Amman 11194 JORDAN Telephone: + 962 6 68 01 39 Fax: + 962 6 68 10 99
KAZMEMST	Committee for Standardization, Metrology and Certification	Pr. Altynsarina 83, 480035 Almaty, KAZAKHSTAN Telephone: +7 327 2 21 08 08 Fax: +7 327 2 28 68 22
KEBS	Kenya Bureau of Standards	Off Mombasa road, Behind Belle Vue Cinema, P.O. Box 54974, Nairobi, KENYA Telephone: +254 2 50 22 10/19 Fax: +254 2 50 32 93 E-mail: kebs@arso.gn.apc.org
KIAA	Industrial Advancement Administration	2, Chungang-dong, Kwachon-city, Kyonggi-do 427-010, REPUBLIC of KOREA Telephone: +82 2 503 79 38 Fax: +82 2 503 79 41
KYRGYZST	State Inspection for Standardization and Metrology [*]	197 Panfilova str., 720040 Bishkek KYRGYZSTAN Telephone: + 7 331 2 26 48 62 Fax: + 7 331 2 26 47 08 E-mail: kmc@infotel.bishkek.su
LIBNOR	Lebanese Standards Institution [*]	Industry Institute Building, Next to Riviera Hotel, Beirut Corniche, P.O. Box 14-6473 Beirut LEBANON Telephone: + 961 1 34 82 19 Fax: + 961 1 42 70 04
LNCSM	Libyan National Centre for Standardization and Metrology	Industrial Research, Centre Building, P.O. Box 5178, Tripoli, LIBYA Telephone: +218 21 499 49 Fax: +218 21 69 00 28
LST	Lithuanian Standards Board [*]	A. Jaksto g. 1/25, Vilnius LITHUANIA Telephone: + 370 2 22 69 62 Fax: + 370 2 22 62 52
LVS	Latvian National Center of Standardization and Metrology [*]	157, Kr. Valdemara Street, 1013 Riga LATVIA Telephone: + 371 2 37 81 65 Fax: + 371 2 36 28 05
MBS	Malta Board of Standards [*]	Department of Industry, Triq il-Kukkanja, Santa Venera CMR 02 MALTA Telephone: + 356 44 62 50 Fax: + 356 44 62 57
MNISM	Mongolian National Institute for Standardization and Metrology	Ulaanbatar-51, MONGOLIAN PEOPLE's REPUBLIC Telephone: +976 1 35 83 49 Fax: +976 1 35 80 32
MSB	Mauritius Standards Bureau	MAURITIUS Telephone: +230 454 19 33 Fax: +230 464 11 44
MSIT	Major State Inspection of Turkmenistan [*]	Seydi, 14, 744000 Ashgabat, TURKMENISTAN Telephone: + 7 363 2 51 14 94 Fax: + 7 363 2 51 04 98

Table 6-4 *(continued)* National standards bodies worldwide

Acronym	Name	Address (headquarters)
MSZT	Magyar Szabványügyi Testület	Üllöi út 25, Pf. 24, H-1450 Budapest 9, HUNGARY Telephone: +36 1 218 30 11 Fax: +36 1 218 51 25
NBSM	Nepal Bureau of Standards and Metrology [*]	P.O. Box 985, Sundhara, Kathmandu NEPAL Telephone: + 977 1 27 26 89 Fax: + 977 1 27 26 89
NC	Oficina Nacional de Normalización	Calle E No.261 entre 11 y 13, Velado, La Habana 10400, CUBA Telephone: +53 7 30 00 22 Fax: +53 7 33 80 48
NISIT	National Institute of Standards and Industrial Technology [*]	P.O. Box 3042, National Capital District, Boroko Papua, NEW GUINEA Telephone: + 675 27 21 02 Fax: + 675 25 87 93
NNI	Nederlands Normalisatie-Instituut	Kalfjeslaan 2,P.O. Box 5059, NL-2600 GB Delft, THE NETHERLANDS Telephone: +31 15 2 69 03 90 Fax: +31 15 2 69 01 90
NSAI	National Standards Authority of Ireland	Glasnevin House, Ballymun Road, Dublin 9, IRELAND Telephone: +353 1 837 01 01 Fax: +353 1 836 98 21
NSF	Norges Standardiserings Forbund	Drammensveien 145, Postbooks 353 Skoyen, N-0212 Oslo, NORWAY Telephone: +47 22 04 92 00 Fax: +47 22 04 92 11 Internet: http://www.standard.no/
ON (ONORM)	Österreichlishes Normungs-institut	Heinestrasse 38, Postfach 130, A-1021 Wien 2, AUSTRIA Telephone: +43 1 213 00 Fax: +43 1 213 00 650 E-mail: iro@tbxa.telecom.at
PKN	Polish Committee for Standardization	ul. Elektoralna, P.O. Box 411, 00-950 Warszawa, POLAND Telephone: +48 22 620 54 34 Fax: +48 22 620 07 41
PSB	Singapore Productivity and Standards Board	1 Science Park Drive, Singapore, 118221, SINGAPORE Telephone: +65 778 77 77 Fax: +65 776 12 80
PSI	Pakistan Standards Institution	39 Garden road, Saddar, Karachi-74400, PAKISTAN Telephone: +92 21 772 95 27 Fax: +92 21 772 81 24
SA (SAA, AS)	Standards Australian Association	1 The Crescent, Homebush- N.S.W. 2140, AUSTRALIA Telephone: +61 2 746 47 00 Fax: +61 2 746 84 50 Internet: http:// www.standards.com.au/ E-mail: intsect@ saa.sa.telememo.au

Table 6-4 *(continued)* National standards bodies worldwide

Acronym	Name	Address (headquarters)
SABS	South African Bureau of Standards	1 Dr Lategan Rd, Groenkloof, Private Bag, X191 Pretoria 0001, SOUTH AFRICA Telephone: +27 12 48 79 11 Fax: +27 12 344 15 68
SARM	Department for Standardization, Metrology and Certification [*]	Komitas Avenue 49/2, 375051 Yerevan, ARMENIA Telephone: + 374 2 23 56 00 Fax: + 374 2 28 56 20 E-mail: sarm@arminco.com
SASMO	Syrian Arab Organization for Standardization and Metrology	P.O. Box 11836, Damascus, SYRIAN ARAB REPUBLIC Telephone: + 963 11 445 05 38 Fax: +963 11 441 39 13
SASO	Saudi Arabian Standards Organization	Imam Saud Bin Abdul Aziz Bin Mohammed Road (West End), P.O. Box 3437 Riyadh 11471, SAUDI ARABIA Telephone: +966 1 452 00 00 Fax: +966 1 452 00 86
SAZ	Standards Association of Zimbabwe	P.O. Box 2259, Harare, ZIMBABWE Telephone: +263 4 88 34 46 Fax: +263 4 88 20 20
SCC (ACNOR)	Standards Council of Canada/ Association Canadienne de Normalisation	45 O'Connor Street, Suite 1200, Ottawa, Ontario, KIP 6N7 CANADA Telephone: +1 613 238 32 22 Fax: +1 613 995 45 64 Internet: http://www.scc.ca/ E-mail: info@scc.ca
SFS	Suomen Standardisoimislütto	P.O. Box 116, FIN-00241 Helsinski, FINLAND Telephone: +358 0 149 93 31 Fax: +358 0 146 49 25 Internet: http://www.sfs.fi/ E-mail: efs@sfs.fi
SII	Standards Institution of Israel	42 Chaim Levanon Sreet, University State, Tel Aviv 69977, ISRAEL Telephone: +972 3 646 51 54 Fax: +972 3 641 96 83 E-mail: standard@netvision.net.il
SIRIM	Standards and Industrial Research Institute of Malaysia	Persiaran Dato'Menteri Section 2, P.O. Box 7035, 40911 Shah Alam, Selangor Darul Ehsan, MALAYSIA Telephone: +60 3 559 26 01 Fax: +60 3 550 80 95
SIS	Standardiseringskommissione n i Sverige	St Eriksgatan 115, Box 6455, S-113 82 Stockholm, SWEDEN Telephone: +46 8 610 30 00 Fax: +46 8 30 77 57 E-mail: info@sis.se
SLBS	Saint Lucia Bureau of Standards [#]	Government Buildings, Block B, 4th floor, John Campton Highway, Castries, SAINT LUCIA Telephone: + 1 809 453 00 49 Fax: + 1 809 453 73 47

Table 6-4 *(continued)* National standards bodies worldwide

Acronym	Name	Address (headquarters)
SLSI	Sri Lanka Standards Institution	53 Dharmapala Mawatha, P.O. Box 17, Colombo 3, SRI LANKA Telephone: +94 1 32 60 51 Fax: +94 1 44 60 18
SMIS	Standards and Metrology Institute	Ministry of Science and Technology, Kotnikova 6, SI-1000 Ljubljana, SLOVENIA Telephone: +386 61 131 23 22 Fax: +386 61 31 48 82 E-mail: ic@usm.mzt.si
SNIMA	Service de Normalisation Industrielle Marocaine	Ministère du Commerce, de l'Industrie et de l'Artisanat, quarier administratif, Rabat Chellah, MOROCCO Telephone: +212 7 76 37 33 Fax: +212 7 76 62 96
SNV	Schweizerische Normen-Vereinigung	Mühlebachstrasse 54, CH-8008 Zurich, SWITZERLAND Telephone: +41 1 254 54 54 Fax: +41 1 254 54 74
SNZ	Standards New Zealand	Standards House, 155 The Terrace, Wellington 6001, NEW ZEALAND Telephone: +64 4 498 59 90 Fax: +64 4 498 59 94
SON	Standards Organization of Nigeria	Federal Secretariat, Phase 1, 9th Floor, Ikoyi, Lagos, NIGERIA Telephone: +234 1 68 26 15 Fax: +234 1 68 18 20
SSUAE	Directorate of Standardization and Metrology [*]	Ministry of Finance and Industry, El Falah Street, P.O. Box 433, Abu Dhabi, UNITED ARAB EMIRATES Telephone: + 971 2 72 60 00 Fax: + 971 2 77 97 71
STRI	Icelandic Council of Standardization	Keldnaholt, IS-112 Reykjavik, ICELAND Telephone: +354 587 70 00 Fax: +354 587 74 09 E-mail: stri@iti.is
SZS	Savezni zavod za standardizaciju	Kneza Milosa 20, Post Pregr. 933, YU-11000 Beograd, YUGOSLAVIA Telephone: +381 11 64 35 57 Fax: +381 11 68 23 82 E-mail: etanasko@ ubbgg.etf.bg.ac.yu
TBS	Tanzania Bureau of Standards	Ubungo Area, Morogoro Road/Sam Nujoma Road, P.O. Box 9524, Dar es Salaam, TANZANIA Telephone: +255 51 4 32 98 Fax: +255 51 4 32 98
TCVN	Directorate for Standards and Quality	Hung dao Street, Hanoi, VIETNAM Telephone: +84 4 26 62 20 Fax: +84 4 26 74 18
TISI	Thai Industrial Standards Institute	Ministry of Industry, Rama VI Street, Bangkok 10400, THAILAND Telephone: +66 2 245 78 02 Fax: +66 2 247 87 41

Table 6-4 *(continued)* National standards bodies worldwide

Acronym	Name	Address (headquarters)
TSE	Türk Standardlari Enstitüsü	Necatibey Cad. 112, Bakanlikdar, 06100 Ankara, TURKEY Telephone: +90 312 417 83 30 Fax: +90 312 425 43 99
TTBS	Trinidad and Tobago Bureau of Standards	#2 Century Drive, Trincity Industrial Estate, Tunapuna, TRINIDAD AND TOBAGO Telephone: +1 809 662 88 27 Fax: +1 809 663 43 35 E-mail: ttbs@opus-networx.com
UNBS	Uganda National Bureau of Standards	P.O. Box 6329, Kampala, UGANDA Telephone: + 256 41 22 23 69 E-mail: unbs@mukla.gn.apc.org
UNI	Ente Nazionale Italiano di Unificazione	Via Battislotti Sassi 11/b, I-20133 Milano, ITALY Telephone: +39 2 70 02 41 Fax: +39 2 70 10 61 06 Internet: http://www.unicei.it/ E-mail: webmaster@uni.unicei.it
UNIT	Instituto Uruguayo de Normas Técnicas	San José 1031 P.7, Galeria Elysée, Montevideo, URUGUAY Telephone: +598 2 91 20 48 Fax: +598 2 92 16 81
UNMS	Slovak Office of Standards, Metrology and Testing	Stefanovicova 3, 814 39 Bratislava, SLOVAKIA Telephone: +42 7 49 10 85 Fax: +42 7 49 10 50
UZGOST	Uzbek State Centre for Standardization, Metrology and Certification	Ulitsa Farobi, 333-A, 700049 Tachkent, UZBEKISTAN Telephone: +7 371 2 46 17 10 Fax: +7 371 2 46 17 11
ZSM	Zavod za standardizacija i metrologija	Ministry of Economy, Samoilova 10, 91000 Skopje, MACEDONIA Telephone: +389 91 22 47 74 Fax: +389 91 11 02 63

6.5.2.2 US Standards and Standardizations Bodies, Technical Associations and Societies

Table 6-5 US standards bodies, technical associations and societies

Acronym	Name	Address (headquarters)
AA	Aluminum Association	Telephone: (202) 862 5100
AEE	Association of Energy Engineers	Telephone: (404) 447 5083
AFS	American Foundrymen's Society	Telephone: (312) 824 0181
AGA	American Gas Association	Internet: http://www.gasweb.org/
AIA	American Insurance Association	85 John Street, New York, New York, 10038, USA
AIChE	American Institute of Chemical Engineers	345, East 47th Street, New York, New York, 10017, USA Telephone: (212) 705 7338

Table 6-5 *(continued)* US standards bodies, technical associations and societies

Acronym	Name	Address (headquarters)
AIP	American Institute of Physics	Internet: http://www.aip.org/
AISI	American Iron and Steel Institute	1000, 16th Street NW, Washington DC, 20036, USA Telephone: (412) 281 6323
ANMC	American National Metric Council	1625, Massachusetts Ave., Washington, DC 20036, USA
ANS	American Nuclear Society	555, N. Kensington Avenue, La Grange Park, Illinois, 60525, USA
APHA	American Public Health Association	1015, 18th Street NW, Washington DC, 20036, USA
API	American Petroleum Institute	1801, K. Street NW, Washington DC, 20006, USA
ASEE	American Society for Engineering Education	Internet: http://www.asee.org/
ASHRAE	American Society of Heating, Refrigerating, and Air Conditioning Engineers	1791 Tulle Circle, N.E. Atlanta, GA 30329, USA Telephone: (404) 636 8400 Fax: (404) 321 5478 Internet: http://www.ashrae.org/
ASM	American Society for Metals	Telephone: (216) 338 5151
ASME	American Society of Mechanical Engineers	United Engineering Center, 345 East 47th Street, New York, New York, 10017, USA Telephone: (212) 705 7722
ASQC	American Society for Quality and Control	Telephone: (414) 272 8575
ASTM	American Society for Testing and Materials	1916, Race Street, Philadelphia, Pa., 19103, USA Telephone: +1 202 862 5100
AWS	American Welding Society	550, NW LeJeune Road, P.O. Box 351040, Miami, Florida, FL-33125, USA Telephone: +1 800 443 9353 Fax: +1 303 443 7559
AWWA	American Water Works Association	Internet: http://www.awwa.org/
CDA	Copper Development Association	2 Greenwich Office, P.O. Box 1840, Greenwich, CT 06836.1840, USA Telephone: +1 230 625 8232
FED	Federal and Military Standards	Telephone: (215) 697 20000
IEEE	Institute of Electrical and Electronic Engineers	Internet: http://www.ieee.org/
IES	Institute of Environmental Science	Telephone: (312) 255 1561
IIE	Institute of Industrial Engineers	Telephone: (404) 449 0460
ISA	Instrument Society of America	400 Stanwix Street, Pittsburg, Pa., 15222 USA
ISS	Iron and Steel Society	Telephone: (412) 776 1535
MTS	Marine Technology Society	Telephone: (202) 775 5966
NACE	National Association of Corrosion Engineers	Telephone: (713) 492 0535

Table 6-5 *(continued)* US standards bodies, technical associations and societies

Acronym	Name	Address (headquarters)
NAPE	National Association of Power Engineers	Telephone: (212) 298 0600
NAPEGG	Association of Professional Engineers, Geologists, and Geophysicists of the Northwest Territories	Telephone: (403) 920 4055
NFPA	National Fire Protection Association	470, Atlantic Avenue, Boston, Massachusetts, 02210, USA
NiDI	Nickel Development Institute	Telephone: (416) 591 7999
NIH	National Institute of Health	
NIST (NBS)	National Institute for Science and Technology (previously the NBS)	Building 225, room B162, Washingto DC, 20234, USA
NSA	National Standards Association	5161, River Road, Washington DC, 20016, USA
PIA	Plastics Institute of America	Telephone: (201) 420 5553
SAE	Society for Automotive Engineers	400, Commonwealth Drive, Warrendale, Pa., 15096.0001, USA Telephone: +1 412 776 4841 Fax: +1 412 776 5760
SAMA	Scientific Apparatus Makers Association	1140, Connecticut Ave. NW Washington, DC 20036, USA
SAME	Society of American Military Engineers	Telephone: (703) 549 3800
SCTE	Society of Carbide and Tool Engineers	Telephone: (216) 338 5151
SPE	Society of Petroleum Engineers	Telephone: (214) 669 3377
STLE	Society of Tribologists and Lubrication Engineers	Telephone: (312) 825 5536
TDA	Titanium Development Association	Telephone: (303) 443 7515
TMS	The Mineral, Metals, and Materials Society	Telephone: (412) 776 9000
USMA	US Metric Association	Sugarloaf Star Route, Boulder, Colorado, 80302, USA
USMB	US Metric Board	1815, N. Lynn Street, Arlington, Va. 22209, USA
USP	US Pharmacopeia	Internet: http.//www.usp.org/
ZI	Zinc Institute	292, Madison Avenue, New York, NY-10017, USA Telephone: +1 212 578 4750 Fax: +1 212 578 4750

.6 Acceleration Due to Gravity at any Latitude and Elevation

φ is the latitude and h the elevation in centimetres the acceleration in SI ıits is given by Helmert's equation:

$$h\ (\mathrm{m.s}^{-2}) = 9.80616.0 - 025928 \cos\varphi + 0.000069 \cos^2 2\varphi - 3.086 \times 10^{-8}$$

.7 International Practical Temperature Scale (IPTS, 1968)

he International Practical Temperature Scale of 1968 (IPTS, 1968) amended ı 1975 is defined by a set of extrapolation based on the following reference :mperatures which are given in degrees Celsius and in kelvins:

Table 6-6 International Practical Temperature Scale of 1968

State of equilibrium	T_{68} (K)	t_{68} (°C)
hydrogen triple point	13.81	−259.34
hydrogen liquid-gas equilibrium at a pressure of 333 330.6 Pa	17.042	−256.108
hydrogen liquid-gas equilibrium (boiling point of hydrogen)	20.28	−252.87
neon liquid-gas equilibrium (boiling point of neon)	27.102	−246.048
oxygen triple point	54.361	−218.789
argon liquid-gas equilibrium (boiling point of argon)	83.798	−189.352
oxygen liquid-gas equilibrium (boiling point of oxygen)	90.188	−182.962
water[16] triple point	273.16	0.01
gallium triple point	302.9169 ± 0.0005	29.7669 ± 0.0005
water liquid-gas equilibrium (boiling point of water)	373.15	100.0
tin solid-liquid equilibrium (freezing point of tin)	505.1181	231.9681
zinc solid-liquid equilibrium (freezing point of zinc)	692.73	419.58
silver solid-liquid equilibrium (freezing point of silver)	1235.08	961.93
gold solid-liquid equilibrium (freezing point of gold)	1337.58	1064.43

16 The water used must have the isotopic composition of sea water.

6.8 French-English Lexicon for Units

Table 6-7 French-English lexicon for units

Unit (UK, US)	translation
acre	acre
area	aire
bag	sac
barleycorn	grain d'orge
barrel	barrel
beer	bière
board	planche
bolt	boulon
bucket	seau
bushel	boisseau
butt	barrique, fût, futaille, tonneau
cable length	longueur de cable
calibre, caliber	calibre
capacity	capacité, volume
chain	chaîne
chaldron	chaudron
chopine	chope
cubic	cube
cubit	coudée
cup	coupe
drachm	petit verre
dram	goutte
dry	sec, sèche
ell	-
fathom	brasse
fluid	fluide
foot	pied
furlong	furlong
gallon	gallon
gill	gill
grain	grain
hay	foin
hock	jarret
hogshead	barrique, muid, tonneau
hundredweight	centpoids
inch	pouce
league	lieue
line	ligne
link	chaînon, maillon
load	charge
mile	mille
minim	goutte
nail	ongle
ounce	once
pace	pas
palm	paume
peck	picotin
perch	perche
pint	pinte
point	point
pound	livre
quart	quart
quarter	quartier
rope	cordon
roquille	roquille
scruple	scripulum
seam	veine (mine)
span	empan
spirit	alcool, spiritueux
square	carré
stone	pierre, pavé
straw	paille
tablespoon	cuillère à soupe
teaspoon	petite cuillère
ton	tonne
township	canton
tumb	tambour
tun	fût
truss	botte
wine	vin
yard	verge

.9 French-English Lexicon for Physical Quantities

he nomenclature for intensive properties according to French and English is e following:

Flux of the quantity *X*: $J_X = \dfrac{\partial X}{A\partial t}$

Specific quantities (= mass basis): the adjective specific in the Engish term for an intensive physical quantity should be avoided if possible and should in all cases be restricted to the meaning divided by mass (mass of the system, if this consists of more than one component or more than one phase). In French, the adjective *massique* is used in the sense of divided by mass to express this concept.

Examples:

Physical quantity	Symbol	Grandeur
specific volume	$v = V/m$	volume massique
specific energy	$e = E/m$	énergie massique
specific heat capacity	$c_p = C_p/m$	capacité thermique massique

- **Molar quantities** (= mole basis): The adjective molar in English which describes an intensive physical quantity should be restricted to the meaning divided by the amount of substance (the amount of substance of the system, if this consists of more than one component or more than one phase).

Examples:

Physical quantity	Symbol	Grandeur
molar volume	$v_m = V/n$	volume molaire
molar energy	$e_m = E/n$	energie molaire
molar heat capacity	$c_{m,p} = C_p/n$	capacité thermique molaire

- **Density quantities**: the English term density for an intensive physical quantity (when it is not modified by the adjective linear or surface) usually implies division by volume for scalar quantities and division by area for a vector quantity denoting flow or flux. In French, the adjectives *volumique*, *surfacique* or *linéique* as appropriate are used with the name of a scalar quantity to express division by volume, area or length, respectively.

Examples:

Physical quantity	Symbol	Grandeur
mass density	$\rho = m/V$	mass volumique
energy density	$e_v = E/V$	energie volumique
current density	$j = i/A$	densité de courant
surface charge density	$\sigma = Q/A$	charge surfacique

Table 6-8 French-English lexicon for physical quantities

English name	Nom Français
abundance	abondance
acoustic impedance	impédance acoustique
area	aire
capacity	capacité
circumference	circonférence
coefficient of diffusion	coefficient de diffusion
coefficient of heat transfer	coefficient de transfert thermique
depth	profondeur
diametre	diamètre
distance	distance
dose equivalent	équivalent de dose
duration	durée
electric charge	charge électrostatique
electric current density	densité de courant électrique
electric current intensity	intensité de courant électrique
electric dipole moment	moment dipolaire électrique
electric field strength	champ électrique
electric inductance	inductance électrique
electric potential	potentiel électrique
electric potential difference	différence de potentiel électrique
electric quadrupole moment	moment quadrupolaire électrique
electric resistance	résistance électrique
electric resistivity	résistivité électrique
electrical conductance	conductance électrique
electrical conductivity	conductivité électrique
electrokinetic potential	potentiel électrocinétique
electromotive force	force électromotrice
energy	énergie
energy density	énergie volumique
energy of electromagnetic radiation	energy d'un rayonnement électromagnétique
entropy	entropie
enzymatic activity	activité enzymatique
exposure	exposition
fluence rate	débit de fluence
force	force
fraction	titre
heat	chaleur
heat insulation coefficient	coefficient d'isolation thermique
high	hauteur
hydrodynamic permeability	permeabilité hydrodynamique
hydrometre degree	degré aréométrique

Table 6-8 *(continued)* French-English lexicon for physical quantities

English name	Nom Français
hydrometer index	indice aréométrique
illuminance	éclairement lumineux
index of kinematic viscosity	indice de viscosité cinématique
interfacial tension	tension interfaciale
irradiance	radiance énergétique
length	longueur
linear mass density	masse linéique
linear momentum, momentum	moment linéaire, impulsion, quantité de mouvement
linear rate of corrosion	vitesse de corrosion linéaire
luminous flux	flux lumineux
luminous intensity	intensité lumineuse
luminous luminance	luminance lumineuse
magnetic dipole moment	moment mangnetique dipolaire
magnetic field strength	champ magnétique
magnetic flux density	densité de flux magnétique
magnetic induction field	champ d'induction magnétique
magnetizability	aimantation
mass	masse
mass flow rate	débit massique
molar density	densité molaire
molar electric charge	charge électrique molaire
molar entropy	entropie molaire
molar heat capacity	capacité thermique molaire
molarity, molar concentration	molarité
moment of a force	moment d'une force
normality	normalité
optical rotatory power	pouvoir rotatoire
period	periode
plane angle	angle plan
polarizability	polarisation
power	puissance
pressure	pression
quadratic moment of a plane area	moment quadratique d'une aire plane
quantity of electricity	quantité d'électricité
radiating power	puissance rayonnée
radioactivity	activité radioactive
radius	rayon
rate of exposure	débit d'exposition
solid angle	angle solide
specific acoustic impedance	impédance acoustique volumique
specific gravity	densité

Table 6-8 *(continued)* French-English lexicon for physical quantities	
English name	**Nom Français**
specific heat capacity	capacité thermique massique
specific length	longueur massique
specific linear mass	masse linéique
specific volume	volume massique
speed	vitesse
stress	contrainte
surface	surface
surface tension	tension superficielle
temperature	température
thermal conductance	conductance thermique
thermal conductivity	conductivité thermique
thermal diffusivity	diffusivité thermique
thickness	épaisseur
time	temps
torque	couple
velocity	célérité
volume	volume
volume flow rate	débit volumique
wave length	longueur d'onde
wave number	nombre d'onde
weight	poids
width, breath	largeur
work	travail
yield, efficiency	rendement

6.10 Old Alchemical Symbols Used in Astronomy, Chemistry, and Biology

Table 6-9 Old alchemical symbols	
Symbol	**Designation (planet, element)**
☉	Sun, gold
☿	Mercury
♀	Venus, copper, female sex
♁	Earth, terra
○, ☽	Moon, silver
♂	Mars, iron, male sex
♃	Jupiter, tin
♄	Saturn, lead
♌	arsenic
△	sulphur

.11 Old Symbols Used in the British System

some of the oldest British textbooks used in pharmacy, it is possible to find d notations for British units of weights and capacity (apothecaries' units). mbols were usually used for writing the quantities of units of the apothecies' system on prescriptions. Occasionally abbreviations were used. These d symbols were taken directly from ancient times (e.g. the Middle Ages), hen alchemy and cabalistic beliefs were one discipline. Some of these mbols used capital letters from the Cyrillic alphabet.

Table 6-10 Obsolete symbols for British apothecaries' units

Old symbol	Unit, name	
℈ ap		scruple apoth.
℈ i, j	1	scruple apoth.
℈ i, ss	3/2	scruple apoth. (semis)
℈ ii, jj	2	scruple apoth.
ʒ ap		drachm apoth.
ʒ i, j	1	drachm apoth.
ʒ i, ss	3/2	drachm apoth. (semis)
ʒ ii, jj	2	drachm apoth.
℥ ap		ounce apoth.
℥ i, j	1	ounce apoth.
℥ i, ss	3/2	ounce apoth. (semis)
℥ ii, jj	2	ounce apoth.
Cong.		gallon (from Latin *Congius*)
O.		pint (from Latin *Octarius*)
f ℥		fluid dram
f ʒ		fluid ounce
♍		minim or drop

7 Bibliography

7.1 Specific References

These references, which contain formulae, definitions, and conversions factors, are books specializing in scientific units and their conversion. They are listed according to their publication dates. For books which were published before 1900 see section 7.4.

1. Hering, C, *Ready Reference Tables. Volume I. Conversion factors of every unit or measure in use based on the accurate legal standard values of the United States. Conveniently arranged for engineers, physicists, students, merchants, etc.*, 1st ed., J. Wiley & Sons, New York, 1904.
2. Guillaume, C-É, and Volet, C, National and local systems of weights and measures, in *International Critical Tables of Numerical Data, Physics, Chemistry and Technology* (vol. 1), EW Washburn (Ed.), McGraw-Hill Book Company, New York, 1926.
3. Zimmerman, OT, and Lavine, I, *Industrial Research Service's Conversion Factors and Tables*, Industrial Research Service, Dover, 1944.
4. Boll, M, *Tables Numériques Universelles des Laboratoires et Bureaux d'Études*, Dunod Éditeur, Paris, 1947.
5. United States. Production and Marketing Administration, *Conversion Factors and Weights and Measures for Agricultural Commodities and Their Products*, Washington, 1952.
6. Zimmerman, OT and Lavine, I, *Industrial Research Service's Conversion Factors and Tables*, 3rd ed., Industrial Research Service, Dover, 1961.
7. *Lexicon of International and National Units*, Elsevier Publishing Company, London, 1964.
8. Anderton, P, and Bigg, PH, *Changing to the Metric system: Conversion Factors, Symbols and Definitions*, 2nd ed. London, H.M.S.O., 1967.
9. Mechtly, EA, *The International System of Units; Physical Constants and Conversion Factors*, Washington, Scientific and Technical Information Division, National Aeronautics and Space Administration Supt. of Docs., U.S. Govt. Print. Off., 1970.
10. Institute of Petroleum, *Recommended SI and Other Metric Units for the Petroleum and Petrochemical Industries, Including Basic Rules of the Oil Companies Materials Association*, Revised ed., Institute of Petroleum, London, 1971.
11. *McGraw-Hill Metrication Manual*, McGraw-Hill, New York, 1972.
12. New Zealand Dept. of Agriculture and the Information Section and Standards Association of New Zealand, *Metrication Handbook: a Guide for N.Z.D.A. Staff. N.Z.D.A. policy, the S.I. system, style, conversion factors & tables*, Wellington, 1972.
13. Sellers, RC, (Ed.), *Basic Training Guide to the New Metrics and SI Units*, RC Sellers & Associates, Floral Park, N.Y., 1972.

14. Singapore Metrication Board, *Conversion Factors and Conversion Tables*, Singa pore, 1971–73.

15. Semioli, WJ, and Schubert, PB, *Conversion Tables for SI Metrication*, Industria Press, New York, 1974.

16. Defix, A, *Mesurages des Volumes des Carburants et Combustibles Liquides* Technip, Paris, 1975.

17. New Zealand Metric Advisory Board, Engineering and Engineering Service Sector, and Australia. Metric Conversion Board, *Metric Conversion Manual fo Engineering Establishments*, Metric Advisory Board, Wellington, 1975.

18. Royal Australian Institute of Architects and Australia. Metric Conversion Board *Metric Data for Building Designers*, Standards Association of Australia, North Sydney, N.S.W, 1975.

19. *Engineering Design Handbook – Metric Conversion Guide*, Publication DARCOM P 706–470, Headquarters, U.S. Army Material Development and Readiness Command, Washington, July 1976.

20. National Research Council. Panel on Metrication in the U.S. Maritime Industry *Maritime Metrication: a Recommended Metric Conversion Plan for the U.S Maritime Industry*, National Academy of Sciences, Washington, 1976.

21. Pedde, LD (Ed.), *Metric Manual*, U.S. Department of the Interior, Bureau of Reclamation, Superintendent of Documents, U.S. Government Printing Office, Washington DC, 1978.

22. Milone, F, and Giacomo, P, (Eds.), *Metrology and Fundamental Constants*, North Holland Publishing, Amsterdam, 1980.

23. Moureau, M, *Guide Pratique pour le Système International d'Unités (SI)*, 2nd ed., Technip, Paris, 1980.

24. *The World Measurement Guide*, The Economist Newspaper, London, 1980.

25. Quatremer, R, and Trotignon, J-P, *Unités et Grandeurs: Système International (SI) Symboles et Grandeurs*, AFNOR-NATHAN, Paris, 1981.

26. National Bureau of Standard (NBS), *The International System of Units (SI)*, Special Publication 330, Superintendent of Documents, U.S. Government Printing Office, Washington DC, 1981.

27. American National Metric Council, *Metric Editorial Guide*, 3rd ed., Bethesda, 1981.

28. International Organization for Standardization (ISO), *ISO Standards Handbook 2, Units of Measurement*, Geneva, 1982.

29. Lukens, RD, Units and Conversion Factors, in *Kirk-Othmer Encyclopedia of Chemical Technology*, Vol. 23, pp 491–502, 3rd ed., John Wiley & Sons, 1983.

30. Defix, A, *Éléments de Métrologie Générale et de Métrologie Légale*, 2nd. ed., Technip, Paris, 1985.

31. Ferrer, JL, *SI Metric Handbook*, Metric Co., 1985.

32. Lafaille, P, *Unités Légales et Facteurs de Conversion*, pp 24–1 à 24–8, Technique de l'ingénieur, Paris, 1986.

33. Lafaille, P., *Unités de Mesures*, pp 23–1 à 23–15, Techniques de l'ingénieur, Paris, 1986.

34. AFNOR, *Grandeurs et Unités*, Edition 1989, AFNOR Publications, Paris, 1989.

35. *The Economist World Measurement Guide*, Economist Books/Random Century, London, 1990.

36. Linderburg, R, *Engineering Unit Conversions*, NACE International, Houston, Texas, 1991.

37. Bureau International des Poids et Mesures, *Le Système International d'Unités (SI)*, 6th ed., BIPM, Pavillon de Breteuil, Sèvres, 1991.

38. Cook, J, *Conversion Factors*, Oxford University Press, 1991.

39. Jerrard, G and McNeill, DB, *Dictionary of Scientific Units*, 6th ed., London, Chapman & Hall, 1992.

40. Hladik, J, *Unités de Mesure: Étalons et Symboles des Grandeurs Physiques*, Masson, Paris, 1992.

41. Laposata, M, *SI Unit Conversion Guide*, New England Journal of Medicine, 1992.

2. Libois, J, *Guide des Unités de Mesures*, De Boeck Université, Brussels, 1993.
3. Finucance, E, *Definitions, Conversions and Calculations for Occupational Safety and Health Professionals*, Lewis, 1993.
4. Wildi, Th., *Metric Units and Conversion Charts: a Metrication Handbook for Engineers, Technologists, and Scientists*, 2nd ed., IEEE, 1995.

.2 References About the Metric System and the SI

45. Great Britain. Ministry of Public Building and Works with Anthony Williams and Burles Component Development Partnership, *Going Metric in the Construction Industry*, London, H.M.S.O., 1967.
46. Kellaway, FW, *Metrication*, Penguin, Harmondsworth, 1968.
47. Vickers, JS, *Making the Most of Metrication*, Gower, London, 1969.
48. Gillot, G and Loftus, P, *Managing Metrication*, Management Quality Systems Ltd., Gravesend, 1969.
49. Peach, J, *Industrial Metrication*, English Universities Press, London, 1970.
50. Ministry of Public Building and Works, *Metric in Practice: Background and General Principles, S.I. Units, Dimensional Coordination*, H.M.S.O., London, 1970.
51. Ministry of Public Building and Works Great Britain. Dept. of the Environment, Dargan Bullivant Associates, *Metric Reference Book*, H.M.S.O., London, 1971.
52. Jefferies, HB, *Metrication in Workshop Practice*, Blackwell, Oxford, 1971.
53. Carey, H, Burge, D and Salt, J, *Going Metric*, Horwitz North Sydney, 1971.
54. Canada. Dept. of Consumer and Corporate Affairs, *Metrication; a Guide for Consumers*, Information Canada, Ottawa, 1972.
55. Cameron, CA, *Going Metric with the US Printing Industry*, Graphic Arts Research Center, Rochester Institute of Technology, Rochester, 1972.
56. Groner, A, George, A, and Boehm, W, *Going Metric*, Amacom, New York, 1973.
57. Deming, R, *Metric Power; Why and How We are Going Metric*, 1st ed., T Nelson, Nashville, 1974.
58. Chalupsky, AB, Crawford, JJ, and Carr, EM, *Going Metric: an analysis of experiences in five nations and their implications for U.S. educational planning: final report*, American Institutes for Research in the Behavioral Sciences, Palo Alto, 1974.
59. Rieger, K, *Going Metric*, 2nd ed., International Correspondence Schools, Scranton, 1975.
60. Lèffin, WW, *Going Metric: Guidelines for the Mathematics Teacher*, National Council of Teachers of Mathematics, Reston, 1975.
61. Brooks, SM, *Going Metric*, AS Barnes, South Brunswick, 1976.
62. Wandmacher, C, *Metric Units in Engineering – Going SI: How to use the new international system of measurement units (SI) to solve standard engineering problems*, Industrial Press, New York, 1978.
63. Richardson, TL, *A Guide to Metrics*, Prakken Publications, Ann Arbor, Mich., 1978.
64. Wandmacher, C, and Johnson, AI, *Metric Units in Engineering – Going SI: How to use the international sytems of measurement units (SI) to solve standard engineering problems*, Revised ed., ASCE Press, New York, 1995.

7.3 References of General Interest

The following references are specialized textbooks and professional handbooks in the main areas of application. These books cover mechanical, electrical, chemical, nuclear, civil, and geological engineering, and also biology, medicine, economics and computer science. They give detailed definitions and descriptions of units. A few of them, especially some of the engineering handbooks, contain conversion tables.

65. Landolt, H, Bornstein, R and Roth, WA (Eds.), *Landolt-Bornstein, Physikalis Chemische Tabellen*, Verlag von Julius Springer, Berlin, 1912.

66. Fabry, C, *Encyclopédie Photométrique, Première section-Généralités: Introduct Générale à la Photométrie*, Editions de la Revue d'Optique Théorique et Inst mentale, Paris, 1927.

67. Perry, RH, *Chemical Engineer's Handbook*, 3rd ed., McGraw-Hill Book Compa New York, 1950.

68. Miner, DF, and Seastone, JB, (Eds.), *Handbook of Engineering Materials*, Jo Wiley & Sons, New York, 1955.

69. Condon, EU, and Odishaw, H, *Handbook of Physics*, McGraw-Hill Book Co pany, New York, 1958.

70. Kaye, GWC, and Laby, TH, *Tables of Physical and Chemical Constants and So Mathematical Functions*, 13th ed., Longman, New York, 1966.

71. Bruhat, G, *Cours de Physique Générale: Mécanique, Électricité, Magnétisr Optique, Thermodynamique*, Masson & Cie, Paris, 1967.

72. Gray, DE, (Ed.), *American Institute of Physics Handbook*, 3rd ed., McGraw-H New York, 1972.

73. Ballantyne, WG, and Lovett, DR, *A Dictionary of Named Effects and Laws Chemistry, Physics and Mathematics*, 3rd ed., Chapman & Hall-Science Pap backs, London, 1972.

74. *British Pharmacopeia 1973*, H.M.S.O., London, 1973.

75. Eshbach, OW, and Souders, M, *Handbook of Engineering Fundamentals*, 3rd John Wiley & Sons, New York, 1974.

76. Lytle, RJ, and Burrows, DT, *American Metric Construction Handbook*, McGra Hill, New York, 1976.

77. Fink, G, and Beaty, W, (Eds.), *Standard Handbook for Electrical Engineers*, 11 ed., McGraw-Hill, New York, 1978.

78. Harris, CM, (Ed.), *Handbook of Noise Control*, 2nd Ed, McGraw-Hill, New Yo 1979.

79. *United States Pharmacopeia XX – The National Formulary XV* – USP conventio Rockville, 1980.

80. Windholz, M (Ed.), *The Merck Index*, 10th ed., Merck & Co, Inc., Rahway, 198

81. Perry, RH, and Green DW (Eds), *Perry's Chemical Engineer's Handbook*, 6th e McGraw-Hill Book Company, New York, 1984.

82. Linden, D, *Handbook of Batteries and Fuel Cells*, McGraw-Hill Book Compan New York, 1984.

83. Granier, R, and Gambini, D-J, *Radiobiologie et Radioprotection Appliquée* Éditions Médicales Internationales/Lavoisier, Paris, 1985.

84. Centre National d'Études Spatiales (CNES) – Conseil International de la Lang Française (CILF), *Dictionnaire de Spatiologie, Sciences et Techniques Spatiale Tome 1: Termes et Définitions*, 2nd ed., CILF-CNES (Eds.), Paris, 1985.

85. Avallone, E, and Baumeister, T, *Mark's Standard Handbook for Mechanic Engineers*, 9th ed., McGraw-Hill, New York, 1987.

86. Mills, I, and Cvitas, T, *Quantities, Units and Symbols in Physical Chemistr* IUPAC, Blackwell Scientific Publications, Oxford, 1988.

87. American Institute of Physics, *A Physicist's Desk Reference, Physics Vade Mecum* 2nd ed., New York, 1989.

88. Weast, RC, (Ed), *CRC Handbook of Chemistry and Physics*, 70th ed., CRC Pres Boca Raton, 1989–1990.

89. McQueen, MC, *SI Unit Pocket Guide*, ASCP Press, Chicago, 1990.

90. *Astronomical Almanack for the Year 1991*, U.S. Government Printing Offic Washington D.C., 1991.

91. Dean, JA, *Lange's Handbook of Chemistry*, 14th ed., McGraw-Hill, 1992.

92. Croft, T, and Summers, W, *American Electrician's Handbook*, 12th ed., McGraw Hill, New York, 1992.

93. Oberg, E, *Machinery's Handbook*, 24th ed., Industrial Press, 1992.

4. Amiss, J, *Machinery's Handbook Guide*, 24th ed., Industrial Press, 1992.

5. Mills, I, and Cvitas, T, *Quantities, Units and Symbols in Physical Chemistry*, 2nd ed., IUPAC, Blackwell Scientific Publications, Oxford, 1993.

6. Dorf, R, (Ed.), *The Electrical Engineering Handbook*, CRC Press, Boca Raton, 1993.

7. Platt, G, *ISA Guide to Measurement Conversions*, Instrument Society of America, Research Triangle Park, 1994.

8. Fink, D, and Beaty, H, *Standard Handbook for Electrical Engineers*, 13th ed., McGraw-Hill, New York, 1994.

9. Parker, S, *Dictionary of Scientific and Technical Terms*, 5th ed., McGraw-Hill, New York, 1994.

0. Lide, D (Ed.), *CRC Handbook of Chemistry and Physics*, 75th ed., CRC Press, Boca Raton, 1994.

01. Kaye, GWC, and Laby, TH, *Tables of Physical and Chemical Constants*, 16th ed., Longman, New York, 1995.

02. Beitz, W, and Kuttner, K-H, *Dubbel's Handbook of Mechanical Engineering*, Springer-Verlag, London, 1995.

.4 Historical References

hese following references are specialized textbooks in historical metrology and the istory of weights and measures since ancient times.

103. Houghton, J, and Bradley, R, *Husbandry and Trade Improved: Being a Collection of Many Valuable Materials Relating to Corn, Cattle, Coals, Hops, Wool, &c.; With a Complete Catalogue of the Several Sorts of Earths, and Their Proper Product . . . as also Full and Exact Histories of Trades, as Malting, Brewing, &c. . . . an Account of the Rivers of England, &c. and How Far They May be Made Navigable; of Weights and Measures . . . the Vegetation of Plants, &c. with Many Other Useful Particulars*, 2nd ed., Woodman and Lyon, London, 1728.

104. Hassler, FR (U.S. Coast and Geodetic Survey and the United States National Bureau of Standards), *Principal Documents Relating to the Survey of the Coast of the United States; and the Construction of Uniform Standards of Weights and Measures for the Custom Houses and States*, J Windt, New York, 1834–36.

105. Aikin, A, and Aikin, CR, *A Dictionary of Chemistry and Mineralogy, with an Account of the Processes Employed in Many of the Most Important Chemical Manufactures* (to which are added a description of chemical apparatus, and various useful tables of weights and measures, chemical instruments), J and A Arch, London, 1807–14.

106. Hussey, R, *An Essay on the Ancient Weights and Money, and the Roman and Greek Liquid Measures, with an Appendix on the Roman and Greek Foot*, S Collingwood, Oxford, 1836.

107. Haswell, CH, *Mensuration and Practical Geometry; Containing Tables of Weights and Measures, Vulgar and Decimal Fractions, Mensuration of Areas, Lines, Surfaces, and Solids . . . To Which is Appended a Treatise on the Carpenter's Slide-rule and Gauging*, Harper & Brothers, New York, 1858.

108. New York (State). Dept. of Weights and Measures, *Report on Weights and Measures in the City of New York*. Office of Superintendent of Weights and Measures. March 11, 1910, J B Lyon Company, State Printers, Albany, 1910.

109. Watson, CM, *British Weights and Measures*, London, 1910.

110. Kettridge JO, *French-English and English-French Dictionary of Technical Terms and Phrases used in Civil, Mechanical, Electrical, and Mining Engineering, and Allied Sciences and Industries, Including Geology, Physical Geography, Petrology, Mineralogy, Crystallography, Metallurgy, Chemistry, Physics, Geometry, Abbreviations and Symbols, Weights and Measures, Compound Conversion Factors, etc., and a Method of Telegraphy*, G. Routledge & Sons Limited, London, 1925.

111. Moreau, H, *Le Système Métrique, des Anciennes Mesures au Système Intern tional d'Unités*, Editions Chiron, Paris, 1977.

112. Connor, RD, *The Weights and Measures of England*, London, Science Museu HMSO, 1987.

113. Edward, ZR, *Revolution in Measurement: Western European Weights and Me sures since the Age of Science*, Memoirs of the American Philosophical Socie Vol. 186, Philadelphia, 1990.

114. Witthoft, H, *Handbuch der Historischen Metrologie*, 4 Vol., Scripta Mercatur Verlag, St. Katharinen, 1991–1993.

114. Hocquet, J-C, *Anciens Systèmes de Poids et Mesures en Occident*, Varioru Reprints, London, 1992.

116. Hocquet, J-C, *Pesi e Misure in Europa nel Medioevo*, in Storia d'Europa, Vol. I Il Medioevo, Einaudi Edizioni, Torino, Italy, 1992.

117. Hocquet, J-C, *La Métrologie Historique*, n° 2972 Collection "Que sais-je? Presses Universitaires de France (PUF), Paris, 1995.

7.5 International and National Standards

- ISO standards (International Organization for Standardization, Geneva, Switzerland

ISO 31–0: 1992	Quantities and units – Part 0: General principles, units and symbols
ISO 31–1: 1992	Quantities and units – Part 1: Space and time
ISO 31–2: 1992	Quantities and units – Part 2: Periodic and related phenomena
ISO 31–3: 1992	Quantities and units – Part 3: Mechanics
ISO 31–4: 1992	Quantities and units – Part 4: Heat
ISO 31–5: 1992	Quantities and units – Part 5: Electricity and magnetism
ISO 31–6: 1992	Quantities and units – Part 6: Light and related electromagnetic radiations
ISO 31–7: 1992	Quantities and units – Part 7: Acoustics
ISO 31–8: 1992	Quantities and units – Part 8: Physical chemistry and molecular physics
ISO 31–9: 1992	Quantities and units – Part 9: Atomic and nuclear physics
ISO 31–10: 1992	Quantities and units – Part 10: Nuclear reactions and ionizing radiations
ISO 31–11: 1992	Quantities and units – Part 11: Mathematical signs and symbols for use in the physical sciences and technology
ISO 31–12: 1992	Quantities and units – Part 12: Characteristic numbers
ISO 31–13: 1992	Quantities and units – Part 13: Solid state physics
ISO 1000: 1992	SI units and recommendations for the use of their multiples and of certain other units

Other ISO standards for quantities and units developed for specific technical fields:

ISO 1151–1: 1988	Flight dynamics – Concepts, quantities and symbols – Part 1: Aircraft motion relative to the air
ISO 1151–2: 1985	Flight dynamics – Concepts, quantities and symbols – Part 2: Motions of the aircraft and the atmosphere relative to the Earth
ISO 1151–3: 1989	Flight dynamics – Concepts, quantities and symbols – Part 3: Derivatives of forces, moments and their coefficients
ISO 1151–4: 1994	Flight dynamics – Concepts, quantities, and symbols – Part 4: Concepts and quantities used in the study of aircraft stability and control
ISO 1151–5: 1987	Flight dynamics – Concepts, quantities and symbols – Part 5: Quantities used inmeasurements
ISO 1151–6: 1982	Terms and symbols for flight dynamics – Part 6: Aircraft geometry
ISO 1151–7: 1985	Flight dynamics – Concepts, quantities and symbols – Part 7: Flight points and flight envelopes

O 1151-8: 1992	Flight dynamics – Concepts, quantities and symbols – Part 8: Concepts and quantities used in the study of the dynamic behaviour of the aircraft
O 1151-9: 1993	Flight dynamics – Concepts, quantities and symbols – Part 9: Models of atmospheric motions along the trajectory of the aircraft
O 3002-1: 1982	Basic quantities in cutting and grinding – Part 1: Geometry of the active part of cutting tools – General terms, reference systems, tool and working angles, chip breakers
O 3002-2: 1982	Basic quantities in cutting and grinding – Part 2: Geometry of the active part of cutting tools – General conversion formulae to relate tool and working angles
SO 3002-3: 1984	Basic quantities in cutting and grinding – Part 3: Geometric and kinematic quantities in cutting
SO 3002-4: 1984	Basic quantities in cutting and grinding – Part 4: Forces, energy, power
SO 3002-5: 1989	Basic quantities in cutting and grinding – Part 5: Basic terminology for grinding processes using grinding wheels
SO 4006: 1991	Measurement of fluid flow in closed conduits – Vocabulary and symbols
SO 4226: 1993	Air quality – General aspects – Units of measurement
SO 7345: 1987	Thermal insulation – Physical quantities and definitions
SO 9288: 1989	Thermal insulation – Heat transfer by radiation – Physical quantities and definitions
ISO 9346: 1987	Thermal insulation – Mass transfer – Physical quantities and definitions
ISO 11145: 1994	Optics and optical instruments – Lasers and laser-related equipment – Vocabulary and symbols

- AFNOR Standards (Association Française de Normalisation, Paris, France)

NF X 02-001 (Août 1985)	Principes généraux concernant les grandeurs, les unités et les symboles – 13 pages (équivalent ISO 31– 1)
NF X 02-002 (Août 1986)	Unités et symboles – Mise en pratique des définitions des principales unités – 7 pages
NF X 02-003 (Août 1985)	Principe de l'écriture des nombres, des grandeurs, des unités et des symboles – 17 pages (équivalent ISO 31– 0)
NF X 02-004 (Août 1985)	Noms et symboles des unités de mesure du système international d'unités (SI) – 5 pages (équivalent ISO 1000)
NF X 02-006 (Août 1985)	Le système international d'unité – Description et règle d'emploi – Choix des multiples et de sous-multiples – 15 pages (équivalent ISO 1000)
NF X 02-010 (Avril 1963)	Sous-multiples décimaux du degré (unité d'angle) – 4 pages
NF X 02-011 (Novembre 1974)	Valeur de la pesanteur terrestre – 2 pages
NF X 02-012 (Avril 1987)	Constantes physiques fondamentales – 8 pages
NF X 02-050 (Janvier 1967)	Principales unités de mesure américaines et britanniques – 14 pages
NF X 02-051 (Août 1985)	Unités de mesure – Facteur de conversion – 13 pages
NF X 02-200 (Août 1985)	Grandeurs et symboles – Liste alphabétique (équivalent ISO 2) – 40 pages
NF X 02-201 (Août 1985)	Grandeurs, unités et symboles d'espace et de temps (équivalent ISO 31-1) – 40 pages
NF X 02-202 (Août 1985)	Grandeurs, unités et symboles de phénomènes périodiques et connexes (équivalent ISO 31-2) – 8 pages
NF X 02-203 (Août 1985)	Grandeurs, unités et symboles de mécanique (équivalent ISO 31-3) – 17 pages
NF X 02-204 (Août 1985)	Grandeurs, unités et symboles de thermique (équivalent ISO 31-4) – 11 pages
NF X 02-205 (Août 1985)	Grandeurs, unités et symboles d'électricité et de magnétisme (équivalent ISO 31-5) – 16 pages
NF X 02-206 (Août 1985)	Grandeurs, unités et symboles des rayonnements électromagnétiques et d'optique (équivalent ISO 31-6) – 21 pages

NF X 02-207 (Août 1985)	Grandeurs, unités et symboles d'acoustique (équivalent ISO 31-7) – 16 pages
NF X 02-208 (Août 1985)	Grandeurs, unités et symboles de chimie physique et physiq moléculaire (équivalent ISO 31-8) – 25 pages
NF X 02-209 (Août 1985)	Grandeurs, unités et symboles de physique atomique et nucléaire (équivalent ISO 31-9) – 19 pages
NF X 02-210 (Août 1985)	Grandeurs, unités et symboles de réactions nucléaires et rayonnements ionisants (équivalent ISO 31-10) – 26 pages
NF X 02-211 (Août 1983)	Signes et symboles mathématiques (équivalent ISO 31-11) – pages
NF X 02-212 (Décembre 1992)	Grandeurs et unités – Nombres caractéristique (équivalent IS 31-12) – 13 pages
NF X 02-213 (Décembre 1992)	Grandeurs et unités – Partie 13: Physique de l'état solide (équivalent ISO 31-13) – 25 pages
NF X 02-300 (Décembre 1992)	Grandeurs et unités – Partie 0: Principes généraux (équivalen ISO 31-0) – 26 pages
NF X 02-301 (Décembre 1992)	Grandeurs et unités – Partie 1: Espace et temps (équivalent IS 31-1) – 17 pages
NF X 02-302 (Décembre 1992)	Grandeurs et unités – Partie 2: Phénomènes périodiques et connexes (équivalent ISO 31-2) – 14 pages
NF X 02-303 (Décembre 1992)	Grandeurs et unités – Partie 3: Mécanique (équivalent ISO 31-3 – 24 pages
NF X 02-304 (Décembre 1992)	Grandeurs et unités – Partie 4: Chaleur (équivalent ISO 31-4) 20 pages
NF X 02-306 (Décembre 1992)	Grandeurs et unités – Partie 6: Lumière et rayonnement électromagnétiques connexes (équivalent ISO 31-6) – 26 pages
NF X 02-307 (Décembre 1992)	Grandeurs et unités – Partie 7: Acoustique (équivalent ISO 31-7 – 18 pages
NF Q 60 002 (Octobre 1975)	Unités typographiques – 7 pages

- IEEE (Institute of Electrical and Electronics Engineers Inc.)

IEEE 268-1982 (1982)	IEEE Standard Metric Practice

- ASTM Standards (Philadelphia, Pa., USA):

ASTM Standard E 380-82 (1982)	ASTM Standard for Metric Practice. Conversion factors are listed both alphabetically and classified by physical quantity.
ASTM Standard E 380-93 (1993)	ASTM Standard Practice for Use of The International System of Units (The Modernized Metric System).

- CSA (ACNOR) Standards (Canadian Standard Association/Association Canadienne de Normalisation, Ontario, CANADA):

CAN3-Z234.1-79 (1979)	Canadian Metric Practice Guide